DYNAMICAL PROCESSES ON COMPLEX NETWORKS

T0273834

The availability of large data sets has allowed researchers to uncover complex properties such as large-scale fluctuations and heterogeneities in many networks, leading to the breakdown of standard theoretical frameworks and models. Until recently these systems were considered as haphazard sets of points and connections. Recent advances have generated a vigorous research effort in understanding the effect of complex connectivity patterns on dynamical phenomena. This book presents a comprehensive account of these effects.

A vast number of systems, from the brain to ecosystems, power grids and the Internet, can be represented as large complex networks. This book will interest graduate students and researchers in many disciplines, from physics and statistical mechanics, to mathematical biology and information science. Its modular approach allows readers to readily access the sections of most interest to them, and complicated maths is avoided so the text can be easily followed by non-experts in the subject.

ALAIN BARRAT is Senior Researcher at the Centre de Physique Théorique (CNRS and Aix-Marseille University, France), and at the Institute for Scientific Interchange in Turin, Italy. His research interests are in the field of out of equilibrium statistical mechanics.

MARC BARTHÉLEMY is Senior Researcher at the Institute for Theoretical Physics (IPhT) at the Commissariat à l'Énergie Atomique et aux Energies Alternatives (CEA), France. His research interests are in the application of statistical physics to complex systems.

ALESSANDRO VESPIGNANI is Sternberg Distinguished Professor in Physics, Computer Science and Health Science at Northeastern University, USA. He is also Fellow of the Institute for Quantitative Social Sciences at Harvard University and Scientific Director of the Institute for Scientific Interchange Foundation in Italy.

DYNAMICAL PROCESSES ON COMPLEX NETWORKS

ALAIN BARRAT
Centre de Physique Théorique, Marseille

MARC BARTHÉLEMY
Commissariat à l'Énergie Atomique et aux Energies Alternatives (CEA)

ALESSANDRO VESPIGNANI
Northeastern University

CAMBRIDGE
UNIVERSITY PRESS

CAMBRIDGE UNIVERSITY PRESS
Cambridge, New York, Melbourne, Madrid, Cape Town,
Singapore, São Paulo, Delhi, Mexico City

Cambridge University Press
The Edinburgh Building, Cambridge CB2 8RU, UK

Published in the United States of America by Cambridge University Press, New York

www.cambridge.org
Information on this title: www.cambridge.org/9781107626256

First published 2008
4th printing 2011
First paperback edition 2013

Printed and Bound in United Kingdom by the MPG Books Group

A catalogue record for this publication is available from the British Library

Library of Congress Cataloguing in Publication data
Barrat, Alain, 1971–
Dynamical processes on complex networks / Alain Barrat,
Marc Barthelemy, Alessandro Vespignani.
p. cm.
Includes bibliographical references and index.
ISBN 978-0-521-87950-7
1. System analysis. 2. Graph theory. 3. Network theory.
I. Barthelemy, Marc, 1965– II. Vespignani, Alessandro, 1965– III. Title.
QA402.B374 2009
511′.5–dc22
2008025427

ISBN 978-0-521-87950-7 Hardback
ISBN 978-1-107-62625-6 Paperback

To Luisa;
To Loulou, Simon, Théo;
To Martina

Contents

Preface

In the past few years, the study of large networked systems has received a boost from the ever-increasing availability of large data sets and computer power for their storage and manipulation. In particular, mapping projects of the World Wide Web and the physical Internet offered the first chance to study the topology of large complex networks. Gradually, other maps followed describing many networks of practical interest in social science, critical infrastructures, and biology. Indeed, large complex networks arise in a vast number of natural and artificial systems. The brain consists of many interconnected neurons; ecosystems consist of species whose interdependency can be mapped into intricate food webs. Social systems may be represented by graphs describing various interactions among individuals. Large networked infrastructures such as power grids and transportation networks are critical to our modern society. Finally, the living cell is not an exception, its organization and function being the outcome of a complex web of interactions among genes, proteins, and other molecules. In the search for the underlying laws governing the dynamics and evolution of these complex systems, researchers have started the systematic analysis and characterization of their network representations. A central result of these activities is that large-scale networks are generally characterized by complex topologies and very heterogeneous structures. These features usually find their signature in connectivity patterns statistically characterized by heavy tails and large fluctuations, scale-free properties, and non-trivial correlations such as high clustering and hierarchical ordering.

The large size and dynamic nature of complex networks has attracted the attention of the statistical physics community. In this context the statistical physics approach has been exploited as a very convenient strategy because of its deep connection with statistical graph theory and the possibility of characterizing emergent macroscopic phenomena in terms of the dynamical evolution of the basic elements of the system. In addition, the large scale analysis of networks appearing in very different fields has provided evidence for the sharing of several asymptotic properties and raised the issue of the supposed emergence in complex networks of general and common organizing principles that go beyond the particulars of the individual system, and are thus amenable to general modeling principles.

The advances in understanding large complex networks have generated increased attention towards the potential implications of their structure for the most important questions concerning the various physical and dynamical processes occurring on top of them. Questions of how pathogens spread in population networks, how blackouts can spread on a nationwide scale, or how efficiently we can search and retrieve data on large information structures are all related to spreading and diffusion phenomena. The resilience and robustness of large infrastructures can be studied by percolation models in which we progressively damage the network. Social behavior may be often modeled through simple dynamical processes and agent-based models. Since the early studies of percolation and spreading processes in complex networks, a theoretical picture has emerged in which many of the standard results were radically altered by the topological fluctuations and the complex features observed in most real-world networks. Indeed, complex properties often imply a virtual infinite heterogeneity of the system and large scale fluctuations extending over several orders of magnitude, generally corresponding to the breakdown of standard theoretical frameworks and models. Therefore, while the definitions of the basic models are expected to remain unchanged, the scientific community has become aware of the need to investigate systematically the impact of the various network characteristics on the basic features of equilibrium and non-equilibrium dynamical processes.

The study of models unfolding on complex networks has generated results of conceptual and practical relevance. The resilience of networks, their vulnerability to attacks, and their synchronization properties are all strongly affected by topological heterogeneities. Consensus formation, disease spreading, and our accessibility to information can benefit from or be impaired by the connectivity pattern of the population or infrastructure we are looking at. Noticeably, all these results have been obtained in different contexts and have relevance across several disciplines including physics, biology, and computer and information science. The purpose of this book is to provide a unified presentation of these results and their impact on our understanding of a wide range of networked systems.

While several techniques and methods have been used in the field, the massive size and heterogeneity of the systems have favored the use of techniques akin to the analysis of non-linear, equilibrium, and non-equilibrium physical systems. As previously mentioned, the statistical physics approach has indeed revealed its convenience and has proven very useful in this context, particularly regarding its ability to shed light on the link between the microscopic dynamical evolution of the basic elements of the system and the emergence of macroscopic phenomena. For this reason we will follow this approach and structure the book according to the general framework provided by statistical physics. In doing this, however, we will make a special effort in defining for each phenomenon or system under study the appropriate language used in the field and offer to the reader a mapping between languages and techniques used in the different disciplines.

For the sake of clarity and readability we have used a modular approach to the material that allows the reader interested in just one phenomenon to refer directly to the corresponding specific chapter. The basic definitions and methods used throughout the book are introduced in the first four chapters that set the stage for the remaining chapters, each one clearly self-contained and addressing a specific class of processes. The material is ordered in a path that goes from equilibrium to non-equilibrium phenomena, in accordance with the general framework of statistical physics and in terms of the homogeneity of the techniques used in each context.

More precisely, we start in Chapter 1 by defining the basics of graph theory and some tools for the statistical characterization and classification of large networks. These tools are used in Chapter 2, where we discuss the complex features which characterize a large number of real-world networks. Chapter 3 is devoted to a brief presentation of the various network modeling techniques and the general approaches used for their analysis. Approximations and computational techniques are discussed as well. Chapter 4 provides an introduction to the basic theoretical concepts and tools needed for the analysis of dynamical processes taking place on networks. All these introductory chapters will allow researchers not familiar with mathematical and computational modeling to get acquainted with the approaches and techniques used in the book.

In Chapter 5, we report the analysis of equilibrium processes and we review the behaviors of basic equilibrium physical models such as the Ising model in complex networks. Chapter 6 addresses the analysis of damage and attack processes in complex networks by mapping those processes into percolation phase transitions. Chapter 7 is devoted to synchronization in coupled systems with complex connectivity patterns. This is a transition chapter in that some of the phenomena analyzed and their respective models can be considered in the equilibrium processes domains while others fall into the non-equilibrium class. The following four chapters of the book are devoted to non-equilibrium processes. One chapter considers search, navigation, and exploration processes of complex networks; the three other chapters concern epidemic spreading, the emergence of social behavior, and traffic avalanche and congestion, respectively. These chapters, therefore, are considering far from equilibrium systems with absorbing states and driven-dissipative dynamics. Finally, Chapter 12 is devoted to a discussion of the recent application of network science to biological systems. A final set of convenient appendices are used to detail very technical calculations or discussions in order to make the reading of the main chapters more accessible to non-expert readers. The postface occupies a special place at the end of the book as it expresses our view on the value of complex network science.

We hope that our work will result in a first coherent and comprehensive exposition of the vast research activity concerning dynamical processes in complex networks. The large number of references and research focus areas that find room in this book make us believe that the present volume will be a convenient entry reference to all researchers and students who consider working in this exciting area of interdisciplinary research. Although this book reviews or mentions more than 600 scientific works, it is impossible to discuss in detail all relevant contributions to the field. We have therefore used our author privilege and made choices based on our perception of what is more relevant to the focus and the structure of the present book. This does not imply that work not reported or cited here is less valuable, and we apologize in advance to all the colleagues who feel that their contributions have been overlooked.

Acknowledgements

There are many people we must thank for their help in the preparation and completion of this book. First of all, we wish to thank all the colleagues who helped us shape our view and knowledge of complex networks through invaluable scientific collaborations: J. I. Alvarez-Hamelin, F. P. Alvarez, L. A. N. Amaral, D. Balcan, A. Baronchelli, M. Boguñá, K. Börner, G. Caldarelli, C. Caretta Cartozo, C. Castellano, F. Cecconi, A. Chessa, E. Chow, V. Colizza, P. Crépey, A. De Montis, L. Dall'Asta, T. Eliassi-Rad, A. Flammini, S. Fortunato, F. Gargiulo, D. Garlaschelli, A. Gautreau, B. Gondran, E. Guichard, S. Havlin, H. Hu, C. Herrmann, W. Ke, E. Kolaczyk, B. Kozma, M. Leone, V. Loreto, F. Menczer, Y. Moreno, M. Nekovee, A. Maritan, M. Marsili, A. Maguitman, M. Meiss, S. Mossa, S. Myers, C. Nardini, M. Nekovee, D. Parisi, R. Pastor-Satorras, R. Percacci, P. Provero, F. Radicchi, J. Ramasco, C. Ricotta, A. Scala, M. Angeles Serrano, H. E. Stanley, A.-J. Valleron, A. Vàzquez, M. Vergassola, F. Viger, D. Vilone, M. Weigt, R. Zecchina, and C.-H. Zhang.

We also acknowledge with pleasure the many discussions with, encouragements, and useful criticisms from: R. Albert, E. Almaas, L. Adamic, A. Arenas, A. L. Barabási, J. Banavar, C. Cattuto, kc claffy, P. De Los Rios, S. Dorogovtsev, A. Erzan, R. Goldstone, S. Havlin, C. Hidalgo, D. Krioukov, H. Orland, N. Martinez, R. May, J. F. Mendes, A. Motter, M. Mungan, M. Newman, A. Pagnani, T. Petermann, D. Rittel, L. Rocha, G. Schehr, S. Schnell, O. Sporns, E. Trizac, Z. Toroczkai, S. Wasserman, F. van Wijland, W. Willinger, L. Yaeger, and S. Zapperi.

Special thanks go to E. Almaas, C. Castellano, V. Colizza, J. Dunne, S. Fortunato, H. Jeong, Y. Moreno, N. Martinez, M. E. J. Newman, M. Angeles Serrano, and O. Sporns for their kind help in retrieving data and adapting figures from their work. A. Baronchelli, C. Castellano, L. Dall'Asta, A. Flammini, S. Fortunato, and R. Pastor-Satorras reviewed earlier versions of this book and made many suggestions for improvement. B. Hook deserves a special acknowledgement for proofreading and editing the first version of the manuscript. Simon Capelin and his outstanding editorial staff at Cambridge University Press were a source of continuous encouragement and help and greatly contributed to make the writing of this book a truly exciting and enjoyable experience. We also thank all those institutions that helped in making this book a reality. The Centre National de la Recherche Scientifique, the Laboratoire de Physique Théorique at the University of Paris-Sud, the Commissariat à l'Energie Atomique, the School of Informatics at Indiana University, and the Institute for Scientific Interchange in Turin have provided us with great environments and the necessary resources to complete this work. We also acknowledge the generous funding support at various stages of our research by the following awards: NSF-IIS0513650, EC-001907 Delis, and CRT-foundation Lagrange laboratory. Finally, A.B. thanks the SNCF for providing fast trains between Paris, Dijon, and Turin which are not always late, M.B. expresses his gratitude to the Cannibale Café which is probably the best coffee-office in Paris, and A.V. thanks Continental Airlines for having power plug-ins in all classes of reservations.

Abbreviations

AS	Autonomous System
BA	Barabási–Albert
BGP	Border Gateway Protocol
CAIDA	Cooperative Association for Internet Data Analysis
CPU	central processing unit
ER	Erdős–Rényi
IATA	International Air Transport Association
DIMES	Distributed Internet Measurements and Simulations
GIN	giant in-component
GOUT	giant out-component
GSCC	giant strongly connected component
GWCC	giant weakly connected component
HOT	Heuristically Optimized Trade-off
ICT	Information and communication technology
LAN	Local Area Network
ME	master equation
MF	mean-field
P2P	peer-to-peer
SCN	scientific collaboration network
SI	Susceptible-Infected model
SIR	Susceptible-Infected-Removed model
SIS	Susceptible-Infected-Susceptible model
TTL	time-to-live
WAN	worldwide airport network
WS	Watts–Strogatz
WWW	World Wide Web

1

Preliminaries: networks and graphs

In this chapter we introduce the reader to the basic definitions of network and graph theory. We define metrics such as the shortest path length, the clustering coefficient, and the degree distribution, which provide a basic characterization of network systems. The large size of many networks makes statistical analysis the proper tool for a useful mathematical characterization of these systems. We therefore describe the many statistical quantities characterizing the structural and hierarchical ordering of networks including multipoint degree correlation functions, clustering spectrum, and several other local and non-local quantities, hierarchical measures and weighted properties.

This chapter will give the reader a crash course on the basic notions of network analysis which are prerequisites for understanding later chapters of the book. Needless to say the expert reader can freely skip this chapter and use it later as a reference if needed.

1.1 What is a network?

In very general terms a network is any system that admits an abstract mathematical representation as a graph whose nodes (vertices) identify the elements of the system and in which the set of connecting links (edges) represent the presence of a relation or interaction among those elements. Clearly such a high level of abstraction generally applies to a wide array of systems. In this sense, networks provide a theoretical framework that allows a convenient conceptual representation of inter-relations in complex systems where the system level characterization implies the mapping of interactions among a large number of individuals.

The study of networks has a long tradition in graph theory, discrete mathematics, sociology, and communication research and has recently infiltrated physics and biology. While each field concerned with networks has introduced, in many cases, its own nomenclature, the rigorous language for the description of networks is found in mathematical graph theory. On the other hand, the study of very large networks has spurred the definitions of new metrics and statistical observables specifically aimed at the study of large-scale systems. In the following we provide an introduction to the basic notions and notations used in network theory and set the cross-disciplinary language that will be used throughout this book.

1.2 Basic concepts in graph theory

Graph theory – a vast field of mathematics – can be traced back to the pioneering work of Leonhard Euler in solving the Königsberg bridges problem (Euler, 1736). Our intention is to select those notions and notations which will be used throughout the rest of this book. The interested reader can find excellent textbooks on graph theory by Bergé (1976), Chartrand and Lesniak (1986), Bollobás (1985, 1998) and Clark and Holton (1991).

1.2.1 Graphs and subgraphs

An undirected graph G is defined by a pair of sets $G = (\mathcal{V}, \mathcal{E})$, where $\mathcal{V}$ is a non-empty countable set of elements, called *vertices* or *nodes*, and $\mathcal{E}$ is a set of *unordered* pairs of different vertices, called *edges* or *links*. Throughout the book we will refer to a vertex by its order i in the set $\mathcal{V}$. The edge (i, j) joins the vertices i and j, which are said to be *adjacent* or *connected*. It is also common to call connected vertices *neighbors* or *nearest neighbors*. The total number of vertices in the graph (the cardinality of the set $\mathcal{V}$) is denoted as N and defines the order of the graph. It is worth remarking that in many biological and physical contexts, N defines the physical size of the network since it identifies the number of distinct elements composing the system. However, in graph theory, the size of the graph is identified by the total number of edges E. Unless specified in the following, we will refer to N as the size of the network.

For a graph of size N, the maximum number of edges is $\binom{N}{2}$. A graph with $E = \binom{N}{2}$, i.e. in which all possible pairs of vertices are joined by edges, is called a *complete N-graph*. Undirected graphs are depicted graphically as a set of dots, representing the vertices, joined by lines between pairs of vertices, representing the corresponding edges.

An interesting class of undirected graph is formed by hierarchical graphs where each edge (known as a child) has exactly one parent (node from which it originates). Such a structure defines a *tree* and if there is a parent node, or *root*, from which the whole structure arises, then it is known as a rooted tree. It is easy to prove that the number of nodes in a tree equals the number of edges plus one, i.e.,

$N = E+1$ and that the deletion of any edge will break a tree into two disconnected trees.

A directed graph D, or digraph, is defined by a non-empty countable set of vertices V and a set of *ordered* pairs of different vertices $\mathcal{E}$ that are called directed edges. In a graphical representation, the directed nature of the edges is depicted by means of an arrow, indicating the direction of each edge. The main difference between directed and undirected graphs is represented in Figure 1.1. In an undirected graph the presence of an edge between vertices i and j connects the vertices in both directions. On the other hand, the presence of an edge from i and j in a directed graph does not necessarily imply the presence of the reverse edge between j and i. This fact has important consequences for the connectedness of a directed graph, as will be discussed in more detail in Section 1.2.2.

From a mathematical point of view, it is convenient to define a graph by means of the *adjacency matrix* $\mathbf{X} = \{x_{ij}\}$. This is a $N \times N$ matrix defined such that

$$x_{ij} = \begin{cases} 1 & \text{if } (i, j) \in \mathcal{E} \\ 0 & \text{if } (i, j) \notin \mathcal{E} \end{cases}. \tag{1.1}$$

For undirected graphs the adjacency matrix is symmetric, $x_{ij} = x_{ji}$, and therefore contains redundant information. For directed graphs, the adjacency matrix is not

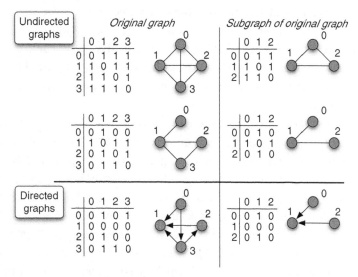

Fig. 1.1. Adjacency matrix and graphical representation of different networks. In the graphical representation of an undirected graph, the dots represent the vertices and pairs of adjacent vertices are connected by a line (edge). In directed graphs, adjacent vertices are connected by arrows, indicating the direction of the corresponding edge.

symmetric. In Figure 1.1 we show the graphical illustrations of different undirected and directed graphs and their corresponding adjacency matrices.

An important feature of many graphs, which helps in dealing with their structure, is their *sparseness*. The number of edges E for a connected graph (i.e., with no disconnected parts) ranges from $N - 1$ to $\binom{N}{2}$. There are different definitions of sparseness, but we will adopt the convention that when the number of edges scales as $E \sim N^\alpha$ with $\alpha < 2$, the graph is said to be *sparse*. In the case where $E \sim N^2$, the corresponding graph is called *dense*. By defining the *connectance* or *density* of a graph as the number of existing edges divided by the maximal possible number of edges $\mathcal{D} = E/[N(N - 1)/2]$, a graph is then sparse if $\mathcal{D} \ll 1$. This feature implies, in the case of large graphs, that the adjacency matrix is mostly defined by zero elements and its complete representation, while costly, does not contain much relevant information. With large graphs, it is customary to represent the graph in the compact form defined by the adjacency lists $\ell(i, v \in \mathcal{V}(i))$, where the set of all neighbors of a fixed vertex i is called the neighborhood (set) of i and is denoted by $\mathcal{V}(i)$. The manipulation of these lists is obviously very convenient in computational applications because they efficiently store large sparse networks.

In many cases, we are also interested in subsets of a graph. A graph $G' = (\mathcal{V}', \mathcal{E}')$ is said to be a *subgraph* of the graph $G = (\mathcal{V}, \mathcal{E})$ if all the vertices in $\mathcal{V}'$ belong to $\mathcal{V}$ and all the edges in $\mathcal{E}'$ belong to $\mathcal{E}$, i.e. $\mathcal{E}' \subset \mathcal{E}$ and $\mathcal{V}' \subset \mathcal{V}$. A *clique* is a complete n-subgraph of size $n < N$. In Figure 1.1 we provide the graphical and adjacency matrix representations of subgraphs in the undirected and directed cases. The abundance of given types of subgraphs and their properties are extremely relevant in the characterization of real networks.[1] Small, statistically significant, coherent subgraphs, called motifs, that contribute to the set-up of networks have been identified as relevant building blocks of network architecture and evolution (see Milo *et al.* [2002] and Chapter 12).

The characterization of local structures is also related to the identification of *communities*. Loosely speaking, communities are identified by subgraphs where nodes are highly interconnected among themselves and poorly connected with nodes outside the subgraph. In this way, different communities can be traced back with respect to varying levels of cohesiveness. In directed networks, edge directionality introduces the possibility of different types of local structures. A possible mathematical way to account for these local cohesive groups consists in examining the number of bipartite cliques present in the graph. A bipartite clique $K_{n,m}$ identifies a group of n nodes, all of which have a direct edge to the same m

---

[1] Various approaches exist to determine the structural equivalence, the automorphic equivalence, or the regular equivalence of subnetworks, and measures for structural similarity comprise correlation coefficients, Euclidean distances, rates of exact matches, etc.

nodes. The presence of subgraphs and communities raises the issue of modularity in networks. Modularity in a network is determined by the existence of specific subgraphs, called *modules* (or communities). Clustering techniques can be employed to determine major clusters. They comprise non-hierarchical methods (e.g., single pass methods or reallocation methods), hierarchical methods (e.g., single-link, complete-link, average-link, centroid-link, Ward), and linkage based methods (we refer the interested reader to the books of Mirkin (1996) and Banks *et al.* (2004) for detailed expositions of clustering methods). Non-hierarchical and hierarchical clustering methods typically work on attribute value information. For example, the similarity of social actors might be judged based on their hobbies, ages, etc. Non-hierarchical clustering typically starts with information on the number of clusters that a data set is expected to have and sorts the data items into clusters such that an optimality criterion is satisfied. Hierarchical clustering algorithms create a hierarchy of clusters grouping similar data items. Connectivity-based approaches exploit the topological information of a network to identify dense subgraphs. They comprise measures such as betweenness centrality of nodes and edges (Girvan and Newman, 2002; Newman, 2006), superparamagnetic clustering (Blatt, Wiseman and Domany, 1996; Domany, 1999), hubs and bridging edges (Jungnickel, 2004), and others. Recently, a series of sophisticated overlapping and non-overlapping clustering methods has been developed, aiming to uncover the modular structure of real networks (Reichardt and Bornholdt, 2004; Palla *et al.*, 2005).

1.2.2 Paths and connectivity

A central issue in the structure of graphs is the *reachability* of vertices, i.e. the possibility of going from one vertex to another following the connections given by the edges in the network. In a connected network every vertex is reachable from any other vertex. The connected components of a graph thus define many properties of its physical structure.

In order to analyze the connectivity properties let us define a *path* $\mathcal{P}_{i_0, i_n}$ in a graph $G = (\mathcal{V}, \mathcal{E})$ as an ordered collection of $n + 1$ vertices $\mathcal{V}_{\mathcal{P}} = \{i_0, i_1, \ldots, i_n\}$ and n edges $\mathcal{E}_{\mathcal{P}} = \{(i_0, i_1), (i_1, i_2), \ldots, (i_{n-1}, i_n)\}$, such that $i_\alpha \in \mathcal{V}$ and $(i_{\alpha-1}, i_\alpha) \in \mathcal{E}$, for all α. The path $\mathcal{P}_{i_0, i_n}$ is said to connect the vertices i_0 and i_n. The *length* of the path $\mathcal{P}_{i_0, i_n}$ is n. The number $\mathcal{N}_{ij}$ of paths of length n between two nodes i and j is given by the ij element of the nth power of the adjacency matrix: $\mathcal{N}_{ij} = (\mathbf{X}^n)_{ij}$.

A *cycle* – sometimes called a *loop* – is a closed path ($i_0 = i_n$) in which all vertices and all edges are distinct. A graph is called *connected* if there exists a path connecting any two vertices in the graph. A *component* $\mathcal{C}$ of a graph is defined as a connected subgraph. Two components $\mathcal{C}_1 = (\mathcal{V}_1, \mathcal{E}_1)$ and $\mathcal{C}_2 = (\mathcal{V}_2, \mathcal{E}_2)$ are disconnected if it is impossible to construct a path $\mathcal{P}_{i,j}$ with $i \in \mathcal{V}_1$ and $j \in \mathcal{V}_2$.

It is clear that for a given number of nodes the number of loops increases with the number of edges. It can easily be shown (Bergé, 1976) that for any graph with p disconnected components, the number of independent loops, or *cyclomatic number*, is given by

$$\Gamma = E - N + p. \tag{1.2}$$

It is easy to check that this relation gives $\Gamma = 0$ for a tree.

A most interesting property of random graphs (Section 3.1) is the distribution of components, and in particular the existence of a *giant component* $\mathcal{G}$, defined as a component whose size scales with the number of vertices of the graph, and therefore diverges in the limit $N \to \infty$. The presence of a giant component implies that a macroscopic fraction of the graph is connected.

The structure of the components of directed graphs is somewhat more complex as the presence of a path from the node i to the node j does not necessarily guarantee the presence of a corresponding path from j to i. Therefore, the definition of a giant component needs to be adapted to this case. In general, the component structure of a directed network can be decomposed into a giant weakly connected component (GWCC), corresponding to the giant component of the same graph in which the edges are considered as undirected, plus a set of smaller disconnected components, as sketched in Figure 1.2. The GWCC is itself composed of several parts because of the directed nature of its edges: (1) the giant strongly connected

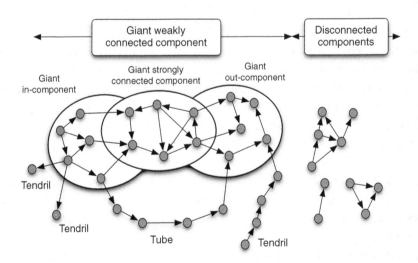

Fig. 1.2. Component structure of a directed graph. Figure adapted from Dorogovtsev *et al.* (2001a).

component (GSCC), in which there is a directed path joining any pair of nodes; (2) the giant in-component (GIN), formed by the nodes from which it is possible to reach the GSCC by means of a directed path; (3) the giant out-component (GOUT), formed by the nodes that can be reached from the GSCC by means of a directed path; and (4) the tendrils containing nodes that cannot reach or be reached by the GSCC (among them, the tubes that connect the GIN and GOUT), which form the rest of the GWCC.

The concept of "path" lies at the basis of the definition of distance among vertices. Indeed, while graphs usually lack a metric, the natural distance measure between two vertices i and j is defined as the number of edges traversed by the shortest connecting path (see Figure 1.3). This distance, equivalent to the chemical distance usually considered in percolation theory (Bunde and Havlin, 1991), is called the *shortest path length* and denoted as ℓ_{ij}. When two vertices belong to two disconnected components of the graph, we define $\ell_{ij} = \infty$. While it is a symmetric quantity for undirected graphs, the shortest path length ℓ_{ij} does not coincide in general with ℓ_{ji} for directed graphs.

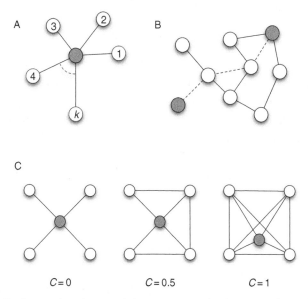

Fig. 1.3. Basic metrics characterizing a vertex i in the network. A, The degree k quantifies the vertex connectivity. B, The shortest path length identifies the minimum connecting path (dashed line) between two different vertices. C, The clustering coefficient provides a measure of the interconnectivity in the vertex's neighborhood. As an example, the central vertex in the figure has a clustering coefficient $C = 1$ if all its neighbors are connected and $C = 0$ if no interconnections are present.

By using the shortest path length as a measure of distance among vertices, it is possible to define the diameter and the typical size of a graph. The *diameter* is traditionally defined as

$$d_G = \max_{i,j} \ell_{ij}. \tag{1.3}$$

Another effective definition of the linear size of the network is the *average shortest path length*,[2] defined as the average value of ℓ_{ij} over all the possible pairs of vertices in the network

$$\langle \ell \rangle = \frac{1}{N(N-1)} \sum_{ij} \ell_{ij}. \tag{1.4}$$

By definition $\langle \ell \rangle \leq d_G$, and in the case of a well-behaved and bounded shortest path length distribution, it is possible to show heuristically that in many cases the two definitions behave in the same way with the network size.

There are also other measures of interest which are related to the characterization of the linear size of a graph. The eccentricity of a vertex i is defined by $ec(i) = \max_{j \neq i} \ell_{ij}$, and the radius of a graph G by $\mathrm{rad}_G = \min_i ec(i)$. These different quantities are not independent and one can prove (Clark and Holton, 1991) that the following inequalities hold for any graph

$$\mathrm{rad}_G \leq d_G \leq 2 \, \mathrm{rad}_G. \tag{1.5}$$

Simple examples of distances in graphs include the complete graph where $\langle \ell \rangle = 1$ and the regular hypercubic lattice in D dimensions composed by N vertices for which the average shortest path length scales as $\langle \ell \rangle \sim N^{1/D}$. In most random graphs (Sections 2.2 and 3.1), the average shortest path length grows logarithmically with the size N, as $(\langle \ell \rangle \sim \log N)$ – a much slower growth than that found in regular hypercubic lattices. The fact that any pair of nodes is connected by a small shortest path constitutes the so-called *small-world effect*.

1.2.3 Degree and centrality measures

When looking at networks, one of the main insights is provided by the importance of their basic elements (Freeman, 1977). The importance of a node or edge is commonly defined as its centrality and this depends on the characteristics or specific properties we are interested in. Various measurements exist to characterize the centrality of a node in a network. Among those characterizations, the most

---

[2] It is worth stressing that the average shortest path length has also been referred to in the physics literature as another definition for the diameter of the graph.

commonly used are the degree centrality, the closeness centrality, or the betweenness centrality of a vertex. Edges are frequently characterized by their betweenness centrality.

Degree centrality

The degree k_i of a vertex i is defined as the number of edges in the graph incident on the vertex i. While this definition is clear for undirected graphs, it needs some refinement for the case of directed graphs. Thus, we define the *in-degree* $k_{in,i}$ of the vertex i as the number of edges arriving at i, while its *out-degree* $k_{out,i}$ is defined as the number of edges departing from i. The degree of a vertex in a directed graph is defined by the sum of the in-degree and the out-degree, $k_i = k_{in,i} + k_{out,i}$. In terms of the adjacency matrix, we can write

$$k_{in,i} = \sum_j x_{ji}, \qquad k_{out,i} = \sum_j x_{ij}. \tag{1.6}$$

For an undirected graph with a symmetric adjacency matrix, $k_{in,i} = k_{out,i}$. The degree of a vertex has an immediate interpretation in terms of centrality quantifying how well an element is connected to other elements in the graph. The *Bonacich power index* takes into account not only the degree of a node but also the degrees of its neighbors.

Closeness centrality

The *closeness centrality* expresses the average distance of a vertex to all others as

$$g_i = \frac{1}{\sum_{j \neq i} \ell_{ij}}. \tag{1.7}$$

This measure gives a large centrality to nodes which have small shortest path distances to the other nodes.

Betweenness centrality

While the previous measures consider nodes which are topologically better connected to the rest of the network, they overlook vertices which may be crucial for connecting different regions of the network by acting as bridges. In order to account quantitatively for the role of such nodes, the concept of betweenness centrality has been introduced (Freeman, 1977; Newman, 2001a): it is defined as the number of shortest paths between pairs of vertices that pass through a given vertex. More precisely, if σ_{hj} is the total number of shortest paths from h to j and $\sigma_{hj}(i)$ is the number of these shortest paths that pass through the vertex i, the betweenness of i is defined as

$$b_i = \sum_{h \neq j \neq i} \frac{\sigma_{hj}(i)}{\sigma_{hj}}. \tag{1.8}$$

A similar quantity, the load or stress centrality, does not discount the multiplicity of equivalent paths and reads as $L_i = \sum_{h \neq j \neq i} \sigma_{hj}(i)$. The above definitions may include a factor $1/2$ to avoid counting each path twice in undirected networks. The calculation of this measure is computationally very expensive. The basic algorithm for its computation would lead to a complexity of order $\mathcal{O}(N^2 E)$, which is prohibitive for large networks. An efficient algorithm to compute betweenness centrality is reported by Brandes (2001) and reduces the complexity to $\mathcal{O}(NE)$ for unweighted networks.

According to these definitions, central nodes are therefore part of more shortest paths within the network than less important nodes. Moreover, the betweenness centrality of a node is often used in transport networks to provide an estimate of the traffic handled by the vertices, assuming that the number of shortest paths is a zero-th order approximation to the frequency of use of a given node. Analogously to the vertex betweenness, the betweenness centrality of edges can be calculated as the number of shortest paths among all possible vertex couples that pass through the given edge. Edges with the maximum score are assumed to be important for the graph to stay interconnected. These high-scoring edges are the "bridges" that interconnect clusters of nodes. Removing them frequently leads to unconnected clusters of nodes. The "bridges" are particularly important for decreasing the average path length among nodes in a network, for speeding up the diffusion of information, or for increasing the size of the part of the network at a given distance from a node. However, networks with many such bridges are more fragile and less clustered.

1.2.4 Clustering

Along with centrality measures, vertices are characterized by the structure of their local neighborhood. The concept of *clustering*[3] of a graph refers to the tendency observed in many natural networks to form cliques in the neighborhood of any given vertex. In this sense, clustering implies the property that, if the vertex i is connected to the vertex j, and at the same time j is connected to l, then with a high probability i is also connected to l. The clustering of an undirected graph can be quantitatively measured by means of the *clustering coefficient* which measures the local group cohesiveness (Watts and Strogatz, 1998). Given a vertex i, the clustering $C(i)$ of a node i is defined as the ratio of the number of links between the neighbors of i and the maximum number of such links. If the degree of node i is k_i and if these nodes have e_i edges between them, we have

$$C(i) = \frac{e_i}{k_i(k_i - 1)/2},$$ (1.9)

---

[3] Also called *transitivity* in the context of sociology (Wasserman and Faust, 1994).

where it is worth remarking that this measure of clustering only has a meaning for $k_i > 1$. For $k_i \leq 1$ we define $C(i) \equiv 0$. Given the definition of e_i, it is easy to check that the number of edges among the neighbors of i can be computed in terms of the adjacency matrix $\mathbf{X}$ as

$$e_i = \frac{1}{2} \sum_{jl} x_{ij} x_{jl} x_{li}. \tag{1.10}$$

In Figure 1.3, we provide an illustration of some simple examples of the clustering of vertices with a given neighborhood. The average clustering coefficient of a graph is simply given by

$$\langle C \rangle = \frac{1}{N} \sum_i C(i). \tag{1.11}$$

It is worth noting that the clustering coefficient has been defined in a number of similar ways, for instance as a function of triples in the network (triples are defined as subgraphs which contain exactly three nodes) and reversing the order of average and division in Equation (1.11)

$$C_\Delta = \frac{3 \times \text{number of fully connected triples}}{\text{number of triples}}, \tag{1.12}$$

where the factor 3 is due to the fact that each triangle is associated with three nodes. This definition corresponds to the concept of the fraction of transitive triples introduced in sociology (Wasserman and Faust, 1994). Different definitions give rise to different values of the clustering coefficient for a given graph. Hence, the comparison of clustering coefficients among different graphs must use the very same measure. In any case, both measures are normalized and bounded to be between 0 and 1.

1.3 Statistical characterization of networks

One of the elements that has fostered the recent development of network science can be found in the recent possibility of systematic gathering and handling of data sets on several large-scale networks. Indeed, in large systems, asymptotic regularities cannot be detected by looking at local elements or properties. In other words, one has to shift the attention to statistical measures that take into account the global behavior of these quantities.

1.3.1 Degree distribution

The degree distribution $P(k)$ of undirected graphs is defined as the probability that any randomly chosen vertex has degree k. It is obtained by constructing the normalized histogram of the degree of the nodes in a network. In the case of directed graphs, one has to consider instead two distributions, the in-degree $P(k_{in})$ and out-degree $P(k_{out})$ distributions, defined as the probability that a randomly chosen vertex has in-degree k_{in} and out-degree k_{out}, respectively. The average degree of an undirected graph is defined as the average value of k over all the vertices in the network,

$$\langle k \rangle = \frac{1}{N} \sum_i k_i = \sum_k k P(k) \equiv \frac{2E}{N}, \tag{1.13}$$

since each edge end contributes to the degree of a vertex. For a directed graph, the average in-degree and out-degree must be equal,

$$\langle k_{in} \rangle = \sum_{k_{in}} k_{in} P(k_{in}) = \langle k_{out} \rangle = \sum_{k_{out}} k_{out} P(k_{out}) \equiv \frac{\langle k \rangle}{2}, \tag{1.14}$$

since an edge departing from any vertex must arrive at another vertex. Analogously to the average degree, it is possible to define the nth moment of the degree distribution,

$$\langle k^n \rangle = \sum_k k^n P(k). \tag{1.15}$$

A *sparse* graph has an average degree that is much smaller than the size of the graph, $\langle k \rangle \ll N$. In the next chapters we will see that the properties of the degree distribution will be crucial to identify different classes of networks.

1.3.2 Betweenness distribution

Analogously to the degree, it is possible to introduce the probability distribution $P_b(b)$ that a vertex has betweenness b, and the average betweenness $\langle b \rangle$ defined as

$$\langle b \rangle = \sum_b b \, P_b(b) \equiv \frac{1}{N} \sum_i b_i. \tag{1.16}$$

For this quantity it is worth showing its relation with the average shortest path length $\langle \ell \rangle$. By simply reordering the sums in the betweenness definition we have

$$\sum_i b_i = \sum_i \sum_{h,j \neq i} \frac{\sigma_{hj}(i)}{\sigma_{hj}} = \sum_{h \neq j} \frac{1}{\sigma_{hj}} \sum_{i \neq h,j} \sigma_{hj}(i). \tag{1.17}$$

A simple topological reasoning gives $\sum_{i \neq h,j} \sigma_{hj}(i) = \sigma_{hj}(\ell_{hj} - 1)$. Plugged into the previous equation, this yields $\sum_i b_i = N(N-1)(\langle \ell \rangle - 1)$, and therefore

$$\langle b \rangle = (N-1)(\langle \ell \rangle - 1). \qquad (1.18)$$

As this formula shows, it is easy to realize that the betweenness usually takes values of the order $\mathcal{O}(N)$ or larger. For instance, in a star graph, formed by $N-1$ vertices with a single edge connected to a central vertex, the betweenness takes a maximum value $(N-1)(N-2)$ at the central vertex (the other peripheral vertices having 0 betweenness). For this reason, in the case of very large graphs ($N \to \infty$) it is sometimes convenient to define a rescaled betweenness $\tilde{b} \equiv N^{-1}b$.

Finally, it is clear that in most cases the larger the degree k_i of a node, the larger its betweenness centrality b_i will be. More quantitatively, it has been observed that for large networks and for large degrees there is a direct association of the form (Goh, Kahng and Kim, 2001, 2003; Barthélemy, 2004)

$$b_i \sim k_i^{\eta}, \qquad (1.19)$$

where η is a positive exponent depending on the network. Such associations are extremely relevant as they correspond to the fact that a large number of shortest paths go through the nodes with large degree (the hubs). These nodes will therefore be visited and discovered easily in processes of network exploration (see Chapter 8). They will also typically see high traffic which may result in congestion (see Chapter 11). Of course, fluctuations are also observed in real and model networks, and small-degree nodes may also have large values of the betweenness centrality if they connect different regions of the network, acting as bridges. Such low-degree nodes with high centrality appear, for instance, in networks where spatial constraints limit the ability of hubs to deploy long links (Guimerà and Amaral, 2004; Barrat, Barthélemy and Vespignani, 2005).

1.3.3 Mixing patterns and degree correlations

As a discriminator of structural ordering of large-scale networks, the attention of the research community has initially been focused on the degree distribution, but it is clear that this function is only one of the many statistics characterizing the structural and hierarchical ordering of a network. In particular, it is likely that nodes do not connect to each other irrespective of their property or type. On the contrary, in many cases it is possible to collect empirical evidence of specific mixing patterns in networks. A typical pattern known in ecology, epidemiology, and social science as "assortative mixing" refers to the tendency of nodes to connect to other nodes with similar properties. This is common to observe in the social context where people prefer to associate with others who share their interests. Interesting observations about assortative mixing by language or race are abundant in the literature. Likewise, it is possible to define a "disassortative mixing" pattern whenever the elements of the network prefer to share edges with those who have a different property or attribute. Mixing patterns have a profound effect on the topological properties of a network as they affect the community formation and the detailed structural arrangements of the connections among nodes.

While mixing patterns can be defined with respect to any type or property of the nodes (Newman, 2003a), in the case of large-scale networks the research community's interest has focused on the mixing by vertex degree. This type of mixing refers to the likelihood that nodes with a given degree connect with nodes with similar degree, and is investigated through the detailed study of multipoint degree correlation functions. Most real networks do exhibit the presence of non-trivial correlations in their degree connectivity patterns. Empirical measurements provide evidence that high or low degree vertices of the network tend, in many cases, to preferentially connect to other vertices with similar degree. In this situation, correlations are named assortative. In contrast, connections in many technological and biological networks are more likely to attach vertices of very different degree. Correlations are then referred to as disassortative. The correlations, although other possibilities could be considered, are characterized through the conditional probabilities $P(k', k'', \ldots, k'^{(n)} \mid k)$ that a vertex of degree k is simultaneously connected to a number n of other vertices with corresponding degrees $k', k'', \ldots, k'^{(n)}$. Such quantities might be the simplest theoretical functions that encode degree correlation information from a local perspective. A network is said to be uncorrelated when the conditional probability is structureless, in which case the only relevant function is just the degree distribution $P(k)$.

In order to characterize correlations, a more compact quantity is given by the Pearson assortativity coefficient r (Newman, 2002a)

$$r = \frac{\sum_e j_e k_e / E - \left[\sum_e (j_e + k_e)/(2E)\right]^2}{\left[\sum_e (j_e^2 + k_e^2)/(2E)\right] - \left[\sum_e (j_e + k_e)/(2E)\right]^2}, \tag{1.20}$$

where j_e and k_e denote the degree of the extremities of edge e and E is the total number of edges. This quantity varies from -1 (disassortative network) up to 1 (perfectly assortative network). However, such a measure can be misleading when a complicated behavior of the correlation functions (non-monotonous behavior) is observed. In this case the Pearson coefficient gives a larger weight to the more abundant degree classes, which in many cases might not express the variations of the correlation function behavior.

More details on the degree correlations are provided by the two-point conditional probability $P(k' \mid k)$ that any edge emitted by a vertex with degree k is connected to a vertex with degree k'. Even in this simple case, however, the direct evaluation of the function from empirical data is a rather cumbersome task. In general, two nodes' degree correlations can be represented as the three-dimensional histograms of $P(k' \mid k)$ or related quantities[4] (Maslov and Sneppen, 2002). On the other hand, such a histogram is highly affected by statistical fluctuations and, thus, it is not a good candidate when the data set is not extremely large and accurate. A more practical quantity in the study of the network structure is given by the average nearest neighbors degree of a vertex i

$$k_{\mathrm{nn},i} = \frac{1}{k_i} \sum_{j \in \mathcal{V}(i)} k_j, \qquad (1.21)$$

where the sum is over the nearest neighbors vertices of i. From this quantity a convenient measure to investigate the behavior of the degree correlation function is obtained by the average degree of the nearest neighbors, $k_{\mathrm{nn}}(k)$, for vertices of degree k (Pastor-Satorras, Vázquez and Vespignani, 2001; Vázquez, Pastor-Satorras and Vespignani, 2002)

$$k_{\mathrm{nn}}(k) = \frac{1}{N_k} \sum_{i/k_i=k} k_{\mathrm{nn},i}, \qquad (1.22)$$

where N_k is the number of nodes of degree k. This last quantity is related to the correlations between the degrees of connected vertices since on average it can be expressed as

$$k_{\mathrm{nn}}(k) = \sum_{k'} k' P(k'|k). \qquad (1.23)$$

If degrees of neighboring vertices are uncorrelated, $P(k'|k)$ is only a function of k' and thus $k_{\mathrm{nn}}(k)$ is a constant. In the presence of correlations, the behavior of $k_{\mathrm{nn}}(k)$ identifies two general classes of networks (see Figure 1.4). If $k_{\mathrm{nn}}(k)$ is an increasing function of k, vertices with high degree have a larger probability of being connected

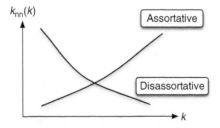

Fig. 1.4. Pictorial representation of the assortative (disassortative) mixing property of networks as indicated by the behavior of the average degree of the nearest neighbors, $k_{\mathrm{nn}}(k)$, for vertices of degree k.

---

[4] For instance, the joint degree distribution $P(k', k)$ that defines the probability that a randomly chosen edge connects two vertices of degrees k and k'.

with large degree vertices. This corresponds to an assortative mixing (Newman, 2002a). On the contrary, a decreasing behavior of $k_{nn}(k)$ defines a disassortative mixing, in the sense that high degree vertices have a majority of neighbors with low degree, while the opposite holds for low degree vertices (Newman, 2002a).

It is important to stress, however, that given a certain degree distribution, a completely degree–degree uncorrelated network with finite size is not always realizable owing to structural constraints. Indeed, any finite-size random network presents a structural cut-off value k_{str} over which the requirement of the lack of dangling edges introduces the presence of multiple and self-connections and/or degree–degree correlations (Boguñá, Pastor-Satorras and Vespignani, 2004; Moreira, Andrade and Amaral, 2002). Networks with bounded degree distributions and finite second moments $\langle k^2 \rangle$ present a maximal degree k_c that is below the structural one k_{str}. However, in networks with heavy-tailed degree distribution (see Section 2.1), this is not always the case and k_{str} is generally smaller than k_c. In this instance, structural degree–degree correlations and higher order effects, such as the emergence of large cliques (Bianconi and Marsili, 2006a; Bianconi and Marsili, 2006b), are present even in maximally random networks.[5] Structural correlations are genuine correlations, which is not surprising since they are just imposed by topological constraints and not by a special ordering or evolutionary principle shaping the network. A more detailed discussion on topological constraints and the properties of random networks can be found in Appendix 1.

In the case of random uncorrelated networks, it is possible to obtain an explicit form for the conditional probability $P(k' \mid k)$. In this case $P(k' \mid k)$ does not depend on k and its functional form in terms of k' can be easily obtained by calculating the probability that any given edge is pointing to a node of degree k'. The probability that one edge is wired to one node of degree k' is just the total number of edges departing from nodes of degree k' divided by the number of all edges departing from nodes of any degree. Since each one of the $N_{k'}$ nodes emanates k' edges, we obtain

$$P_{unc}(k' \mid k) = \frac{k' N_{k'}}{\sum_{k''} k'' N_{k''}}, \tag{1.24}$$

where considering that $P(k) = N_k/N$, finally yields

$$P_{unc}(k' \mid k) = \frac{1}{\langle k \rangle} k' P(k'). \tag{1.25}$$

This expression states that even in an uncorrelated network, the probability that any edge points to a node of a given degree k' is not uniform but proportional to

---

[5] Operatively, the maximally random network can be thought of as the stationary ensemble of networks visited by a process that, at any time step, randomly selects a couple of links of the original network and exchanges two of their ending points (automatically preserving the degree distribution).

the degree itself. In other words, by following any edge at random it is more likely you will end up in a node with large degree: the more connected you are, the easier it is to find you. This result is particularly relevant in the case of networks where degree values may be very different, and will prove to be particularly useful in several analytical calculations where the heterogeneous nature of the network is affecting equilibrium and non-equilibrium processes even in the uncorrelated case. The behavior of $k_{nn}(k)$ in the uncorrelated case can be easily derived from Equation (1.25) and reads as

$$k_{nn}^{unc}(k) = \frac{\langle k^2 \rangle}{\langle k \rangle}. \tag{1.26}$$

As we stated previously, in a random uncorrelated network the average nearest neighbor degree does not depend on k and has a constant value determined by the two first moments of the degree distribution.

1.3.4 Clustering spectrum

Correlations among three vertices can be measured by means of the probability $P(k', k'' \mid k)$ that a vertex of degree k is simultaneously connected to two vertices with degrees k' and k''. As previously indicated, the conditional probabilities $P(k', k'' \mid k)$ or $P(k'' \mid k)$ are difficult to estimate directly from real data, so other assessments have been proposed. All of them are based on the concept of clustering coefficient $C(i)$ as expressed in Equation (1.9), which refers to the tendency to form triangles in the network (see Section 1.2.4). A measure customarily used in graph characterization is the average clustering coefficient $\langle C \rangle = N^{-1} \sum_i C(i)$ which expresses the statistical level of cohesiveness measuring the global density of interconnected vertex triplets in the network. Although statistical scalar measures are helpful as a first indication of clustering, it is always more informative to work with quantities that explicitly depend on the degree. As in the case of two vertices' correlations, a uniparametric function is defined by the average clustering coefficient $C(k)$ restricted to classes of vertices with degree k (Vázquez *et al.*, 2002; Ravasz *et al.*, 2002)

$$C(k) = \frac{1}{N_k} \sum_{i/k_i=k} C(i) \tag{1.27}$$

where N_k is the number of vertices with degree k. A functional dependence of the local clustering on the degree can be attributed to the presence of a complex structure in the three-vertex correlation pattern. Indeed, it has been observed that $C(k)$ exhibits, in many cases, a non-trivial dependence on k that is supposed to partly encode the hierarchical structure of the network (Vázquez *et al.*, 2002; Ravasz *et al.*, 2002).

1.3.5 Rich-club phenomenon

Several other statistical measures have been defined in the case of large-scale networks as simple proxies for their architectures and many of them are specifically devised for certain types of graphs, as seen in the density of bipartite cliques $K_{n,m}$ in directed graphs. Analogously, the "rich-club" phenomenon has been discussed in both social and computer sciences (de Solla Price, 1986; Wasserman and Faust, 1994; Zhou and Mondragon, 2004; Pastor-Satorras and Vespignani, 2004), and refers to the tendency of high degree nodes, the hubs of the network (the *rich nodes*), to be very well connected to each other, forming well-interconnected subgraphs (*clubs*) more easily than low degree nodes. Zhou and Mondragon (2004) have introduced a quantitative measure of this tendency through the rich-club coefficient, expressed as

$$\phi(k) = \frac{2E_{>k}}{N_{>k}(N_{>k} - 1)}, \qquad (1.28)$$

where $E_{>k}$ is the number of edges among the $N_{>k}$ nodes with degree larger than k, and $N_{>k}(N_{>k} - 1)/2$ represents the maximum possible number of edges among these $N_{>k}$ nodes, so that $\phi(k)$ measures the fraction of edges actually connecting those nodes out of the maximum number of edges they might possibly share. A growing behavior as a function of k indicates that high degree nodes tend to be increasingly connected among themselves.[6] On the other hand, a monotonic increase of $\phi(k)$ does not necessarily imply the presence of a rich-club organizing principle. Indeed, even in the case of the Erdős–Rényi graph – a completely random network – there is an increasing rich-club coefficient. This implies that the increase of $\phi(k)$ is a natural consequence of the fact that vertices with large degree have a larger probability of sharing edges than low degree vertices. This feature is therefore imposed by construction and does not represent a signature of any particular organizing principle or structure. For this reason, it is appropriate to measure the rich-club phenomenon as $\rho_{ran}(k) = \phi(k)/\phi_{ran}(k)$, where $\phi_{ran}(k)$ is the rich-club coefficient of the maximally random network with the same degree distribution $P(k)$ as the network under study (Colizza *et al.*, 2006b). In this case an actual rich-club ordering is denoted by a ratio $\rho_{ran}(k) > 1$. In other words, $\rho_{ran}(k)$ is a

---

[6] It is also worth stressing that the rich-club phenomenon is not trivially related to the mixing properties of networks described in Section 1.3.3, which permit the distinction between assortative networks, where large degree nodes preferentially attach to large degree nodes, and disassortative networks, showing the opposite tendency (Colizza *et al.*, 2006b).

normalized measure which discounts the structural correlations due to unavoidable connectivity constraints, providing a better discrimination of the actual presence of the rich-club phenomenon due to the ordering principles shaping the network. This example makes explicit the need to consider the appropriate null hypotheses when measuring correlations or statistical properties. Such properties might either be simply inherent to the connectivity constraints present in networks, or be the signature of real ordering principles and structural properties due to other reasons.

1.4 Weighted networks

Along with a complex topological structure, real networks display a large hetero-geneity in the capacity and intensity of their connections: the weights of the edges. While the topological properties of a graph are encoded in the adjacency matrix x_{ij}, weighted networks are similarly described by a matrix w_{ij} specifying the weight on the edge connecting the vertices i and j ($w_{ij} = 0$ if the nodes i and j are not connected). The weight w_{ij} may assume any value and usually represents a phys-ical property of the edge: capacity, bandwidth, traffic. A very significant measure of a network's properties in terms of the actual weights is also obtained by looking at the vertex *strength* s_i defined as (Yook *et al.*, 2001; Barrat *et al.* 2004a)

$$s_i = \sum_{j \in \mathcal{V}(i)} w_{ij}, \tag{1.29}$$

where the sum runs over the set $\mathcal{V}(i)$ of neighbors of i. The strength of a node integrates the information about its degree and the importance of the weights of its links and can be considered as the natural generalization of the degree. When the weights are independent of the topology, the strength typically grows linearly with the degree, i.e. with the number of terms in the sum (1.29): $s \simeq \langle w \rangle k$ where $\langle w \rangle$ is the average weight. In the presence of correlations we obtain in general $s \simeq A k^\beta$ with $\beta = 1$ and $A \neq \langle w \rangle$, or $\beta > 1$. Statistical measures for weighted networks are readily provided by the probability distributions $P(w)$ and $P(s)$ that any given edge and node have weight w and strength s, respectively.

In general, topological measures do not take into account that some edges are more important than others. This can easily be understood with the simple exam-ple of a network in which the weights of all edges forming triples of interconnected vertices are extremely small. Even for a large clustering coefficient, it is clear that these triples have a minor role in the network's dynamics and organization, and the network's clustering properties are definitely overestimated by a simple topo-logical analysis. Similarly, high degree vertices could be connected to a majority of low degree vertices while concentrating the largest fraction of their strength only on the vertices with high degree. In this case the topological information would point to disassortative properties while the network could be considered as effectively assortative, since the more relevant edges in terms of weight are linking high degree vertices. In order to solve these incongruities, it is possible to introduce specific definitions that explicitly consider the weights of the links and combine the topological information with the weight distribution of the network. Many different clustering coefficient definitions have been introduced in the literature (Barrat *et al.*, 2004; Onnela *et al.*, 2005; Serrano, Boguñá and Pastor-Satorras, 2006; Saramäki *et al.*, 2007). A convenient *weighted clustering coefficient* is defined as

$$C^{\mathrm{w}}(i) = \frac{1}{s_i(k_i - 1)} \sum_{j,h} \frac{(w_{ij} + w_{ih})}{2} x_{ij} x_{ih} x_{jh}. \tag{1.30}$$

The quantity $C^{\mathrm{w}}(i)$ is a count of the weight of the two participating edges of the vertex i for each triple formed in the neighborhood of i. This definition not only considers the number of closed triangles in the neighborhood of a vertex but also considers their total relative edge weights with respect to the vertex's strength. The factor $s_i(k_i - 1)$ is a normalization factor that ensures that $0 \leq C^{\mathrm{w}}(i) \leq 1$. Consistently, the $C^{\mathrm{w}}(i)$ definition recovers the topological clustering coefficient in the case that $w_{ij} = \text{const}$. It is customary to define C^{w} and $C^{\mathrm{w}}(k)$ as the weighted clustering coefficient averaged over all vertices of the network and over all vertices with degree k, respectively. For a large randomized network (without any correlations between weights and topology), it is easy to see that $C^{\mathrm{w}} = C$ and $C^{\mathrm{w}}(k) = C(k)$. In real weighted networks, however, we can face two opposite cases. If $C^{\mathrm{w}} > C$, we observe a network in which the interconnected triples are more likely formed by edges with larger weights (see Figure 1.5). In contrast, $C^{\mathrm{w}} < C$ signals a network in which the topological clustering is generated by edges with low weight. In the latter, it is explicit that the clustering has a minor effect in the organization of the network since the largest part of the interactions (traffic, frequency of the relations, etc...) occurs on edges not belonging to interconnected triples. In order to obtain a more detailed knowledge of the structure of the network, the variations of $C^{w}(k)$ with respect to the degree class k can be analyzed and compared with those of $C(k)$.

Similarly, the *weighted average nearest neighbors degree* is defined as (Barrat *et al.*, 2004a)

$$k^{\mathrm{w}}_{\mathrm{nn},i} = \frac{1}{s_i} \sum_{j=1}^{N} x_{ij} w_{ij} k_j. \tag{1.31}$$

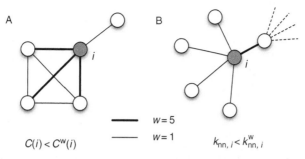

$C(i) < C^{w}(i)$ $w=5$ ⎯⎯ $w=1$ ⎯⎯ $k_{nn,\,i} < k^{w}_{nn,\,i}$

Fig. 1.5. A, The weighted clustering of a node i is larger than its topological coun-
terpart if the weights of the links (i, j) are concentrated on the cliques in which i
participates. B, Example of a node i with small average nearest neighbors degree
but large *weighted* average nearest neighbors degree: i is mostly connected to
low-degree nodes but the link with largest weight points towards a well-connected
hub.

This quantity (see also Serrano *et al.*, 2006) is a simple generalization of the
average nearest neighbors degree (Pastor-Satorras *et al.*, 2001)

$$k_{nn,i} = \frac{1}{k_i} \sum_{j=1}^{N} x_{ij} k_j \qquad (1.32)$$

and performs a local weighted average of the nearest neighbor degree according to
the normalized weight of the connecting edges, w_{ij}/s_i. This definition implies that
$k^{w}_{nn,i} > k_{nn,i}$ if the edges with the larger weights point to the neighbors with larger
degree and $k^{w}_{nn,i} < k_{nn,i}$ for the opposite (see Figure 1.5). Thus, $k^{w}_{nn,i}$ measures
the effective *affinity* to connect with high or low degree neighbors according to
the magnitude of the actual interactions. Also, the behavior of the function $k^{w}_{nn}(k)$
(defined as the average of $k^{w}_{nn,i}$ over all vertices with degree k) marks the weighted
assortative or disassortative properties considering the actual interactions among
the system's elements.

A final note should be made concerning the local heterogeneities introduced
by weights. The strength of a node i is the sum of the weights of all links in
which i participates. The same strength can, however, be obtained with very different configurations: the weights w_{ij} may be either of the same order s_i/k_i or
heterogeneously distributed among the links. For example, the most heterogeneous situation is obtained when one weight dominates over all the others. A
simple way to measure this "disparity" is given by the quantity Y_2 introduced in
other contexts (Herfindal, 1959; Hirschman, 1964; Derrida and Flyvbjerg, 1987;
Barthélemy, Gondran and Guichard, 2003).

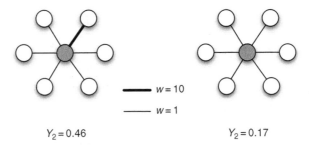

$Y_2 = 0.46$ $Y_2 = 0.17$

Fig. 1.6. If there is a small number of important connections around a node, the quantity Y_2 is of order $1/m$ with m of order unity. In contrast, if all the connections are of the same order, Y_2 is small and of order $1/k$ where k is the node's degree.

$$Y_2(i) = \sum_{j \in \mathcal{V}(i)} \left[\frac{w_{ij}}{s_i} \right]^2. \tag{1.33}$$

If all weights are of the same order then $Y_2 \sim 1/k_i$ (for $k_i \gg 1$) and if a small number of weights dominate then Y_2 is of the order $1/m$ with m of order unity (see Figure 1.6).

A similar finding is encoded in the local entropy, defined for nodes of degree larger than 2 as

$$f(i) = -\frac{1}{\ln k_i} \sum_{j \in \mathcal{V}(i)} \frac{w_{ij}}{s_i} \ln \left[\frac{w_{ij}}{s_i} \right]. \tag{1.34}$$

This quantity goes from 0 if the strength of i is fully concentrated on one link to the maximal value 1 for homogeneous weights: it can thus be used as an alternative or complement to the disparity Y_2 to investigate the local heterogeneity of the weights.

It is important to stress regarding weighted networks that we have emphasized some of the measures that are customarily used to analyze large-scale complex systems. In particular, most of them have been adopted in the context of the recent analysis of infrastructure and communication networks, and will be used to evaluate the effect of heterogeneities and complexity in dynamical phenomena affected by the weighted nature of the connections. Several other quantities related to weighted properties are, however, defined and used in graph theory. In particular, weighted distances between two nodes are defined by assigning an effective length (depending on the weight) to each edge and summing these lengths along the path followed. The minimum spanning tree corresponds to the way of connecting all the vertices together with the smallest possible global weight. In standard graph theory, mostly *flows* have been studied, which represent a particular case of weighted networks where the weights must satisfy the conservation equation at each node, stating that the total amount of flow into a node is equal to the amount of flow going out of it. Although this problem is very important in many instances (such as electrical networks, fluids in pipes, etc.), the conservation equation is not satisfied in all real-world networks and we refer the reader interested in flow problems (such as the maximum flow) to graph theory books (Clark and Holton, 1991; Cormen *et al.*, 2003; Jungnickel, 2004).

2

Networks and complexity

Undeniably, the visualizations of the Internet or the airport network convey the notion of intricate, in some cases haphazard, systems of a very complicated nature. *Complexity*, however, is not the same as the addition of complicated features. Despite the fact that there is no unique and commonly accepted definition of complexity – it is indeed very unlikely to find two scientists sharing the same definition of *complex system* – we discuss from what perspectives many real-world networks can be considered as complex systems, and what are the peculiar features signaling this occurrence. In this chapter we review the basic topological and dynamical features that characterize real-world networks and we attempt to categorize networks into a few broad classes according to their observed statistical properties. In particular, self-organized dynamical evolution and the emergence of the small-world and scale-free properties of many networks are discussed as prominent concepts which have led to a paradigm shift in which the dynamics of networks have become a central issue in their characterization as well as in their modeling (which will be discussed in the next chapter). We do not aim, however, at an exhaustive exposition of the theory and modeling of complex networked structures since, as of today, there are reference textbooks on these topics such as those by Dorogovtsev and Mendes (2003), Pastor-Satorras and Vespignani (2004), and Caldarelli (2007), along with journal reviews by Albert and Barabási (2002), Dorogovtsev and Mendes (2002), Newman (2003c), and Boccaletti *et al.* ([2006]).

2.1 Real-world systems

In recent times, the increased power of computers and the informatics revolution have made possible the systematic gathering and handling of data sets on several large-scale networks, allowing detailed analysis of their structural and functional properties. As a first guide to the classification of the data obtained, we can provide a rudimentary taxonomy of real-world networks. Two main different classes are infrastructure systems, and natural or living systems. Each of these classes can be further divided into different subgroups. Natural systems networks can be differentiated into the subgroups of biology, social systems, food webs and ecosystems, among others. For instance, biological networks refer to the complicated sets of interactions among genes, proteins and molecular processes which regulate biological life, while social networks describe the relations between individuals such as family links, friendships, and work relations. In particular, leading sociologists refer to our societies as networked societies and even if their analysis has been largely focused on small and medium-scale networks, social network studies play a key role in introducing basic definitions and quantities of network science. Indeed, biological and social networks are prominent research topics and much of network science is influenced by what has happened in these fields recently.

In turning our attention to infrastructure networks we can readily separate two main subcategories. The first contains virtual or cyber networks. These networks exist and operate in the digital world of cyberspace. The second subcategory includes physical systems such as energy and transportation networks. This is a rough classification since there are many interrelations and interdependencies existing among physical infrastructure networks, as well as between physical and digital networks. In the Internet, for instance, the cyber features are mixed with the physical features. The physical Internet is composed of physical objects such as routers – the main computers which allow us to communicate – and transmission lines, the cables which connect the various computers. On top of this physical layer lies a virtual world made of software that may define different networks, such as the World Wide Web (WWW), email networks, and Peer-to-Peer networks. These networks are the information transfer channels for hundreds of millions of users and, like the physical Internet, have grown to become enormous and intricate networks as the result of a self-organized growing process. Their dynamics are the outcomes of the interactions among the many individuals forming the various communities, and therefore are mixtures of complex socio-technical aspects. Infrastructure networks represent a combination of social, economic, and technological processes. Further examples can be found in the worldwide airport network (WAN) and power distribution networks where physical and technological constraints cooperate with social, demographic, and economic factors.

2.1.1 Networks everywhere

The various systems mentioned so far are characterized by the very different nature of their constitutive elements. It is therefore appropriate to offer a list of some network data sets prototypically considered in the literature, making clear to which property and elements their graph representation refers.

Social networks

The science of social networks is one of the pillars upon which the whole field of network science has been built. Since the early works of Moreno (1934) and the definition of the sociogram, social networks have been the object of constant analysis and study. Social networks represent the individuals of the population as nodes and the social ties or relations among individuals as links between these nodes. The links therefore may refer to very different attributes such as friendship among classmates, sexual relations among adults, or just the belonging to common institutions or work teams (collaborative interactions). The importance of these networks goes beyond social sciences and affects our understanding of a variety of processes ranging from the spreading of sexually transmitted diseases to the emergence of consensus and knowledge diffusion in different kinds of organizations and social structures.

Recently, the recording of social interactions and data in electronic format has made available data sets of unprecedented size. The e-mail exchanges in large corporate organizations and academic institutions make tracking social interactions among thousands of individuals possible in a precise and quantitative way (Ebel, Mielsch and Bornholdt, 2002; Newman, Forrest and Balthrop, 2002). Habits and shared interests may be inferred from web visits and file sharing. Professional communities have been analyzed from wide databases such as complex collaboration networks. Examples are the already classic collaboration network of film actors (Watts and Strogatz, 1998; Barabási and Albert 1999; Amaral *et al.*, 2000; Ramasco, Dorogovtsev and Pastor-Satorras, 2004), the company directors network (Newman, Strogatz and Watts, 2001; Davis, Yoo and Baker, 2003; Battiston and Catanzaro, 2004) and the network of co-authorship among scientists (Newman, 2001a; 2001b; 2001c). In these last examples we encounter bipartite networks (Wasserman and Faust, 1994), although the one mode projection is often used so that members are tied through common participation in one or more films, boards of directors, or academic papers. As an example, we report in Figure 2.1 an illustration of a construction of the scientific collaboration network (SCN). According to this construction, the authors are the nodes of the network and share an edge if they co-authored a paper. Similarly, for the actors' collaboration network, two actors are linked if they have co-starred in a movie. More information can be projected in the graph by weighting the intensity of the collaboration. A convenient definition of the weight is introduced by Newman (2001b), who considers that the intensity w_{ij} of the interaction between two collaborators i and j is given by

$$w_{ij} = \sum_p \frac{\delta_i^p \delta_j^p}{n_p - 1},$$
(2.1)

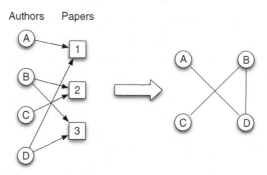

Fig. 2.1. A bipartite graph is defined as a graph whose nodes can be divided into two separate sets or classes such that every edge links a node of one set to a node of the other set. For a co-authorship network, authors and papers form two distinct classes and the edges representing the authorship just go from authors to papers. The one mode projection defines a unipartite graph, where the nodes of one class share an edge if they are connected to a common node of the other class. The one mode projection on the author set defines the co-authorship network in which authors share an edge if they co-authored at least one paper.

where the index p runs over all co-authored papers, n_p is the number of authors of the paper p, and δ_i^p is 1 if author i has contributed to paper p, and 0 otherwise. The strength of the interactions is therefore large for collaborators having many papers in common but the contribution to the weight introduced by any given paper is inversely proportional to the number of authors. Similarly, other definitions of weight may be introduced in order to measure the impact of collaborations in the scientific community. For instance, Börner *et al.* (2004) consider that the weight of each edge is also a function of the number of citations of each paper, providing a quantitative assessment of the impact of co-authorship teams. This is an example of social science merging with bibliometrics, another discipline which has recently benefited from the impact of the e-revolution and text digitalization on gathering large-scale network data sets. Just to cite a paramount example, the networks of citations (there is a directed link between two papers if one cites the other) among scientific papers of several databases for journals such as the *Proceedings of the National Academy of Sciences* (*PNAS*) or *Physical Review* contain thousands of nodes (Redner, 2005).

Transportation networks

Transportation systems such as roads, highways, rails, or airlines are crucial in our modern societies and will become even more important in a world where more than 50% of the population lives in urban areas.[1] The importance of this subject

---

[1] Source: United Nations, population division http://www.unpopulation.org

justifies the huge literature published on these systems for at least the past 70 years. Studies cover a broad range of approaches from applied engineering to mathematic works on users' equilibrium (for an introductory book on the subject, see Sheffi [1985]). Geographers, regional science experts, and economists have long observed the existence of structures such as hierarchies or particular subgraphs present in the topology (see for example Yerra and Levinson [2005] and references therein). In the air traffic network, for instance, O'Kelly (1998) has discussed the existence of hubs and of particular motifs. Hierarchies present in a system of cities were also noticed a long time ago, and Christaller (1966) proposed his famous central place theory in which heterogeneous distributions of facilities and transportation costs are the cause of the emergence of hierarchies. More recently, Fujita, Krugman and Venables (1999) have proposed a model which explains the hierarchical formation of cities in terms of decentralized market processes. These different models are, however, very theoretical, and more thorough comparison with empirical data is needed at this stage. Only recently, large-scale data and extensive statistics have opened the path to better characterization of these networks and their interaction with economic and geographic features. In this context, the network representation operates at different scales and different representative granularity levels are considered.

The TRANSIMS project characterizes human flows at the urban level (Chowell *et al.*, 2003; Eubank *et al.*, 2004). This study concerns the network of locations in the city of Portland, Oregon, where nodes are city locations including homes, offices, shops, and recreational areas. The edges among locations represent the flow of individuals going at a certain time from a location to another. In Figure 2.2 we illustrate the network construction and we report an example of the type of data set used by Chowell, Hyman, Eubank and Castillo-Chavez (2003). Other relevant networks studied at this scale are subways (Latora and Marchiori, 2002), and roads and city streets (Cardillo *et al.*, 2006; Buhl *et al.*, 2006; Crucitti, Latora and Porta, 2006; Scellato *et al.*, 2006; Kalapala *et al.*, 2006).

At a slightly coarser scale, it is possible to map into a weighted network the commuting patterns among urban areas, municipalities, and counties (Montis *et al.*, 2007), as well as railway systems (Sen *et al.*, 2003). Finally, on the global scale, several studies have provided extensive analysis of the complete worldwide airport network and the relative traffic flows (Barrat *et al.*, 2004a; Guimerà and Amaral, 2004; Guimerà *et al.*, 2005): this transportation system can be represented as a weighted graph where vertices denote airports, and weighted edges account for the passenger flows between the airports. The graphs resulting from large-scale transportation networks define the basic structure of metapopulation or patch models used in studies of epidemic spread. In this representation the

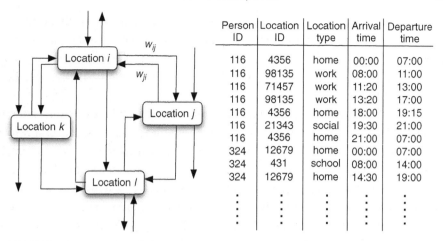

Person ID	Location ID	Location type	Arrival time	Departure time
116	4356	home	00:00	07:00
116	98135	work	08:00	11:00
116	71457	work	11:20	13:00
116	98135	work	13:20	17:00
116	4356	home	18:00	19:15
116	21343	social	19:30	21:00
116	4356	home	21:00	07:00
324	12679	home	00:00	07:00
324	431	school	08:00	14:00
324	12679	home	14:30	19:00
⋮	⋮	⋮	⋮	⋮

Fig. 2.2. Construction of the network between the various locations in the city of Portland as reported by Chowell *et al.* (2003). The table is a small sample of the data set, obtained through a TRANSIMS simulation, which records for each individual the successive locations visited during a day, with the arrival and departure time. The network between locations is constructed from this data set: an edge exists between location i and j whenever an individual has gone directly from i to j, and the weight w_{ij} gives the daily traffic between i and j.

network's nodes refer to different populations such as urban areas or cities, and an edge between two nodes denotes an exchange of individuals between the two populations (see also Chapter 9). These networks are usually directed and weighted in that the number of individuals going from one population to the other is not necessarily symmetric and the weight of each edge quantifies the number of exchanged individuals. In general, transportation networks naturally define such networked structures, with the edges' weights denoting the traffic flows (commuters, passengers etc.) between the different locations, and network theory appears as the natural tool for the study of these systems.

Internet

Characterizing how routers, computers, and physical links interconnect with each other in the global Internet is very difficult because of several key features of the network. One main problem is related to its exponential growth rate. The Internet, in fact, has already increased by five orders of magnitude since its birth. A second difficulty is the intrinsic *heterogeneity* of the Internet which is composed of networks engineered with considerable technical and administrative diversity. The different networking technologies are merged together by the TCP/IP architecture which provides connectivity but does not imply a uniform behavior. Also very important is the fact that the Internet is a *self-organizing* system whose properties cannot be traced back to any global blueprint or chart. This means that routers and links are added by competing entities according to local economic and technical constraints, leading to a very intricate physical structure that does not comply with any globally optimized plan (Pastor-Satorras and Vespignani, 2004; Crovella and Krishnamurthy, 2006).

The combination of all these factors results in a general lack of understanding about the large-scale topology of the Internet, but in recent years, several research groups have started to deploy technologies and infrastructures in order to obtain a more global picture of this network. Efforts to obtain such maps have been focused essentially on two levels. First, the inference of router adjacencies amounts to a measure of the internet router (IR) level graph. The second mapping level concerns the Autonomous System (AS) graph of the Internet, referring to autonomously administered *domains* which to a first approximation correspond to internet service providers and organizations. Therefore, internet maps are usually viewed as undirected graphs in which vertices represent routers or Autonomous Systems and edges (links) represent the physical connections between them. Although these two graph representations are related, it is clear that they describe the Internet at rather different scales (see Figure 2.3). In fact, the collection of Autonomous Systems and inter-domain routing systems defines a coarse-grained picture of the Internet in which each AS groups many routers together, and links are the aggregations of all the individual connections between the routers of the corresponding AS.

Internet connectivity information at the level of Autonomous Systems can be retrieved by the inspection of routing tables and paths stored in each router (passive measurements) or by direct exploration of the network with a software probe (active measurements). Based on the first strategy, the *Oregon Route Views* project (RV) provides maps of the AS graph obtained from the knowledge of the routing

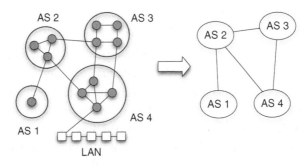

Fig. 2.3. Different granularity representations of the Internet. The hosts (squares) are included in the Local Area Networks (LAN) that connect to routers (shaded circles) and are excluded from the maps. Each Autonomous System is composed of many routers.

tables of several Border Gateway Protocol (BGP) peers.[2] This project has been running since 1997 and is one of the first projects to provide regular snapshots of the Internet's evolution. On the other hand, the most famous large infrastructure for active measurement has been implemented by the skitter project at CAIDA.[3] This project deployed several strategically placed probing monitors devoted to Internet mapping and measurement (Huffaker *et al.*, 2002b). All the data are then centrally collected and merged in order to obtain large-scale Internet maps that minimize measurement biases. A different active strategy, the Distributed Internet Measurements and Simulations (DIMES) project (Shavitt and Shir, 2005), considers a distributed measurement infrastructure, based on the deployment of thousands of lightweight measurement agents around the globe. Several other projects have focused attention on different levels such as the maps of specific Internet providers or Internet regions.

World Wide Web

The World Wide Web (WWW) is probably the most famous virtual network living on the physical structure of the Internet. It is nowadays considered as a critical infrastructure in the sense that it has acquired a central role in the everyday functioning of our society, from purchasing airplane tickets to business teleconferences. The Web is so successful that its rapid and unregulated growth has led to a huge and complex network for which it is extremely difficult to estimate the total number of web pages, if possible at all.

The experiments aimed at studying the Web's graph structure are based on Web *crawlers* which explore connectivity properties by following the links found on each page. In practice, crawlers are special programs that, starting from a source page, detect and store all the links they encounter, then follow them to build up a set of pages reachable from the starting one. This process is then repeated for all pages retrieved, obtaining a second layer of vertices and so on, iterating for as many possible layers as allowed by the available storage capacity and CPU time. From the collected data it is then possible to reconstruct a graph representation of the WWW by identifying vertices with web pages and edges with the connecting hyperlinks.

Large crawls of the WWW have been constantly gathered for several years because of the importance of indexing and storing web pages for search engine effectiveness. This data availability has made the WWW the starting place for the study of large-scale networks (Albert, Jeong and Barabási, 1999; Broder

---

[2] University of Oregon Route Views Project. http://www.routeviews.org/
[3] Cooperative Association for Internet Data Analysis; http://www.caida.org/tools/measurement/skitter/router_topology/

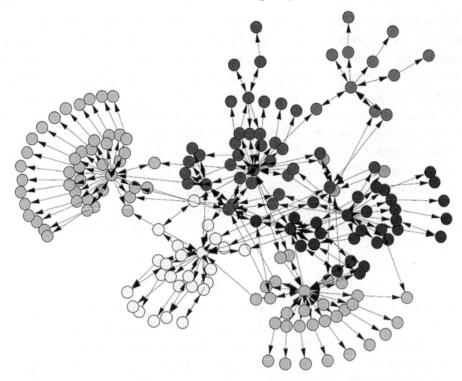

Fig. 2.4. Pages of a website and the directed hyperlinks between them. The graph represents 180 pages from the website of a large corporation. Different shades identify topics as identified with a community detection algorithm. Reprinted with permission from Newman and Girvan (2004). Copyright 2004 by the American Physical Society.

et al., 2000; Adamic and Huberman, 2001). The sheer size of the web graphs has also made possible the investigation of large-scale statistical properties and has led to the development of new measurement tools aimed at characterizing graphs with 10^7–10^8 vertices. In addition, the web graph is a prototypical *directed* graph, in which edges connect ordered pairs of vertices (see Chapter 1). Crawls count the number of each web page's outgoing hyperlinks, but in principle we know nothing about the incoming hyperlinks from other pages (see Figure 2.4). We can follow hyperlinks to reach pointed pages, but we cannot navigate backwards through the incoming hyperlinks. From this perspective, web graphs are often the gold standard used to address the discussion of paths, directionality, and component structure in large-scale directed networks.

Biological networks

It is fair to say that one of the reasons for the explosion of interest in networks lies in the possibility, due to high throughput experiments, of gathering large data collections on the interactions or relations of entire proteomes or genomes, also called the "omic" revolution. Networks now pervade the biological world at various levels ranging from the microscopic realm of biological chemistry, genetics, and proteomics to the large scale of food webs.

At the microscopic level, many relevant aspects of biological complexity are encapsulated in the structure and dynamics of the networks emerging at different organizational levels, ranging from intracellular biochemical pathways to genetic regulatory networks or protein interaction networks (Barabási and Oltvai, 2004; Alon, 2003). A prominent example in this area is provided by protein interaction networks (PIN) of various organisms which can be mathematically represented as graphs with nodes representing proteins and edges connecting pairs of interacting proteins (see Chapter 12). The importance of microscopic biological networks is clearly related to the biological significance of the network's topology, and analysis in this direction has indeed pointed out correlation signatures between gene knock-out lethality and the connectivity of the encoded protein, negative correlation between the evolution rate of a protein and its connectivity, and functional constraints in protein complexes.

At a larger scale, biological networks can describe individuals' interactions in various animal and human populations. In this area, biology may overlap with social science. A typical example is given by the network describing the web of sexual relations that is both of interest from a social point of view and of great concern in the epidemiology of sexually transmitted diseases (Liljeros *et al.*, 2001; Schneeberger *et al.*, 2004).

Finally, at the very large scale we find the networks describing the food web of entire ecosystems. Roughly speaking, food webs describe which species eat which other species. In this perspective, nodes represent species and links are antagonistic trophic interactions of the predator–prey type (see Dunne, Williams and Martinez [2002a]; Montoya, Pimm and Solé [2006] and Chapter 12).

2.1.2 Measurements and biases

In the previous section and throughout the rest of the book we have made a particular effort to discuss and cite some of the most recent network data sets available in the literature. Data gathering projects, however, are continuously making public new data on large-scale networks and very likely larger and more accurate samples than those mentioned here will be available by the time this book is published. In addition the list of systems for which it is possible to analyze network data is continually enlarging. Genomic and proteomic data of new organisms are constantly added to the various community data repositories. New and very large maps of the WWW are acquired along with a better knowledge of the physical Internet.

In addition, many other networks related to the cyberworld are explored, such as Peer-to-Peer or email networks. Data on large-scale infrastructures such as transportation, power-grid, freight and economic networks are also constantly enlarging our inventory of network data.

On the other hand, approaching network data requires great caution and a critical perspective informed about all the details of the data gathering process. Indeed, for many systems the acquisition of the complete network structure is impossible owing to time, resource, or technical constraints. In this case, network sampling techniques are applied to acquire the most reliable data set and minimize the error or biases introduced in the measurement. The aim is to obtain network samples that exhibit reliable statistical properties resembling those of the entire network. Network sampling and the relative discussion of statistical reliability of the data sets are therefore a major issue in the area of network theory, unfortunately not always carefully scrutinized (Willinger *et al.*, 2002).

Examples of sampling issues can be found in all of the areas discussed previously. Crawling strategies to gather WWW data rely on exhaustive searches by following hyperlinks. Internet exploration consists of sending probes along the computer connections and storing the physical paths of these probes. These techniques can be applied recursively or repeated from different vantage points in order to maximize the discovered portion of the network. In any case, it is impossible to know the actual fraction of elements sampled in the network (Broido and claffy, 2001; Barford *et al.*, 2001; Qian *et al.*, 2002; Huffaker *et al.*, 2000; Huffaker *et al.*, 2002a; 2002b). In biological networks, the sources of biases lie in the intrinsic experimental error that may lead to the presence of a false positive or negative on the presence of a node or edge. A typical example can be found in high throughput techniques in biological network measurements such as experiments for detecting protein interactions (Bader and Hogue, 2002; Deane *et al.*, 2002). For these reasons, a large number of model-based techniques, such as probabilistic sampling design (Frank, 2004) developed in statistics, provide guidance in the selection of the initial data sets. In addition, the explosion in data gathering has spurred several studies devoted to the bias contained in specific, large-scale sampling of information networks or biological experiments (Lakhina *et al.*, 2002; Petermann and De Los Rios, 2004a; Clauset and Moore, 2005; Dall'Asta *et al.*, 2005, 2006a; Viger *et al.*, 2007).

2.2 Network classes

The extreme differences in the nature of the elements forming the networks considered so far might lead to the conclusion that they share few, if any, features.

The only clear commonalities are found in the seemingly intricate haphazard set of points and connections that their graph representations produce, and in their very large size. It is the latter characteristic that, making them amenable to large-scale statistical analysis, allowed the scientific community to uncover shared properties and ubiquitous patterns which can be expressed in clear mathematical terms.

2.2.1 Small-world yet clustered

Let us first consider the size of networks and the distance among elements. Even though graphs usually lack a metric, it is possible to define the distance between any pair of vertices as the number of edges traversed by the shortest connecting path (see Chapter 1). A first general evidence points toward the so-called small-world phenomenon. This concept describes in simple words that it is possible to go from one vertex to any other in the system passing through a very small number of intermediate vertices. The small-world effect, sometimes referred to as the "six degrees of separation" phenomenon, has been popularized in the sociological context by Milgram (1967) who has shown that a short number of acquaintances (on average six) is enough to create a connection between any two people chosen at random in the United States. Since then, the small-world effect has been identified as a common feature in a wide array of networks, in particular infrastructure networks where the small average distance is crucially important to speed up communications. For instance, if the Internet had the shape of a regular grid, its characteristic distance would scale with the number of nodes N as $\langle \ell \rangle \sim N^{1/2}$; with the present Internet size, each Internet Protocol packet would pass through 10^3 or more routers, drastically depleting communication capabilities. The small-world property is therefore implicitly enforced in the network architecture which incorporates hubs and backbones connecting different regional networks, thus strongly decreasing the value of $\langle \ell \rangle$.

To be more precise, the small-world property refers to networks in which $\langle \ell \rangle$ scales logarithmically, or more slowly, with the number of vertices. In many cases, data on the same network at different sizes are not available. The small-world property is then measured by inspecting the behavior of the quantity $M(\ell)$ defined as the average number of nodes within a distance less than or equal to ℓ from any given vertex. While in regular lattices $M(\ell)$ is expected to increase as a power law with the distance ℓ, in small-world networks this quantity follows an exponential or faster increase. In Figure 2.5 we report this quantity for several prototypical networks. At first sight the small-world feature appears a very peculiar property. By itself, however, it is not the signature of any special organizing principle, and it finds its explanation in the basic evidence that randomness appears as a major

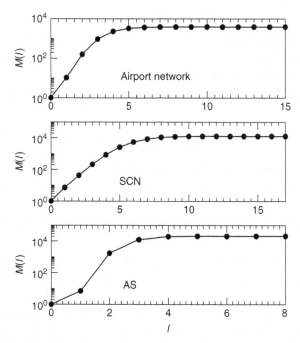

Fig. 2.5. Hop-plot for several prototypical networks. In all cases the number of nodes at distance l or less grows exponentially fast as shown by the behavior on a linear–log scale, before saturating because of the finite size of the network. The data sets considered are, from top to bottom: the worldwide airport network (data from the International Air Transport Association IATA, see Barrat *et al.* (2004a) and http://www.iata.org), the scientific collaboration network (SCN), see Newman (2001a; 2001b; 2001c) and http://www-personal.umich.edu/~mejn/netdata/, and the map of the Autonomous Systems of the Internet (AS) as obtained by the DIMES project (http://www.netdimes.org).

ingredient in the shaping of large-scale networks. Indeed, by looking at the intricacy and apparent absence of order of networks, the simplest assumption one can make is that the creation of a connection between two elements is a random event determined by the sum of a very large number of unpredictable circumstances. This is the guiding principle that defines the Erdős–Rényi model and the general formulation of random graph models and algorithms presented in Chapter 3. In brief, all of these models assume that the presence of an edge between two vertices is a random process occurring with the same probability, independent of the vertex characteristics. This simple construction suffices in yielding small-world graphs. It can be shown rigorously that the average shortest path distance between any two vertices in the graph increases just as the logarithm of the number N of vertices considered. In other words, the small-world feature can be explained with the simple inclusion of randomness. This readily explains the ubiquity of this property since in all natural systems we have to allow for the presence of some level of noise and stochastic behavior in the dynamics at the origin of the networks.

More interesting evidence is that, in many social and technological networks, the small-world effect goes along with a high level of clustering (Watts and Strogatz, 1998; Watts, 1999). The clustering coefficient characterizes the local cohesiveness of the networks as the tendency to form groups of interconnected elements (see Chapter 1). It is easy to perceive that a random graph model cannot achieve high clustering in that no organizing principle is driving the formation of groups of interconnected elements. Interconnections happen only by chance, randomness being the only force shaping the network, and therefore in large random graphs the clustering coefficient becomes very small (see Chapter 3). In other words, random graphs feature the small-world effect but are not clustered, while regular grids tend to be clustered but are not small-world. This means that the high clustering is a memory of a grid-like ordering arrangement that is not readily understandable in a purely random construction. This puzzle has been addressed by the famous small-world model of Watts and Strogatz (1998) (see Chapter 3) which is able to capture both properties at the same time.

2.2.2 Heterogeneity and heavy tails

The evidence for the presence of more subtle structural organizations in real networks is confirmed by the statistical analysis of centrality measures. The functional form of the statistical distributions characterizing large-scale networks defines two broad network classes. The first refers to the so-called statistically *homogeneous* networks. The distributions characterizing the degree, betweenness, and weighted quantities have functional forms with fast decaying or "light" tails such as Gaussian or Poisson distributions. The second class concerns networks with statistically *heterogeneous* connectivity and weight patterns usually corresponding to skewed and heavy-tailed distributions. In order to better understand the basic difference between these two classes, let us focus for the moment on the degree distribution $P(k)$. The evidence for the high level of heterogeneity of many networks is simply provided by the fact that many vertices have just a few connections, while a few hubs collect hundreds or even thousands of edges. For instance, this feature is easily seen in the airport networks (Figure 2.6), showing the "hub" policy that almost all airlines have adopted since deregulation in 1978. The same arrangement can easily be perceived in many other networks where the presence of "hubs" is a natural consequence of different factors such as popularity, strategies,

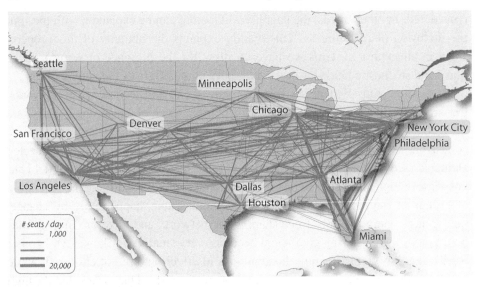

Fig. 2.6. Main air travel routes for the United States. Only the connections with more than 1000 available seats per day are shown. The presence of hubs is clearly visible. Figure courtesy of V. Colizza.

and optimization. For instance, in the WWW some pages become hugely popular and are pointed to by thousands of other pages, while in general most documents are almost unknown.

The presence of hubs and connectivity ordering turns out to have a more striking manifestation than initially thought, yielding in many cases a degree distribution $P(k)$ with heavy tails (Barabási and Albert, 1999). In Figure 2.7 we show the degree distribution resulting from the analysis of several real-world networks. In all cases the distribution is skewed and highly variable in the sense that degrees vary over a broad range, spanning several orders of magnitude. This behavior is very different from the case of the bell-shaped, exponentially decaying distributions and in several cases the heavy tail can be approximated by a power-law decay $P(k) \sim k^{-\gamma}$,[4] which results in a linear behavior on the double logarithmic scale. In distributions with heavy tails, vertices with degree much larger than the average $\langle k \rangle$ are found with a non-negligible probability. In other words, the average behavior of the system is not typical. In the distributions shown in Figure 2.7, the vertices will often have a small degree, but there is an appreciable probability of finding vertices with very large degree values. Yet all intermediate values are present and the average degree does not represent any special value for the distribution. This is

---

[4] Power-law distributions are in many cases also referred to as Pareto distributions.

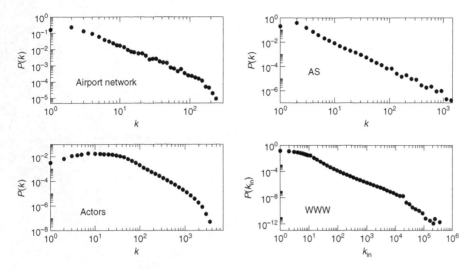

Fig. 2.7. Degree distribution $P(k)$ of real-world networks described in Section 2.1. Worldwide airport network (IATA data, see `http://www.iata.org/`); actors' collaboration network (Internet Movie database at `http://www.imdb.com/` and `http://www.nd.edu/~networks/resources.htm`); map of the Autonomous Systems of the Internet obtained by the DIMES mapping project (`http://www.netdimes.org`); map of the WWW collected in 2003 by the WebBase project (`http://dbpubs.stanford.edu:8091/~testbed/doc2/WebBase/`). For the WWW we report the distribution of the in-degree k_in. Data courtesy of M. A. Serrano.

in strong contrast to the democratic perspective representation offered by homogeneous networks with bell-shaped distributions and fast decaying tails. The average value here is very close to the maximum of the distribution which corresponds to the most probable value in the system. The contrast between these types of distributions is illustrated in Figure 2.8, where we compare a Poisson and a power-law distribution with the same average degree.

In more mathematical terms, the heavy-tail property translates to a very large level of degree fluctuations. The significance of the heterogeneity contained in heavy-tailed distributions can be understood by looking at the first two moments of the distribution. We can easily compute the average value that the degree assumes in the network as

$$\langle k \rangle = \int_m^\infty k P(k) \, \mathrm{d}k, \tag{2.2}$$

where $m \geq 1$ is the lowest possible degree in the network. Here for the sake of simplicity we consider that k is a continuous variable but the same results hold in the discrete case, where the integral is replaced by a discrete sum. By computing the above integral, it is possible to observe that in any distribution with a power-law

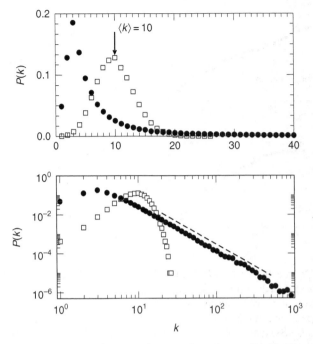

Fig. 2.8. Comparison of a Poisson and power-law degree distribution on a linear scale plot (top) and a double logarithmic plot (bottom). The two distributions have the same average degree $\langle k \rangle = 10$. The dashed line in the bottom figure corresponds to the power law $k^{-\gamma}$, where $\gamma = 2.3$.

tail with exponent $2 < \gamma \leq 3$, the average degree is well defined and bounded. On the other hand, a measure of the typical error we make if we assume that $\langle k \rangle$ is the typical degree value of a vertex is given by the normalized variance of the distribution $\sigma^2 / \langle k \rangle^2$ that expresses the statistical fluctuations present in our system. The variance $\sigma^2 = \langle k^2 \rangle - \langle k \rangle^2$ is dominated by the second moment of the distribution

$$\langle k^2 \rangle \sim \int k^2 P(k) \, dk \sim \int_m^{k_c} k^{2-\gamma} \, dk \sim k_c^{3-\gamma}. \qquad (2.3)$$

In the asymptotic limit of infinite network sizes, the cut-off k_c corresponding to the largest possible degree value diverges ($k_c \to \infty$) (see Appendix 1), so that $\langle k^2 \rangle \to \infty$: fluctuations are unbounded and depend on the system size.[5] The absence of any intrinsic scale for the fluctuations implies that the average value is not a characteristic scale for the system. In other words, we observe a *scale-free* network as far as the degree of the vertices is concerned. This reasoning can

---

[5] In many cases we will refer to the infinite size limit as the *thermodynamic limit*.

be extended to values of $\gamma \leq 2$, since in this case even the first moment is unbounded.

The absence of an intrinsic characteristic scale in a power-law distribution is also reflected in the self-similarity properties of such a distribution; i.e. it looks the same at all length scales. This means that if we look at the distribution of degrees by using a coarser scale in which $k \rightarrow \lambda k$, with λ representing a magnification/reduction factor, the distribution would still have the same form. This is not the case if a well-defined characteristic length is present in the system. From the previous discussion, it is also possible to provide a heuristic characterization of the level of heterogeneity of networks by defining the parameter

$$\kappa = \frac{\langle k^2 \rangle}{\langle k \rangle}. \tag{2.4}$$

Indeed, fluctuations are denoted by the normalized variance which can be expressed as $\kappa / \langle k \rangle - 1$, and scale-free networks are characterized by $\kappa \rightarrow \infty$, whereas homogeneous networks have $\kappa \sim \langle k \rangle$. For this reason, we will generally refer to scale-free networks as all networks with heterogeneity parameter $\kappa \gg \langle k \rangle$.[6] We will see in the following chapters that κ is a key parameter for all properties and physical processes that are affected by the degree fluctuations. It is also important to note that, in the case of uncorrelated networks, $\kappa = k_{nn}(k)$ (see Equation (1.26)), so that there is a link between the divergence of fluctuations and the average degree of nearest neighbors.

The evidence of heavy-tail distributions is not found only in the connectivity properties. In general, betweenness distributions are also broad and exhibit scale-free behavior. Even more striking is the evidence for the heavy-tail character of weight and strength distributions with values spanning up to eight or nine orders of magnitude. In Figures 2.9 and 2.10 we report some examples of the betweenness, weight, and strength distributions observed in real networks data. It is worth commenting at this point that several studies have been devoted to whether the observed distributions can or cannot be approximated by a power-law behavior. A thorough discussion of the issues found in the measurement and characterization of power-law behavior is contained in the extensive review by Newman (2005b). While the presence or absence of power-law functional form is a well-formulated statistical question, it is clear that we are now describing real-world systems, for which in most cases the statistical properties are affected by noise, finite size, and other truncation effects. The presence of such effects, however, should not be considered a surprise. For instance, the heavy-tail truncation is the natural effect of the upper limit of the distribution, which must be necessarily present in every real world system. Indeed, bounded power laws (power-law distributions with a cut-off)

---

[6] Obviously, in the real world κ cannot be infinite since it is limited by the network size.

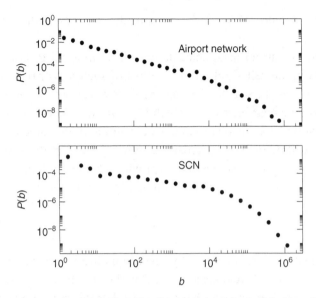

Fig. 2.9. Betweenness centrality distributions in the worldwide airport network (top, http://www.iata.org) and in the scientific collaboration
network (SCN, bottom, http://www-personal.umich.edu/~mejn/
netdata/).

are generally observed in real networks (Amaral *et al.*, 2000) and different mechanisms have been proposed to account for the presence of large degree truncations.
In such a context, fitting to a power-law form can yield different results, depending on the range of values actually considered for the fit. In this sense, the exact
value of the exponents or the precise analytical form of the observed behavior is of
secondary importance, at least given the amount of noise often encountered in the
data sets. In some other situations the behavior is surely more complicated than a
pure power law, exponential cut-offs, or other functional forms being present. The
crucial issue is that the observation of heavy-tailed, highly variable distributions
provides statistical fluctuations which are extremely large and therefore cannot be
neglected.

From this perspective, scale-free networks refer to all those systems in which
fluctuations are orders of magnitude larger than expected values. Table 2.1 summa-
rizes the numerical properties of some of the heavy-tailed probability distributions
analyzed so far. The scale-free behavior is especially clear from the values of the
heterogeneity parameter κ and the wide ranges spanned by the variables. The differences introduced by large fluctuations in many properties of the graph are too
significant to be ignored: as we will see in the next chapters they have an enormous
impact in the properties of the dynamical and physical processes occurring on these
networks.

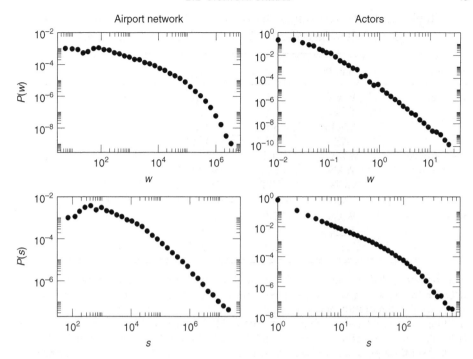

Fig. 2.10. Weight (top row) and strength (bottom row) distributions in the worldwide airport network (left column) and in the actors' collaboration network (right column). In the airport network (`http://www.iata.org`), the weight of a link corresponds to the average number of available seats per year between the two airports connected by this link (Barrat *et al.*, 2004a). The strength therefore represents the average number of passengers (traffic) handled by an airport. For the actors, collaboration network (`http://www.imdb.com/`) the weight of a link connecting two actors *i* and *j* is given by Equation (2.1): the contribution of each co-starred movie is inversely proportional to the number of actors in the movie. The strength of an actor corresponds therefore to the number of movies in which she/he has appeared. All these quantities are clearly broad in distribution.

2.2.3 Higher order statistical properties of networks

A full account of the structural and hierarchical ordering calls for the analysis and study of higher order statistical indicators and other local and non-local graph measures. The list of interesting features and their specific measures is obviously strongly dependent on the specific network analyzed. Any study of real-world networks cannot neglect the specific nature of the system in the choice of the relevant measures and quantities to analyze. While we do not want to provide an extensive discussion of each specific network, it is worth turning our attention to the multipoint degree correlation functions and the clustering properties. These quantities, as will be apparent in the next chapters, characterize

Table 2.1 Numerical values characterizing the probability distributions described in this chapter, for various real-world networks.

Variable x	Sample	$\langle x \rangle$	x_{max}	2σ	$\kappa = \langle x^2 \rangle / \langle x \rangle$
Degree k	WAN	9.7	318	41.4	53.8
	SCN	6.3	97	12.8	12.8
	Actors	80	3956	328	418
	AS	5.3	1351	70.8	242
In-degree k_{in}	WWW	24.1	378 875	843.2	7414.9
Betweenness b	WAN	6655	929 110	67 912	179 924
	SCN	37 087	3 098 700	235 584	411 208
	AS	14 531	9.3×10^6	439 566	3.3×10^6
Strength s	WAN	725 495	5.4×10^7	6×10^6	1.3×10^7
	SCN	3.6	91	9.4	9.8
	Actors	3.9	645	21	32

Networks included are the worldwide airport network (WAN), the scientific collaboration network (SCN, see http://www-personal.umich.edu/~mejn/ netdata/), the Actors' collaboration network (http://www.imdb.com/), the map of the Internet at the AS level, obtained by the DIMES project in May 2005 (http://www.netdimes.org), and the map of the WWW collected in 2003 by the WebBase project (http://dbpubs.stanford.edu: 8091/~testbed/ doc2/WebBase/). The quantity x_{max} is the maximum value of the variable observed in the sample. The parameter $\kappa = \langle x^2 \rangle / \langle x \rangle$ and the mean square root deviation $\sigma = \sqrt{\langle x^2 \rangle - \langle x \rangle^2}$ estimate the level of fluctuations in the sample. The quantity 2σ is usually considered as a 95% confidence interval in measurements. It is also possible to appreciate that all heavy-tailed distributions show a maximum value of the variable $x_{max} \gg \langle x \rangle$.

structural properties which are affecting the dynamical processes occurring on the network.

In Figure 2.11 we report the correlation and clustering spectra of several real-world networks. The figure clearly shows that most networks exhibit non-trivial spectra with a high level of variability. The correlation spectrum generally has one of the two well-defined mixing patterns presented in the previous chapter. This has led to the definition of the two broad classes of assortative and disassortative networks depending on whether their average nearest neighbors degree is an increasing or decreasing function of k (see Chapter 1). In most cases the clustering spectra also indicate significant variability of the cohesiveness properties as a function of the degree. These features have been extensively used in the attempt to formulate a conceptual and modeling understanding of the hierarchical and modular properties of networks (Ravasz *et al.*, 2002; Ravasz and Barabási, 2003). While many problems related to the structural ordering and hierarchical arrangement of large-scale networks remain open, the data of Figure 2.11 strengthen the evidence

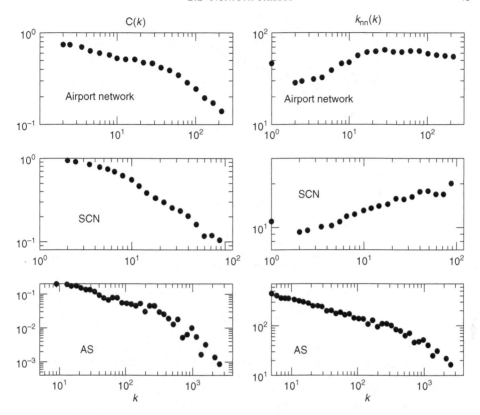

Fig. 2.11. Clustering spectrum (left column) and degree correlation spectrum (right column) for various real-world networks. The networks considered are, from top to bottom, the worldwide airport network (`http://www.iata.org`), the scientific collaboration network (SCN, see `http://www-personal.umich.edu/~mejn/netdata/`), and the map of the Internet at the AS level, obtained by the DIMES project (`http://www.netdimes.org`).

for considerable heterogeneity of networks and the lack of a typical scale in their description. At each degree value, different statistical correlation and clustering properties are found. These properties are highly variable, defining a *continuum* of hierarchical levels, non-trivially interconnected. In particular, there is no possibility of defining any degree range representing a *characteristic* hierarchical level.

A final note concerns the correlations usually present among weighted and topological properties. It appears in many cases that the weights of edges and the strengths of vertices are not trivially related to the connectivity properties of the graph. For instance, a good indication is provided by the dependence of the strength s_i on the degree k_i of vertices. In Figure 2.12 we show as an example the average

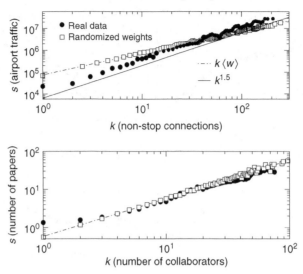

Fig. 2.12. Average strength $s(k)$ as a function of the degree k in the worldwide airport network (top, `http://www.iata.org`) and the scientific collaboration network (bottom, `http://www-personal. umich.edu/~mejn/netdata/`). In the airport network, the average strength represents the average number of passengers (traffic) handled by an airport with k direct connections. For the scientific collaboration network it gives the average number of papers of an author with k collaborators. We report the behavior obtained for both the real weighted networks and their randomized versions, which are generated by a random re-shuffling of the actual weights on the existing topology of the network.

strength $s(k)$ of vertices with degree k in the scientific collaboration network and in the worldwide airport network (Barrat *et al.*, 2004a). It is no surprise that in both cases the strength grows as a function of the degree. On the other hand, in the absence of correlations between the weight of edges and the degree of vertices, the strength of a vertex would be on average simply proportional to its degree. While this is the case for the scientific collaboration network, in general it is possible to observe a non-linear behavior of the form $s(k) \sim k^\beta$, with exponent $\beta \neq 1$, as in the worldwide airport network where $\beta \simeq 1.5$. Such behavior implies that the weights of edges belonging to highly connected vertices tend to have higher values than the ones corresponding to an uncorrelated assignment of weights. The analysis of the weighted quantities and the study of the correlations between weights and topology offer a complementary perspective on the structural organization of the network that might be undetected by quantities based only on topological information. In particular the presence of non-linear associations between topological and weighted quantities indicates a non-trivial interplay between the dynamical and structural properties of networks. Also, in this case, the evidence for heterogeneity, variability and the absence of typical scale provide the signatures that generally characterize complex systems.

2.3 The complicated and the complex

In order to understand where complex networks can be found, and why they are defined as "complex," it is necessary to clarify the distinction between what is "complex" and what is merely complicated. This distinction is a critical one because the characteristic features and the behavior of complex systems in general differ significantly from those of merely complicated systems (Amaral and Barthélemy, 2003; Amaral and Ottino, 2004; Barabási, 2005).

The intricate appearance of large-scale graphs naturally evokes the idea of complicated systems in which a large number of components work together to perform a function. On the other hand, a computer or an airplane is also a very complicated system made by the assembly of millions of elements. Even a standard house is generally made of more than 20,000 different pieces, all of them performing different functions and tasks that follow a precise project. It is thus natural to ask what could distinguish complex systems from complicated ones. While a precise definition would certainly be very subjective, we can identify a few basic features typically characterizing complex systems.

A first point which generally characterizes complex systems is that they are emergent phenomena in the sense that they are the spontaneous outcome of the interactions among the many constituent units. In other words, complex systems are not engineered systems put in place according to a definite blueprint. Indeed, loosely speaking, complex systems consist of a large number of elements capable of interacting with each other and their environment in order to organize in specific emergent structures. In this perspective, another characteristic of complex systems is that decomposing the system and studying each subpart in isolation does not allow an understanding of the whole system and its dynamics, since the self-organization principles reside mainly in the collective and unsupervised dynamics of the many elements. It is easy to realize that the WWW, the Internet, and the airport network are all systems which grow in time by following complicated dynamical rules but without a global supervision or blueprint. The same can be said for many social and biological networks. All of these networks are self-organizing systems, which at the end of their evolution show an emergent architecture with unexpected properties and regularities. For example, in many cases complex systems can adapt to, evolve, and resist random removals of their components. It is clear that random removal of components of a computer will rapidly lead to malfunction or even to complete failure. In contrast, this is not the case for complex systems as illustrated, for instance, by the Internet, for which the failure of a few routers will not prevent its normal functioning at the global level.

In simple terms, the Internet is a physical system that can be defined as a collection of independently administered computer networks, each one of them (providers, academic and governmental institutions, private companies, etc.) having its own administration, rules, and policies. There is no central authority overseeing the growth of this network-of-networks, where new connection lines (links) and computers (nodes) are being added on a daily basis. It is clear that the Internet is subject to technical constraints and, to a certain extent, engineered when we go down to the level of Local Area Networks and within single administered domains. On the other hand, the appearance and disappearance of computers and internet providers did not follow any global blueprint and was dictated by complicated dynamics in which economic, technical, and social reasons all played a role, along with the random variable of many individual decisions. In Figure 2.13 we show the monthly number of new and deleted vertices (in the Internet at the AS level) from November 1998 to November 2000 (Qian *et al.*, 2002). The plot clearly indicates that the overall growth of the Internet is the outcome of the net balance of a birth/death dynamic involving large fractions of the system. Despite this complex dynamic and the growth of vertices and edges, the statistical behavior in time of the various metrics characterizing internet graphs shows much smaller variations (Pastor-Satorras and Vespignani, 2004). The Internet has self-organized itself in a growing structure in which the complex statistical properties have reached a stationary state.

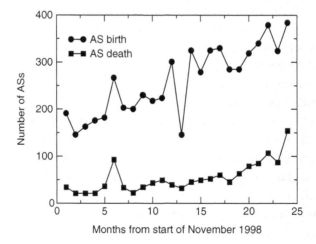

Fig. 2.13. Monthly number of new and dead autonomous systems in the period November 1998 to November 2000. Data from Qian *et al.* (2002).

Another main feature characterizing many complex systems is the presence of complications on all scales possible within the physical constraints of the system. In other words, when facing complex systems we are in the presence of structures whose fluctuations and heterogeneities extend and are repeated at all scales of the system. A typical example of this complexity is represented by fractal objects in which the same level of details and complications appears at whatever resolution we observe the object. Another clear example is provided by critical phenomena where infinitesimal localized perturbations can trigger macroscopic rearrangements across the entire system. Indeed, the long range correlations among the elements of the system may generate cascades of microscopic events disrupting the system at all scales. In the case of networks, the all-scale complication is statistically encoded in the heavy-tail distributions characterizing the structural properties. The larger the size of a system, the larger its heterogeneity and the variability of its properties. The absence of a typical degree or strength is, indeed, an indication that fluctuations extend over all the orders of magnitude allowed by the network size. Moreover, many other properties also change continuously over several orders of magnitude, with virtually infinite fluctuations. It is then impossible to define a typical scale in which an average description would be reliable.

As we have seen previously, heterogeneity and heavy-tail properties appear to be common characteristics of a large number of real-world networks, along with other complex topological features, such as hierarchies and communities. Analogously, most of these networks are dynamically evolving in time according to local processes that define the collective properties of the system. The evidence that a complex topology is the ubiquitous outcome of the evolution of networks can be hardly considered as incidental. It raises the question of the existence of some general organizing principles that might explain the emergence of this architecture in very different contexts. This consideration leads naturally to a shift of modeling focus to place more emphasis on the microscopic processes that govern the appearance and disappearance of vertices and links. As we will see in the next chapter, this approach has triggered the development of new classes of models which aim to predict the large-scale properties and behavior of the system from the dynamical interactions among its constituent units.

3

Network models

In this chapter we present a review of network models that will be used to study dynamical processes in the context of computational approaches. These different models will also help to determine the influence that specific network features have on the various phenomena that will be considered in the next chapters. To this end, we will discuss the different modeling approaches and put the activity focused on each of the different dynamical models into the proper perspective. Particular emphasis will be devoted to models which have been developed as theoretical examples of the different specific classes of real-world networks empirically observed.

3.1 Randomness and network models

Static random graph models and topology generators such as the paradigmatic Erdős–Rényi model (Erdős and Rényi, 1959; 1960; 1961) and the network generation algorithm of Molloy and Reed (1995) are the simplest network models to include stochasticity as an essential element. They are characterized by an absolute lack of knowledge of the principles that guide the creation of edges between nodes. Lacking any information, the simplest assumption one can make is to connect pairs of nodes at random with a given connection probability p. In its original formulation, an Erdős–Rényi graph $G_{N,E}$ is constructed starting from a set of N different vertices which are joined by E edges whose ends are selected at random among the N vertices. A variation of this model (Gilbert, 1959) is the graph $G_{N,p}$ constructed from a set of N different vertices in which each of the $N(N-1)/2$ possible edges is present with probability p (the *connection probability*) and absent with probability $1 - p$.

The relation between them is straightforward: in the latter case, the probability, when constructing a graph $G_{N,p}$, of obtaining a particular graph $G_{N,E}$ with N vertices and exactly E edges is

$$P(G_{N,E}) = p^E(1-p)^{\frac{1}{2}N(N-1)-E}. \tag{3.1}$$

Since each edge can be either present or absent, the ensemble of graphs $G_{N,p}$ contains $2^{N(N-1)/2}$ elements, and many of their properties can be easily derived. For example, in order to compute the average degree, we observe that the average number of edges generated in the construction of the graph is $\langle E \rangle = \frac{1}{2}N(N-1)p$. Since each edge contributes to the degree of two vertices, we obtain

$$\langle k \rangle = \frac{2\langle E \rangle}{N} = (N-1)p \simeq Np, \tag{3.2}$$

where the last equality is valid for large N. The two ensembles of graphs $G_{N,p}$ and $G_{N,E}$ are in fact statistically equivalent when N goes to infinity with

$$pN(N-1)/2 = E.$$

From the previous equation we observe that, for any finite p, the average degree diverges with the number of vertices in the graph. Since real-world graphs are most often characterized by a finite average degree, in many cases it is a natural choice to consider the behavior of the model for a wiring probability that decreases with N; i.e. $p(N) = \langle k \rangle / N$. The average degree of the random graph is also a determinant parameter in establishing the connectivity structure of the resulting network. If $\langle k \rangle < 1$ the network is composed of many small subgraphs that are not interconnected. For $\langle k \rangle = 1$, we observe a phase transition equivalent to the percolation transition in infinite dimension and for $\langle k \rangle > 1$ a giant component emerges with size proportional to the number of vertices in the network. A more detailed account of the component structure of random graphs can be found in Chapter 6.

In order to obtain the degree distribution $P(k)$, we notice that, in a graph with wiring probability p, the probability of creating a vertex of degree k is equal to the probability that it is connected to k other vertices and not connected to the remaining $N - 1 - k$ vertices. Since the establishment of each edge is an independent event, this probability is simply given by the binomial distribution

$$P(k) = \binom{N-1}{k} p^k (1-p)^{N-1-k}. \tag{3.3}$$

In the limit of large N and for $pN = \langle k \rangle$ constant, the binomial distribution can be approximated by the Poisson distribution (Gnedenko, 1962)

$$P(k) = e^{-\langle k \rangle} \frac{\langle k \rangle^k}{k!}, \tag{3.4}$$

recovering the result obtained from more rigorous arguments by Bollobás (1981). The most characteristic trait of the degree distribution of the Erdős–Rényi model is that it decays *exponentially* for large k, allowing only very small degree fluctuations. The Erdős–Rényi model represents, in this sense, the prototypical example of a statistically *homogeneous* random graph, in which, for the purpose of the large-scale characterization of the network, the degree of the different vertices can be approximately considered as uniform and equal to the average degree, $k \simeq \langle k \rangle$.

In the Erdős–Rényi model it is also possible to derive easily the clustering and small-world properties. The clustering coefficient $\langle C \rangle$ of the Erdős–Rényi model follows from the independence of the connections. For any vertex, the probability that any two of its neighbors are also connected to each other is given by the connection probability p. Therefore the average clustering coefficient is equal to

$$\langle C \rangle = p = \frac{\langle k \rangle}{N}. \tag{3.5}$$

From the previous expression it is easy to conclude that the clustering coefficient of the Erdős–Rényi model, at fixed $\langle k \rangle$, *decreases* with the graph size, and approaches zero in the limit of an infinitely large network. In other words, the absence of cohesive ordering is implicit in the complete randomness of the model. For a connected network of average degree $\langle k \rangle$, the average number of neighbors at distance 1 of any vertex i is $\langle k \rangle$. If the position of the edges is completely random and the effect of cycles is neglected, the number of neighbors at a distance d can be approximated by $\langle k \rangle^d$. Let us define r_G such that $\langle k \rangle^{r_G} \simeq N$. Since the quantity $\langle k \rangle^d$ grows exponentially fast with d, an overwhelming majority of vertices are at a distance of order r_G from the vertex i. We can thus approximate the average shortest path length $\langle \ell \rangle$ by r_G and we obtain[1]

$$\langle \ell \rangle \simeq \frac{\log N}{\log \langle k \rangle}. \tag{3.6}$$

This approximate estimate can be proved rigorously (Bollobás, 1981), showing that the Erdős–Rényi model exhibits an average shortest path length $\langle \ell \rangle$ that scales logarithmically with the graph size N. This scaling behavior is the signature of the *small-world* effect observed in many complex networks.

3.1.1 Generalized random graphs

The random graph paradigm of the Erdős–Rényi model can be extended to accommodate the construction of generalized random graphs with a predefined degree distribution – not necessarily Poisson – that are otherwise random in the assignment of the edges' end-points. This procedure, first proposed by Bender and Canfield

---

[1] It must be noted that the present result is not valid if one considers averages over a structure that is wholly tree-like. In this case, the existence of one-dimensional-like trees leads to the result that $\langle \ell \rangle$ usually scales as a power of N (Burda, Correia and Krzywicki, 2001).

(1978), and later developed in several works (Molloy and Reed, 1995; Molloy and Reed, 1998; Aiello, Chung and Lu, 2001), consists of assigning the graph a fixed degree sequence $\{k_i\}$, $i = 1, \ldots, N$, such that the ith vertex has degree k_i, then distributing the end-points of the edges among the vertices according to their respective degrees. This procedure generates graphs which are in all respects random,[2] with the imposed degree distribution $P(k)$. The inherent randomness of these networks allows the analytical calculation of properties such as the clustering coefficient and the average shortest path in the general case of random uncorrelated graphs with given degree distribution $P(k)$, as detailed in Appendix 1. For instance, the general expression for the average clustering coefficient reads as

$$\langle C \rangle = \frac{1}{N} \frac{(\langle k^2 \rangle - \langle k \rangle)^2}{\langle k \rangle^3}. \tag{3.7}$$

This implies that the clustering properties have an intrinsic dependence on the degree distribution moments. In the limit of infinite size $N \to \infty$, the clustering coefficient is null as expected in a completely random graph. However, in finite size networks with large fluctuations of the degree distribution, the clustering coefficient could have much larger values than in the case of the Poissonian Erdős–Rényi graph because of the large values attained by $\langle k^2 \rangle$.

A generalization of Equation (3.6) for the scaling of the average shortest path of the graph can similarly be obtained and reads (Appendix 1)

$$\langle \ell \rangle \approx 1 + \frac{\log[N/\langle k \rangle]}{\log[(\langle k^2 \rangle - \langle k \rangle)/\langle k \rangle]}. \tag{3.8}$$

Small-world properties are thus confirmed for general degree distribution $P(k)$, and exist simply because of the randomness of the graph.

As we have just seen, the random graph framework readily explains the presence of small-world properties and provides a convenient first order approximation for the modeling of a wide range of networks. It is therefore not surprising that, in view of their simplicity and elegance, these graphs have been used as the central paradigm of network modeling for almost four decades.

3.1.2 Fitness or "hidden variables" models

In the Erdős–Rényi and the generalized random graph models, the probability of connecting two nodes is independent of the nodes themselves. In certain real situations, it is however reasonable to think that two nodes will be connected depending on some of their intrinsic properties such as social status, information content, or

---

[2] Correlations are generated because of finite size effects and may not be negligible in some cases (see Section 3.4).

friendship. In order to explore this idea, Söderberg (2002), and Caldarelli *et al.*
(2002), have introduced a network model in which each node i ($i = 1, \ldots, N$) is
assigned a "fitness" described by a random real variable x_i distributed according to
a certain probability distribution $\rho(x)$, or by a discrete "type" (Söderberg, 2002).
Each pair of nodes (i, j) is then connected with a probability depending on the
fitnesses or types of the two nodes, $p_{ij} = f(x_i, x_j)$, where f is a given function.
In the case where $f = $ const., the present model is equivalent to an Erdős–Rényi
graph. The expected degree of a node with fitness x is given by

$$k(x) = N \int_0^\infty f(x, y)\rho(y)\mathrm{d}y \equiv N F(x), \tag{3.9}$$

and the degree distribution can then be deduced by

$$P(k) = \int \mathrm{d}x \rho(x)\delta[k - k(x)]$$

$$= \rho\left[F^{-1}\left(\frac{k}{N}\right)\right]\frac{\mathrm{d}}{\mathrm{d}k}F^{-1}\left(\frac{k}{N}\right), \tag{3.10}$$

where we have supposed that $F(x)$ is a monotonic function of x. From this last
equation, one can see that if the fitness distribution is a power law and F is for
example linear the resulting network will be scale-free. This property, however,
does not yield an explanation of the presence of scale-free degree distributions in
real complex networks since the use of a power-law $\rho(x)$ is not a priori justified.
A more surprising result appears if we choose a peaked fitness distribution of the
exponential type ($\rho(x) \sim \mathrm{e}^{-x}$) and for the function f a threshold function of the
form

$$f(x_i, x_j) = \theta\left[x_i + x_j - z(N)\right], \tag{3.11}$$

where θ is the Heaviside function and $z(N)$ is a threshold depending in general on
the size N. Caldarelli *et al.* (2002) have shown that the degree distribution is then
a power law of the form $P(k) \sim k^{-2}$. This result implies that for static networks a
peaked distribution of fitnesses can generate scale-free networks.[3] Moreover, both
the assortativity $k_{nn}(k)$ and the clustering spectrum $C(k)$ behave as power laws.
Interestingly, Boguñá, Pastor-Satorras (2003) show that this framework can be generalized into a class of models in which nodes are tagged by "hidden" variables
that completely determine the topological structure of the network through their
probability distribution and the probability of connecting pairs of vertices.

---

[3] Geographical constraints may easily be introduced as well, see e.g. Masuda, Miwa and Konno (2005)

3.1.3 The Watts–Strogatz model

In random graph models the clustering coefficient is determined by the imposed degree distribution and vanishes in the limit of very large graphs. The empirical observation of a very large clustering coefficient in many real-world networks is therefore a conceptual challenge that spurred the definition of models in which it is possible to tune $\langle C \rangle$ to any desired value. Inspired by the fact that many social networks (Milgram, 1967; Wasserman and Faust, 1994) are highly clustered while at the same time exhibiting a small average distance between vertices, Watts and Strogatz (1998) have proposed a model that interpolates between ordered lattices and purely random networks (which possess a small average path length).

The original Watts and Strogatz model starts with a ring of N vertices in which each vertex (i_1, i_2, etc.) is symmetrically connected to its $2m$ nearest neighbors (m vertices clockwise and counterclockwise of i_1, then i_2, etc. as shown in Figure 3.1 for $p = 0$). Then, for every vertex, each edge connected to a clockwise neighbor is rewired with probability p, and preserved with probability $1 - p$. The rewiring connects the edge's end-point to a randomly chosen vertex, avoiding self-connections, and typically creates *shortcuts* between distant parts of the ring. The parameter p therefore tunes the level of randomness present in the graph, keeping the number of edges constant. With this construction, after the rewiring process, a graph with average degree $\langle k \rangle = 2m$ is obtained. It is however worth noting that even in the limit $p \to 1$, the fact that each vertex keeps a minimum degree m induces some memory of the generating procedure so that the network is not locally equivalent to an Erdős–Rényi graph (Barrat and Weigt, 2000). The degree distribution of the Watts–Strogatz model can be computed analytically (Barrat and Weigt, 2000) and reads

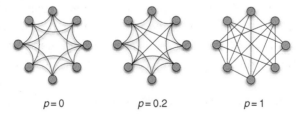

$p = 0$ $p = 0.2$ $p = 1$

Fig. 3.1. Construction leading to the Watts–Strogatz model. We start with $N = 8$ nodes, each one connected to its four nearest neighbors. By increasing p, an increasing number of edges is rewired. Rewired edges are represented as straight arcs. At $p = 1$ all edges have been rewired. Adapted from Watts and Strogatz (1998).

$$P(k) = \sum_{n=0}^{\min(k-m,m)} \binom{m}{n}(1-p)^n p^{m-n} \frac{(pm)^{k-m-n}}{(k-m-n)!}e^{-pm}, \quad \text{for } k \geq m. \quad (3.12)$$

In the limit of $p \to 1$ the above expression reduces to

$$P(k) = \frac{m^{k-m}}{(k-m)!}e^{-m}, \quad (3.13)$$

which is a Poisson distribution for the variable $k' = k - m$, with average value $\langle k' \rangle = m$.

While the degree distribution has essentially the same features as a homogeneous random graph, the parameter p has strong effects on the clustering coefficient and the average shortest path length. When $p = 0$ the number of connections among the neighbors of each node is $3m(m - 1)/2$, while the total possible number of connections is $2m(2m - 1)/2$. This yields a clustering coefficient $\langle C \rangle = 3(m - 1)/2(2m - 1)$. At the same time, the shortest path length scales as in the one-dimensional regular grid, i.e. $\langle \ell \rangle \sim N$. This picture changes dramatically as soon as the rewiring probability is switched on. For very small p the resulting network has a full memory of a regular lattice and consequently a high $\langle C \rangle$. In particular, Barrat and Weigt (2000) derived the dependence of the clustering coefficient defined as the fraction of transitive triples, obtaining

$$\langle C(p) \rangle \simeq \frac{3(m - 1)}{2(2m - 1)}(1 - p)^3. \quad (3.14)$$

On the other hand, even at small p, the appearance of shortcuts between distant vertices in the lattice dramatically reduces the average shortest path length. For $p \to 1$ the network eventually becomes a randomized graph, with a logarithmically small $\langle \ell \rangle$ and a vanishing clustering coefficient. Watts and Strogatz (1998) focused on the transition between these two regimes (shown in Figure 3.2), noting that in a wide range of $p \ll 1$, the shortest path length, after decreasing abruptly, almost reaches the value corresponding to a random graph, while the clustering coefficient remains almost constant and close to that of the original ordered lattice. Therefore, a broad region of the parameter space exists in which it is possible to find graphs with a large $\langle C \rangle$ and a small $\langle \ell \rangle$, as observed in most natural networks.

Interestingly, the smallest value of p at which the small-world behavior sets in is related to the size of the network, as can be seen from the following scaling form for the average shortest path (Barthélemy and Amaral, 1999b)

$$\langle \ell \rangle \sim N^* F\left(\frac{N}{N^*}\right), \quad (3.15)$$

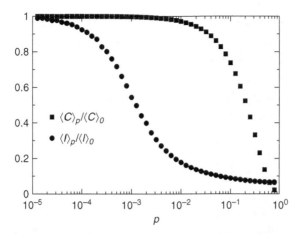

Fig. 3.2. Normalized clustering coefficient $\langle C \rangle_p / \langle C \rangle_0$ (squares) and average shortest path length $\langle \ell \rangle_p / \langle \ell \rangle_0$ (circles) as a function of the rewiring probability p for the Watts–Strogatz model. The results correspond to networks of size $N = 1000$ and average degree $\langle k \rangle = 10$, and are averaged over 1000 different realizations.

where the scaling function has the limit behaviors $F(x \ll 1) \sim x$ and $F(x \gg 1) \sim \log x$. We thus have a crossover from a lattice-like behavior to a small-world one and the crossover size N^* scales with p as $N^* \sim 1/p$ (Barthélemy and Amaral, 1999a; Barrat and Weigt, 2000). This behavior can easily be understood qualitatively. The typical size N^* of the regions between shortcuts is given by the total number of vertices N divided by the average number pN of shortcuts present in the graph, leading to $N^* \sim 1/p$. If the characteristic size N^* of these regions is much smaller than the size of the graph, enough shortcuts connect distant regions on the ring, producing the desired small-world effect. This immediately tells us that if $p \gg 1/N$ the average shortest path is going to be very small. Since, on the other hand, a large clustering coefficient is obtained for $p \ll 1$, the whole parameter region $N^{-1} \ll p \ll 1$ yields networks sharing both small-world properties and a high clustering coefficient. Notably, in the case of very large graphs, $N \to \infty$, even a very small amount of randomness is then sufficient to produce the small-world effect.

The Watts–Strogatz model represents an important development in the modeling of social networks and many other systems (Strogatz, 2000a) since it allows the tuning of the clustering coefficient within the framework of static random graph theory. In addition it can explain the high clustering coefficients observed in real networks as the memory of an initial ordered structure that has been reshaped by some stochastic element. Finally, as the original model displays a Poisson degree distribution, several variations have been proposed in the literature in order to make room within the Watts–Strogatz construction for arbitrary degree distributions. Several models have therefore been defined in order to obtain highly clustered graphs with heavy-tailed degree distributions and non-trivial correlation properties (Dorogovtsev, Mendes and Samukhin, 2001b; Davidsen, Ebel and Bornholdt, 2002; Holme and Kim, 2002a; Warren, Sander and Sokolov, 2002; Rozenfeld *et al.*, 2002; Chung and Lu, 2004; Volz, 2004; Andersen, Chung and Lu, 2005; Serrano and Boguñá, 2005).

3.2 Exponential random graphs

By looking at the basic definition of the Erdős–Rényi graph, it is possible to see that, by drawing edges at random with a given probability, there is a total of $\binom{N(N-1)/2}{E}$ possible different graphs, which form a probability ensemble in which each graph has the same likelihood. In this respect, the previous construction resembles the *microcanonical ensemble* in classical equilibrium statistical mechanics (Pathria, 1996; Burda *et al.*, 2001).

This idea can be better formalized by the well-established group of models represented by the exponential random graph family[4] largely studied in social network analysis (Holland and Leinhardt, 1981; Frank and Strauss, 1986), and more recently cast in the general framework of the equilibrium statistical mechanics of networks. This modeling approach considers the adjacency matrix – also called the sociomatrix in the social network literature – $\mathbf{X} = \{x_{ij}\}$ characterizing the graph of size N as a random matrix whose realization occurs with a probability $P(\mathbf{X})$ defined in the sample space of all possible graphs. In this hyper-dimensional space, each coordinate represents the possible values that each degree of freedom (in the case of the graph the variables x_{ij}) may have. Each point in this space thus specifies the values of the microscopic variables defining a realization of the network. The exponential random graphs family defined in social network analysis assumes that the general probability distribution in the sample space has the form

$$P(\mathbf{X}) = \frac{\exp\left[\sum_i \theta_i z_i(\mathbf{X})\right]}{\kappa(\{\theta_i\})}, \tag{3.16}$$

where $\{\theta_i\}$ is a set of model parameters and $z_i(\mathbf{X})$ is a set of the network's statistical observables. The statistics $z_i(\mathbf{X})$ are numerous and range from the very simple average degree of the graph $\langle k \rangle$ (equivalently the total number of edges $E = N\langle k \rangle/2$) to complete degree sequences and even probability distributions of attributes. The function $\kappa(\{\theta_i\})$ ensures the correct normalization (the sum of $P(\mathbf{X})$ over all possible graphs $\mathbf{X}$ allowed in the sample space is equal to 1). Once the relevant statistics

---

[4] In the statistical and social science literature these models are also referred to as Logit models, p^*-models, and Markov random graphs.

and assumptions are included in the model, the parameters θ_i have to be estimated by comparison with the real data. This has spurred the development of a wide array of techniques such as pseudo-likelihood estimation and Monte Carlo maximum likelihood estimation (Strauss and Ikeda, 1990; Wasserman and Pattison, 1996). This framework has been adapted to introduce the dynamical evolution of edges, given a fixed number of nodes (Sanil, Banks and Carley, 1995; Banks and Carley, 1996), and numerical techniques for the estimation of the modeling distribution parameters have been put forward in computational tools (Snijders, 2001).

The form of Equation (3.16) has deep connections with the basic principles of equilibrium statistical physics. It is indeed possible to show in a very simple way that exponential random graph models are equivalent to the statistical mechanics of Boltzmann and Gibbs for networks (Burda *et al.*, 2001; Berg and Lässig, 2002; Burda and Krzywicki, 2003; Dorogovtsev, Mendes and Samukhin, 2003; Farkas *et al.*, 2004; Park and Newman, 2004). Equilibrium statistical physics assumes that the probability for the system (in our case the network) to be in a specific configuration $\mathbf{X}$ is given by the distribution $P(\mathbf{X})$ that maximizes the Gibbs entropy

$$S[P] = -\sum_{\mathbf{X}} P(\mathbf{X}) \ln P(\mathbf{X}), \tag{3.17}$$

where the sum is over all possible stochastic realizations allowed. The entropy is a measure of the disorder encoded in the probability distribution and it is reasonable to expect that, at equilibrium, statistical disorder is maximal. It is worth stressing that equilibrium here does not imply a static vision and only refers to a situation in which the probability distribution describing the possible states is not biased or constrained to be on a specific subset of the allowed states. In the context of physical systems, this equilibrium assumption can be more formally stated and related to their microscopic dynamics. The maximization of the entropy is moreover constrained by a certain number of statistical observables $z_i(\mathbf{X})$ for which one assumes one has statistical estimates

$$\langle z_i \rangle = \sum_{\mathbf{X}} P(\mathbf{X}) z_i(\mathbf{X}), \tag{3.18}$$

and by the normalization condition $\sum_{\mathbf{X}} P(\mathbf{X}) = 1$. The maximization of the entropy function is made by introducing a Lagrange multiplier α_i for each constraint $\langle z_i \rangle$, with α_0 being the multiplier relative to the normalization condition. The distribution has therefore to satisfy the equation

$$\frac{\delta}{\delta P(\mathbf{X})} \left[S[P] + \alpha_0 \left(1 - \sum_{\mathbf{Y}} P(\mathbf{Y}) \right) + \sum_i \alpha_i \left(\langle z_i \rangle - \sum_{\mathbf{Y}} P(\mathbf{Y}) z_i(\mathbf{Y}) \right) \right] = 0. \tag{3.19}$$

This functional derivative yields the condition, valid for all possible realizations $\mathbf{X}$,

$$\ln P(\mathbf{X}) + 1 + \alpha_0 + \sum_i \alpha_i z_i(\mathbf{X}) = 0, \tag{3.20}$$

which leads to the solution

$$P(\mathbf{X}) = \frac{\exp\left[-\sum_i \alpha_i z_i(\mathbf{X})\right]}{Z(\{\alpha_i\})}, \tag{3.21}$$

where the normalization condition imposes

$$Z(\{\alpha_i\}) = e^{\alpha_0+1} = \sum_{\mathbf{X}} e^{-\sum_i \alpha_i z_i(\mathbf{X})}. \tag{3.22}$$

Finally, the explicit values of the parameters α_i are found by imposing the self-consistent condition on the statistical observables for all the observables z_i used in the model construction:

$$\langle z_i \rangle = \sum_{\mathbf{X}} z_i(\mathbf{X}) \frac{\exp\left[-\sum_j \alpha_j z_j(\mathbf{X})\right]}{Z(\{\alpha_j\})}. \tag{3.23}$$

The simple substitution $\theta_i = -\alpha_i$ and $\kappa(\{\theta_i\}) = Z(\{-\theta_i\})$ readily yields a probability distribution $P(\mathbf{X})$ identical to the distribution of the exponential random graph family. Generally, the function $H(\mathbf{X}) = \sum_i \alpha_i z_i(\mathbf{X})$ defining the statistical weight is called the Hamiltonian of the system and the function Z defines the partition function. Analogies can be pushed further with different statistical constraints corresponding to different statistical ensembles in the statistical mechanics definition. Moreover, it is possible to show that this formalism also contains random graphs such as the Erdős–Rényi one (Park and Newman, 2004). For instance, the random graph family $G_{N,p}$ can be recovered by imposing, as a constraint, the corresponding value of the number of links. The exponential random graph family used in statistics is therefore the distribution form corresponding to the equilibrium ensembles of statistical mechanics developed in physics and can be considered as the statistical mechanics of Gibbs for networks.

3.3 Evolving networks and the non-equilibrium approach

The modeling approaches introduced in the previous section are focused on the stationary properties of the network for which they derive the probability distribution in phase space. However, many physical, social, and biological systems are the result of microscopic dynamical processes determining the occurrence of the various configurations. The creation of a social relation, the introduction of a hyperlink to a web page, and the peering of two Internet service providers are dynamical events based on local interactions among individuals that shape the evolution of the network. Non-equilibrium systems may still have a stationary state in which the probability distribution is time-independent, but in this case we find an overwhelming number of constraints that must be considered for the entropy maximization, rendering its computation infeasible. In the non-equilibrium situation, the system might favor a particular region of the phase space depending on the initial conditions or on a specific dynamical drift. In this case, unless we know exactly all the constraints on the system evolution, for instance the exact partition of accessible configurations depending on the initial conditions, it is impossible to find a meaningful solution. This is also the case for networks with a continuously increasing number of nodes and edges, whose phase space dimensionality is continuously enlarging. In many of these cases, it is more convenient to rely on approaches dealing directly with the dynamical evolution of the network. To this end, we have to introduce the time variable and the probability of a particular network realization $\mathbf{X}$ at time t given by the distribution $P(\mathbf{X}, t)$. The temporal evolution of the probability distribution is generally expressed in the form of a master equation (Dorogovtsev, Mendes and Samukhin, 2000; Krapivsky and Redner, 2001). This is a linear differential equation for the probability that any network, owing to the microscopic dynamics, is in the configuration $\mathbf{X}$. In this case a stochastic description is applied by introducing the rates $r_{\mathbf{X} \rightarrow \mathbf{Y}}$ that express the transition from the realization $\mathbf{X}$ to the realization $\mathbf{Y}$.[5] If we assume that the process has no time memory, we have a Markovian process, and the temporal change of $P(\mathbf{X}, t)$ obeys a master equation of the form

$$\partial_t P(\mathbf{X}, t) = \sum_{\mathbf{Y} \neq \mathbf{X}} [P(\mathbf{Y}, t) r_{\mathbf{Y} \rightarrow \mathbf{X}} - P(\mathbf{X}, t) r_{\mathbf{X} \rightarrow \mathbf{Y}}]. \qquad (3.24)$$

The probability distribution $P(\mathbf{X}, t)$ is normalized and it is trivial to see that the master equation (3.24) preserves this normalization.

In the master equation approach, it is crucial to consider transition rates or probabilities r reflecting the actual dynamics of the system under consideration. In the modeling, the attention thus shifts from the statistical quantities describing the system to the dynamical laws governing its evolution. In the absence of any details on such laws, the dynamical approach is often a difficult exercise in which rough assumptions and often uncontrolled approximations have to be made. On the other hand, it has the advantage of being more intuitive and suitable to large-scale computer simulations and theoretical discussions. In practice, the master equation can

---

[5] In the case of dynamics in discrete time $t \rightarrow t + 1$, one assigns transition probabilities $p_{\mathbf{X} \rightarrow \mathbf{Y}}$ instead of transition rates.

be exactly solved in only a small number of specific cases and it is more practical to work with a specific projection of the probability distribution, depending
on the quantity of interest, such as the degree distribution or any other statistical
observables in the network.

To provide a specific example, let us consider a continuously growing network
in which new nodes appear and wiring processes such as edge adding, removal
and rewiring take place (Krapivsky and Redner, 2003b). Each time a new node
enters the network, it establishes m new reciprocal edges with nodes already existing in the network. For the sake of simplicity let us assume that once an edge is
established it will not be rewired. In this case, the simplest quantity we might be
interested in is the degree distribution specified by the number N_k of vertices with
degree k. The master equation for such a growing scheme is simply given by

$$\partial_t N_k = r_{k-1 \to k} N_{k-1} - r_{k \to k+1} N_k + \delta_{k,m}. \tag{3.25}$$

The first term on the right corresponds to processes in which a vertex with $k - 1$
links is connected to the new vertex, thus entering the class of degree k and yielding
an increase of the number N_k. The second term corresponds to vertices with degree
k that acquire a new edge, with, as a consequence, a decrease of N_k. Finally the last
term, given by a Kronecker symbol with values $\delta_{k,m} = 1$ if $k = m$, and $\delta_{k,m} = 0$
otherwise, corresponds to the entry of the new vertex into the ensemble of vertices
with degree m. The solution of this evolution equation therefore depends on the
rates $r_{k-1 \to k}$ and $r_{k \to k+1}$ that specify the network dynamics. In various specific
cases, it can be obtained explicitly, allowing the calculation of many quantities of
interest.

In the case of continuously growing networks the natural time scale for the network's evolution is given by its size N. In this way, the time is measured with
respect to the number of vertices added to the graph, resulting in the definition
$t = N - m_0$, m_0 being the size of the initial core of vertices from which the growth
process starts. Therefore, each time step corresponds to the addition of a new vertex
that establishes a number of connections (edges) with already existing vertices following a given set of dynamical rules. A full description of the system is achieved
through the probability $p(k, s, t)$ that a vertex introduced at time s has degree k at
the time $t \geq s$. Once the probability $p(k, s, t)$ is known, we can obtain the degree
distribution at time t (i.e. for a network of size $N = t + m_0$) using the expression

$$P(k, t) = \frac{1}{t + m_0} \sum_{s=0}^{t} p(k, s, t). \tag{3.26}$$

In this case a simple projection is provided by the study of the average degree value
$k_s(t) = \sum_{k=0}^{\infty} k p(k, s, t)$ of the sth vertex at time t. Let us consider for the sake

of simplicity that the only relevant characteristic of a node which will determine its further evolution is given by its degree (see Section 3.4 for references on more involved approaches). The dynamical rate equation governing the evolution of $k_s(t)$ can then be formally obtained by considering that the degree growth rate of the sth vertex will increase proportionally to the probability $\Pi[k_s(t)]$ that an edge is attached to it.[6]

In the simple case that edges only come from the newborn vertices, the rate equation reads

$$\frac{\partial k_s(t)}{\partial t} = m\Pi[k_s(t)], \qquad (3.27)$$

where the proportionality factor m indicates the number of edges emanating from every new vertex. This equation is moreover constrained by the boundary condition $k_s(s) = m$, meaning that at the time of its introduction, each vertex has an initial degree m. In this formulation, all the dynamical information is contained in the probability or growth kernel $\Pi[k_s(t)]$, which defines the properties of each particular model.

The projection could consider other quantities such as the number of vertices $N(k|\ell)$ with degree k which share an edge with a vertex of degree ℓ, etc. Higher order statistics clearly yield increasingly involved evolution equations. Similarly, the dynamics might be complicated further by the introduction of more refined processes such as edge removal, rewiring, inheritance, and vertices' disappearance. Space and other attributes for the network formation have also been considered in various forms. For a review of dynamical models we refer the reader to Albert and Barabási (2002); Dorogovtsev and Mendes (2003); Pastor-Satorras and Vespignani (2004); Newman (2003b); Boccaletti *et al.* ([2006]); Caldarelli (2007) and references therein.

It is clear from the previous discussion that the dynamical approach is potentially risky and unless we have precise experimental information on the dynamics, it will not allow quantitatively accurate predictions. Moreover it does not provide any systematic theoretical framework, with each model focusing on some specific features of the system of interest. On the other hand, the study of the dynamics is well suited to identify general growing mechanisms from seemingly very different dynamical rules. In this perspective, the dynamical approach appears as the right tool to capture the emergent behavior and complex features of networks. The emphasis is indeed on the evolutionary mechanisms that generate the observed topological properties, which become a by-product of the system's dynamics. The similarities are evident with the statistical physics approach to complex phenomena

---

[6] For the sake of analytical simplicity one often considers the degree k and the time t as continuous variables. (Barabási, Albert and Jeong, 1999; Dorogovtsev *et al.*, 2000).

that aims to predict the large-scale emergent properties of a system by studying the collective dynamics of its constituents.

3.3.1 The preferential attachment class of models

The dynamical approach to networks is exemplified by the Barabási–Albert class of models (Barabási and Albert, 1999) which provides an example of the emergence of networks with heavy-tailed degree distributions in terms of the elementary process governing the wiring of new vertices joining the network. The insight behind this approach is the fact that in most real networks, new edges are not located at random but tend to connect to vertices which already have a large number of connections (a large degree). For example, very popular web pages are likely to receive new incoming edges from new websites. Similarly, a person with a large number of social relations is more likely to acquire new friends than somebody who has just a close and small circle of friends. In the Internet, new service providers aim at optimizing their centrality by peering with well-connected existing providers. This class of systems might therefore be described by models based on such a preferential attachment mechanism – also known as the rich-get-richer phenomenon, the Matthew effect (Merton, 1968), the Gibrat principle (Simon, 1955), or cumulative advantage (de Solla Price, 1976).

Barabási and Albert (1999) have combined the preferential attachment condition with the growing nature of many networks by defining a simple model based on the following two rules:

Growth: The network starts with a small core of m_0 connected vertices. At every time step we add a new vertex, with m edges ($m < m_0$) connected to old vertices in the system.

Preferential attachment: Each new edge is connected to the old sth vertex with a probability proportional to its degree k_s.

These rules are used to define a class of dynamical algorithms that, starting from a connected initial core, generates connected graphs with fixed average degree $\langle k \rangle = 2m$ (Barabási and Albert, 1999; Barabási *et al.*, 1999; Bollobás *et al.*, 2001). These algorithms can be easily implemented in computer simulations, and Figure 3.3 represents a typical graph of size $N = 200$ and average degree 6 ($m = 3$).

The numerical simulations indicate that the graphs generated with the Barabási–Albert algorithms spontaneously evolve into a stationary power-law degree distribution with the form $P(k) \sim k^{-3}$ (see Figure 3.4). The model can be further explored by using the dynamical approach outlined in the previous section. The preferential attachment mechanism can be easily cast in mathematical form since

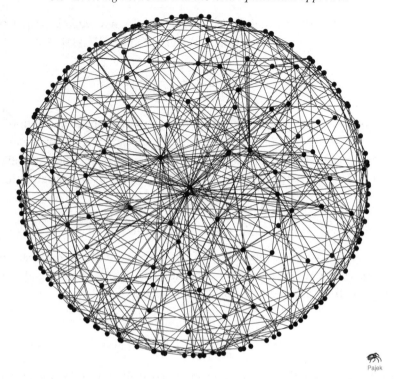

Fig. 3.3. Typical Barabási–Albert network of size $N = 200$ and average degree $\langle k \rangle = 6$. Higher degree nodes are at the center of the graph. The figure is generated with the Pajek package for large network analysis, `http://vlado.fmf.uni-lj.si/pub/networks/pajek/`.

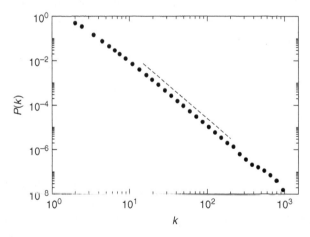

Fig. 3.4. Degree distribution of a Barabási–Albert network of size $N = 10^5$ with average degree $\langle k \rangle = 4$ in double logarithmic scale. The dashed line corresponds to a power-law behavior $P(k) \sim k^{-\gamma}$ with exponent $\gamma = 3$.

it states that the probability that the vertex s acquires a new edge is proportional to its degree, obtaining the explicit form of the growth rate $\Pi[k_s(t)]$

$$\Pi[k_s(t)] = \frac{k_s(t)}{\sum_j k_j(t)}, \tag{3.28}$$

where the denominator is the required normalization factor; i.e. the sum of all the degrees of the vertices in the network. Since each new edge contributes with a factor 2 to the total degree, and at time t we have added tm edges, the evolution equation (3.27) for k_s takes the form, within the continuous k approximation,

$$\frac{\partial k_s(t)}{\partial t} = \frac{mk_s(t)}{2mt + m_0\langle k\rangle_0}, \tag{3.29}$$

where $\langle k\rangle_0$ is the average connectivity of the initial core of m_0 vertices. This differential equation with the boundary condition $k_s(s) = m$ can be readily solved, yielding in the limit of large networks $(t, s \gg m\langle k\rangle_0)$,

$$k_s(t) \simeq m\left(\frac{t}{s}\right)^{1/2}. \tag{3.30}$$

By considering the continuum limit, the degree distribution is obtained as

$$P(k, t) = \frac{1}{t + m_0}\int_0^t \delta[k - k_s(t)]\, ds \equiv -\frac{1}{t + m_0}\left(\frac{\partial k_s(t)}{\partial s}\right)^{-1}\Bigg|_{s=s(k,t)}, \tag{3.31}$$

where $\delta(k - k_s(t))$ is the Dirac delta function and $s(k, t)$ is the solution of the implicit equation $k = k_s(t)$. In the Barabási–Albert model the solution of the previous expression yields

$$P(k, t) = 2m^2\frac{t + (m_0/2m)\langle k\rangle_0}{t + m_0}k^{-3}, \tag{3.32}$$

and gives in the limit of large sizes $t \to \infty$ the solution (valid for any $\langle k\rangle_0$),

$$P(k) = 2m^2 k^{-3}. \tag{3.33}$$

For the Barabási–Albert class of models it is also possible to obtain analytic expressions providing the clustering coefficient and the average shortest path scaling. More precisely, the average clustering coefficient reads (Klemm and Eguíluz, 2002a; Szabó, Alava and Kertész, 2003; Barrat and Pastor-Satorras, 2005)

$$\langle C\rangle_N = \frac{m}{8N}(\ln N)^2, \tag{3.34}$$

and the average shortest path length scales as (Bollobás and Riordan, 2003; Cohen and Havlin, 2003)

$$\langle \ell \rangle \sim \frac{\log(N)}{\log \log(N)}. \tag{3.35}$$

In some situations, it can be reasonable to assume that besides degrees, nodes have other attributes which can render them more "attractive." Bianconi and Barabási (2001) have proposed that for these cases, all the properties (other than the degree) that modify the preferential attachment probability can be encoded in a *fitness* η chosen randomly according to some distribution $\rho(\eta)$. Note that fitness is considered here in a dynamical perspective in contrast with the quantities used in the static fitness model described in Section 3.1.2. In such a modified preferential attachment framework, the probability that a new edge will connect to node s is then given by

$$\Pi_s = \frac{\eta_s k_s}{\sum_j \eta_j k_j}. \tag{3.36}$$

Very fit nodes can thus attract new edges despite a moderate value of their degree and this new "fittest-get-richer" process is thus superimposed on the degree-driven attachment mechanism. Analytical insights into this model can be obtained by using an approximation for the sum $\sum_j \eta_j k_j$, and in the continuous time and degree limits, the degree distribution of the resulting network reads as

$$P(k) = \frac{c}{m} \int d\eta \frac{\rho(\eta)}{\eta} \left(\frac{k}{m} \right)^{-(1+c/\eta)}, \tag{3.37}$$

where the constant c depends on the fitness distribution (Bianconi and Barabási, 2001). In the case of a uniform distribution of fitness in the interval $[0, 1]$, this leads to $P(k) \sim k^{-\gamma}$ for $k \to \infty$ with $\gamma \approx 2.255$. Remarkably enough, this result demonstrates that a small amount of randomness in fitness can lead to growing scale-free networks with a non-trivial exponent for the degree distribution.

The success of the models based on the preferential attachment mechanism resides in the simple explanation, through a basic dynamical principle, of the emergence of graphs with a power-law degree distribution and small-world properties. The importance of the preferential attachment has even been reinforced recently. Fortunato, Flammini and Menczer (2006a) discuss the attachment mechanism for different quantities such as degree, age, etc., and argue that the relative values of nodes' properties matter more than their absolute values. In other words, the relevant quantity in a preferential attachment mechanism is the rank of each node according to a certain attribute. Such information is often easier to estimate than the absolute value of a node's attribute. For example, even if we are not usually able to quantify precisely the wealth of different people, we can propose a ranking of the richest individuals. Following this idea, Fortunato *et al.* (2006a) study the general case of the attachment probability of a new node to an old vertex s given by the form

$$\Pi_s = \frac{R_s^{-\alpha}}{\sum_j R_j^{-\alpha}}, \tag{3.38}$$

where R_s denotes the rank of the node s for some specific attribute and where α is a positive parameter. In this case, the degree distribution is given by $P(k) \sim k^{-\gamma}$ with

$$\gamma = 1 + \frac{1}{\alpha}. \tag{3.39}$$

This result points to the fact that preferential attachment mechanisms in growing networks lead to scale-free networks, even in the absence of a complete knowledge of the values of the nodes' attributes.

The Barabási–Albert class of models represents a very simple implementation of the preferential attachment mechanism and is not intended to be a realistic model of any real-world network. Rather, it is a zeroth order conceptual model which can be used as the paradigm for much more realistic models taking into account the particular processes in the system under consideration. For this reason, after the introduction of the Barabási–Albert model, a large number of other network models, considering a wide array of degree-based growth mechanisms, have been proposed, incorporating various ingredients in order to account for power-law degree distributions with a connectivity exponent $2 < \gamma < 3$, local geographical factors, rewiring among existing nodes, or age effects. We refer the reader to the reviews by Dorogovtsev and Mendes (2003) and Newman (2003b), and to the references therein to get a flavor of the work done in this direction.

3.3.2 Copy and duplication models

One of the strengths of the dynamical approach is that the presence of similarities in the dynamical microscopic mechanisms allows the explanation of shared properties and similarities in very different systems. In particular, there is a large class of models that at first sight seem completely unrelated to the preferential attachment mechanism, but in which a closer look reveals the existence of this mechanism in disguise. Such a result represents an important step in the basic understanding of the microscopic origin of the preferential mechanism in many real-world systems.

The class of copying models relies on the plausible assumption that new elements arriving in a system have the tendency to copy the choice of already present elements. This mechanism was first proposed in the context of WWW simulations for the generation of skewed degree distributions (Kleinberg *et al.*, 1999; Kumar

et al., 2000). The chief consideration for this mechanism is that new pages dedi-
cated to a certain thematic area copy hyperlinks from existing pages with similar
content. This has been translated into a directed growing model in which, at each
time step, a new vertex (web page) is added to the network and a corresponding
prototype vertex is selected at random among those already existing. Each new
vertex emits $m \geq 1$ new outgoing edges initially pointing towards vertices pointed
by the prototype vertex. At this point a *copy factor* α (constant for all new vertices) is introduced. With probability $1 - \alpha$ each new edge is retained as it is; with
probability α it is rewired towards a randomly chosen vertex of the network. The
copy factor introduces the possibility that the new vertex does not just copy all its
edges from the prototype vertex, since the web page author might find other interesting pages in the network by a random exploration. A pictorial illustration of the
copying model dynamics is provided in Figure 3.5.

This model can be easily studied analytically and numerically: the in-degree
distribution is power-law distributed and the copy factor acts as a tuning parameter
for the degree exponent of the model. These results can be obtained by writing the
basic evolution equation for the model. Let us focus here on a generic vertex of
the network and calculate its probability of receiving an edge during the addition
of a new vertex. For each of the m edges of the new vertex, there is a probability
α to rewire it to another vertex chosen at random (uniformly for example). Thus
any vertex has a probability α/N of receiving an edge, where N is the size of the
network. With probability $1 - \alpha$, on the other hand, the vertex which is pointed
to by one of the edges of the prototype vertex is selected. The probability that any
given vertex s is pointed to by this edge is given by the ratio between the number
of incoming edges of that vertex and the total number of edges, i.e. $k_{\mathrm{in},s}/(mt)$. This
second process increases the probability that high degree vertices will receive new
incoming edges, and in the limit of large network sizes the mean-field evolution for
the copying model can be written in the form of the usual growth rate equation as

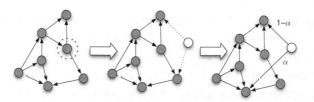

Fig. 3.5. Illustration of the rules of the copying model. A prototype vertex is
selected at random (circled by a dashed line) and a new vertex is created with
edges pointing to the out-neighbors of the prototype. Each new edge is kept with
probability $1 - \alpha$, and rewired to a randomly chosen vertex with probability α.

$$\frac{\partial k_{in,s}(t)}{\partial t} = m \left[\frac{\alpha}{t} + (1-\alpha)\frac{k_{in,s}(t)}{mt} \right], \tag{3.40}$$

where $N \simeq t$ for large linearly growing networks. Therefore, through its local dynamical rules, the copying model produces effective preferential attachment growth dynamics. This is a striking result, since the model is defined on the very simple assumption of selecting a prototype vertex, without any knowledge of the popularity or the degree importance of the vertex. The copying model thus offers a microscopic explanation for the preferential attachment mechanism that was just used as a phenomenological law in other models. The solution of the above equation, with the boundary condition $k_{in,s}(s) = 0$, yields the in-degree distribution

$$P(k_{in}) \sim (k_0 + k_{in})^{-(2-\alpha)/(1-\alpha)}, \tag{3.41}$$

where $k_0 = \alpha m/(1-\alpha)$ is an offset constant, confirming the presence of heavy tails when a preferential attachment mechanism is governing the network growth.

Although the copy model was first used in the context of the World Wide Web, similar copying mechanisms have also been used in models for the citation network (Krapivsky and Redner, 2005) and the evolution of gene and protein interaction networks (Solé *et al.*, 2002; Vázquez *et al.*, 2003; Wagner, 2003). Indeed, the genomes of most organisms are generally thought to evolve through the duplication of genes and subsequent diversification in order to perform different biochemical roles. Genome evolution by duplication/divergence corresponds to the evolution of the protein–protein interaction network (interactome) whose nodes represent the proteins expressed by the genes. The process of gene duplication can be translated, in terms of protein–protein interaction networks, into the duplication of a node (protein) sharing the same interacting partners as its ancestor, while divergence mechanisms lead to the loss or gain of interactions (see also Chapter 12 on biological networks).

3.3.3 Trade-off and optimization models

Despite the complexity and evolving characters of most networks, it is clear that a series of technical constraints and engineering principles are acting at least on the local properties of networks. Just to give an example, it is clear that within the administrative boundaries, Internet service providers decide upon the topology structure of the network and in addition experience the constraints imposed by the capacity of industrial routers. From this perspective, the network topology might be viewed as emerging from deterministic design decisions that seek to optimize certain domain-specific and network characteristics. Models adopting this perspective are based on the competition between a global optimization of the system and some local constraints. In other cases, non-trivial collective behavior arises from conflict among the various local expectations. This is, for instance, the case of the heuristically optimized trade-off (HOT) model introduced by Fabrikant, Koutsoupias and Papadimitriou (2002), successively applied to Internet router level topology (Li *et al.*, 2004b). This model suggests that the network emerges through the optimization of conflicting objectives pursued in its set-up. As a practical implementation of these ideas, the HOT Internet model is a growing model in which at every time step a new vertex is added to the network and placed in a random position on the unit square. The new vertex i is connected with an edge to the vertex j that minimizes the function $\Psi(i, j) = \alpha(N)d_{\mathrm{E}}(i, j) + \phi(j)$, where $d_{\mathrm{E}}(i, j)$ is the Euclidean distance between vertices i and j, $\alpha(N)$ is a constant that depends on the final size of the network, and $\phi(j)$ is a measure of the centrality of the vertex j. In the original paper, several centrality measures based on the shortest path length (in terms of hops on the network) were used. It is clear that in this model each new element seeks to lower the costs of establishing the physical connection by reducing as much as possible the Euclidean distance, while trying at the same time to be "centrally located" in the network, thus reducing the hop distance to other vertices. Several other variations of this model have been proposed, some reproducing specific network architecture, others accounting for the degree heterogeneity (Li *et al.*, 2004b; Alvarez-Hamelin and Schabanel, 2004; Mahadevan *et al.*, 2006). Another important aspect of these models is that randomness plays only a marginal role, since the choice of the connection (i, j) is deterministic once i has been added to the network. In principle this is different from stochastic evolution models where randomness models a set of external factors that acts as a noise on the system. The network then emerges as a result of the contrast between such randomness and the preference function which is encoded in the form of the attachment probability. On closer examination, however, HOT models and preferential-attachment-like models are found to lie on common ground. For instance, the original HOT model of Fabrikant *et al.* (2002) minimizes the average distance from the attachment node to the rest of the network. Since this distance directly depends on the degree of the node (Dorogovtsev, Mendes and Oliveira, 2006; Hołyst *et al.*, 2005), the model can actually be reduced to a form of the preferential attachment model. In other words, the HOT model somehow considers the same ingredients that stand at the basis of degree-driven models, although cast in a different dynamical rule. Analogously, the introduction of more involved generating rules in stochastic evolving networks may effectively account for design principles or constraints of increasing complexity and often competing amongst themselves. Such considerations clearly lead to a convergence of modeling perspectives between the stochastic evolution and the optimization approaches.

While the trade-off paradigm balances an optimization principle with some opposing local constraints, a wide class of models just considers the emergence of networks from purely *global* optimization principles. For instance, variational approaches have been used in practical problems by road traffic engineers (Wardrop, 1952) and problems both of optimal traffic on a network (Ahuja, Magnanti and Orlin, 1993) and of optimal networks (Jungnickel, 2004) have a long tradition in mathematics and physics. Recently, it has been shown that optimal networks are relevant in a large variety of systems such as the mammalian circulatory system (Mahon and Bonner, 1983), food webs (Garlaschelli, Caldarelli and Pietronero, 2003), general transportation networks (Banavar, Maritan and Rinaldo, 1999), metabolic rates (West, Brown and Enquist, 1997), river networks (Maritan *et al.*, 1996), gas pipelines or train tracks (Gastner and Newman, 2006), and the air-travel network (Barthélemy and Flammini, 2006). In these studies, the nodes of the network are embedded in a Euclidean space so that the degree is almost always limited and the connections restricted to "neighbors" only. A second recently investigated broad class of optimal networks where spatial constraints are absent has shown that optimizing both the average shortest path and the total length can lead to small-world networks (Mathias and Gopal, 2001). More generally, degree correlations (Berg and Lässig, 2002) or scale-free features (Valverde, Ferrer i Cancho and Solé, 2002) can emerge from an optimization process. Valverde *et al.* (2002) have shown that the minimization of the average shortest path and the link density leads to a variety of networks including exponential-like graphs and scale-free networks. The minimization of search costs leads, on the other hand, either to star-like or to homogeneous networks (Guimerà *et al.*, 2002b). Finally, the interplay between obtaining short routes and little congestion leads to a variety of network structures when the imposed average degree is varied (Colizza *et al.*, 2004). Even if the philosophy behind these studies differs from the usual approach in complex systems, namely that a complex behavior emerges from a local mechanism and interactions between a large set of units, they show that global optimization could in certain cases be a relevant mechanism in the formation of complex networks.

3.4 Modeling higher order statistics and other attributes

As we have seen in the previous chapter, real-world networks are not just characterized by the degree distribution or the small-world properties. Non-trivial association properties, degree correlations, and other complex features all contribute to define the network structure. In many cases, such features are precisely the elements that allow a meaningful model validation. In the study of the dynamical processes occurring on networks, on the other hand, we might be interested in generating networks with specific correlations or clustering properties whose effects we want to study. In addition, we are generally interested in associating these properties with a specific degree distribution or growth behavior. This still constitutes an open field of study (Boguñá and Pastor-Satorras, 2003; Serrano and Boguñá, 2005; Catanzaro, Boguñá and Pastor-Satorras, 2005), since it is not completely known how the various topological properties are related, or if it is possible to single out a family of metrics defining all others (Mahadevan *et al.*, 2006). On the other hand, basic degree correlation properties are inherent even in simple random and evolving models just because of the constraints imposed by the generating algorithms. For instance, it is possible to show that disassortative correlations are inherent in heavy-tailed random graphs unless very specific precautions are taken in the network's construction. The origin of the problem lies in the fact that the completely random wiring process may generate multiple edges among the same nodes. If the degree distribution imposed by construction has a finite second moment $\langle k^2 \rangle$, the fraction of multiple edges resulting from the construction process vanishes in the thermodynamic limit and, as a consequence, they can be neglected. In contrast, in scale-free degree distributions with exponent $2 < \gamma \leq 3$, the weight of these multiple edges with respect to the overall number of edges cannot be ignored since they are not evenly distributed among all the degree classes. In the thermodynamic limit, a finite fraction of multiple edges will remain among high degree vertices, and imposing restrictions on the algorithm to avoid those occurrences induces disassortative correlations (Boguñá *et al.*, 2004; Maslov, Sneppen and Zaliznyak, 2004; Catanzaro *et al.*, 2005). Analogously, dynamically evolving networks with scale-free degree distributions also spontaneously generate disassortative correlations. The first theoretical derivation of this result (Krapivsky and Redner, 2001) was obtained by calculating the number of nodes of degree k attached to an ancestor node of degree k'. In the framework of the rate equation approach, this joint distribution does not factorize so that correlations exist (Dorogovtsev and Mendes, 2002; Park and Newman, 2003; 2004; Szabó *et al.*, 2003; Barrat and Pastor-Satorras, 2005). For the average degree of nearest neighbors, it is found that in the large k limit

$$k_{nn}(k) \sim N^{(3-\gamma)/(\gamma-1)} k^{-(3-\gamma)}, \qquad (3.42)$$

for $\gamma < 3$. Correlations between two vertices are therefore disassortative by construction, and characterized by a power-law decay. It is also interesting to note that for $\gamma = 3$, the $k_{nn}(k)$ function converges to a constant value independent of k and proportional to $\ln N$: the Barabási-Albert model lacks appreciable correlations between degrees of neighboring vertices.

While in this chapter we have mainly discussed the modeling of the topological properties of networks, the need for more specific and data-driven models has led to the formulation of models in which many other features characterizing real-world structures are introduced. Among those, spatial characteristics and vertex attributes have been plugged into both static and growing network models (Medina, Matt and Byers, 2000; Manna and Sen, 2002; Xulvi-Brunet and Sokolov, 2002; Yook, Jeong and Barabási, 2002; Barthélemy, 2003; Barrat *et al.*, 2005). Weighted properties have also been introduced in modeling attempts in order to understand the interplay and feedback between topology and traffic in complex networks (Yook *et al.*, 2001; Barrat, Barthélemy and Vespignani, 2004b; Barrat, Barthélemy and Vespignani, 2004c; Wang and Zhang, 2004; Barrat *et al.*, 2005; Bianconi, 2005; Dorogovtsev and Mendes, 2005; Wang *et al.*, 2005; Almaas, Krapivsky and Redner, 2005). Finally, various social, economic, and demographic factors have been considered in realistic approaches to the Internet and large-scale infrastructures (Chang, Jamin and Willinger, 2003; Alderson, 2004; Guimerà and Amaral, 2004; Ramasco *et al.*, 2004; Serrano, Boguñá and Díaz-Guilera, 2005; Dimitropoulos *et al.*, 2005; Chang, Jamin and Willinger, 2006; Mahadevan *et al.*, 2006). These works pair with the important observation that many networks have a considerable level of engineering acting at the local and intermediate scale level (Li *et al.*, 2004b; Doyle *et al.*, 2005). This corresponds to particular structural arrangements and connectivity properties requiring specific modeling approaches that cannot be captured with general models, which generally work on simple statistical assumptions. While these elements are very important in the faithful representation of networked systems, they introduce a level of specificity that in most cases does not allow general conclusions to be drawn. For this reason, in the rest of the book we will mainly use models that abstract very general properties whose effect on dynamical processes can be considered as general results, valid in a wide range of systems. On the other hand it is worth stressing that any realistic or detailed study of dynamical processes in specific networks cannot neglect the introduction of domain-specific features and details that might play an important role in the system's behavior.

3.5 Modeling frameworks and model validation

The considerations of the previous sections provide the proper context for a discussion of the various modeling frameworks. The exponential random models, which are akin to the statistical mechanics of networks, are built on very solid statistical foundations and have been mathematically and conceptually developed for many years. On the other hand, they are less intuitive and in many practical instances present technical problems out of our reach. This is the case for a network whose size is rapidly changing and for many non-equilibrium situations. In these cases, the dynamical approach, even if based on uncontrolled assumptions, is the only viable option. This is particularly true if we want to study very large-scale networks, for which it is nowadays possible to rely on large-scale computer simulations based on the microscopic dynamics of the system's elements. In many ways, the recent explosion in dynamical modeling approaches is a consequence of the informatics revolution over recent years. The advent of high-throughput biological experiments, and the possibility of gathering and handling massive data sets on large information structures, and of tracking the relations and behaviors of millions of individuals, have challenged the community to characterize and to model networks of unprecedented sizes. For instance, at the moment, internet data contain more than 10^4 service providers and 10^5 routers, and keeps track of the behavior of 10^7 to 10^8 users, and available data set sizes are continuously increasing. WWW crawls offer maps of the Web with more than 10^8 nodes. In addition, networks with similar sizes and dynamical characteristics are gathered every day for communication infrastructure such as mobile telephone and ad hoc networks, transportation networks, digital documents, etc. Finally, in biology we are witnessing a change of paradigm with an increasing focus on the so-called system's biology and the many large interaction networks that may be measured by taking advantage of high-throughput experiments. Most importantly, the dynamical features of these systems cannot be neglected since we are dealing in general with networks growing exponentially in time because of their intrinsic dynamics. In this context, dynamical modeling offers an intuitive grasp of the main mechanisms at play in network formation and allows the construction of basic generators that capture the rapidly evolving dynamics of these systems. In other words, the dynamical modeling approach suits the need to understand the evolution and non-equilibrium properties of these networks and might be easily plugged into large-scale numerical simulations which allow the generation of synthetic networks of 10^6 nodes with reasonable computational effort.

The availability of large-scale data sets and the feasibility of large-scale simulations also result in new conceptual questions. While networks may be extremely different from each other in their functions and attributes, the large-scale analysis of their fabric has provided evidence for the ubiquity of several asymptotic properties (see Chapter 2) and raised the issue of the emergence of general and common self-organizing principles. This question is akin to the issue of "universality" addressed in statistical physics and in particular in phase transitions. Universality refers to the fact that very different physical systems (such as a ferromagnet or a liquid), are governed by the same set of statistical laws for their large-scale properties at specific points of their phase diagram. While the nature of these systems might be very different, the shared similarities of the microscopic interactions give rise to the same macroscopic behavior. This behavior is an emergent phenomenon and does not reveal any characteristic scale, i.e., the large-scale properties look the same for all systems. The statistical physics approach has been exploited as a very convenient tool because of the possibility of characterizing emergent macroscopic phenomena in terms of the dynamical evolution of the microstates of the various systems.

At this point, however, it is worth remarking that universality does not imply equivalence. Universality refers only to large-scale statistical properties or correlation functions. Naturally, at the local level each system will be described by different properties and details. Universality focuses on specific large-scale properties and looks for possible similarities in the dynamical mechanisms responsible for creating these features across different systems. It is related to the identification of general classes of complex networks whose macroscopic properties are dictated by suitable evolution rules for the corresponding microstates, but this identification is representing a scientific challenge.

In such a context, the issue of model validation, the falsification of assumptions, and the reliability of the statistical models are major questions for which it is difficult to provide a general answer. First of all, models always contain hypotheses which are mainly dictated by our interests. While dynamic modeling is well suited for large-scale and evolutionary properties, statistical modeling based on maximum entropy considerations is more suitable to take fully into account the statistical observables at hand. Indeed, the data set properties play a different role in the two approaches. By using the exponential random graph modeling, the statistical measures obtained from the data are used to obtain the parameters of the model via fitting techniques or via the exact solution. The obtained distribution will then provide statistical predictions that can be tested through new measurements on the data set. In the dynamical approach, the network is defined by local dynamical rules that in general are not related to the statistical observables. The data set properties are not an input of the model and are instead used in the model validation process.

In most cases one needs to identify suitable approximations, since it is rarely possible to consider all the properties of a network. Depending on which question one wants to tackle, different models will be considered and different quantities will become prominent or neglected. In this sense, most models are incomplete and address only a limited set of questions. In particular, it is important to distinguish between models aimed at predicting the behavior of a system in a realistic way and models whose goals are to understand the basic mechanisms governing the behavior of the system. This discussion is the subject of many debates in the complex systems community, and as we will see in the next chapter, many of these considerations can be exported to the discussion of models concerning the dynamics occurring on networks.

4

Introduction to dynamical processes: theory and simulation

The present chapter is intended to provide a short introduction to the theory and modeling of equilibrium and non-equilibrium processes on networks and to define the basic modeling approaches and techniques used in the theory of dynamical processes. In particular, we define the master equation formalism and distinguish between equilibrium and non-equilibrium phenomena. Unfortunately, while the master equation allows for important conceptual distinction and categorization, its complete solution is hardly achievable even for very simple dynamical processes. For this reason we introduce the reader to techniques such as mean-field and continuous deterministic approximations, which usually represent viable approaches to understand basic features of the process under study. We also discuss Monte Carlo and agent-based modeling approaches that are generally implemented in large-scale numerical simulation methods.

These different theoretical methods help to define a general framework to demonstrate how the microscopic interactions between the elements of the system lead to cooperative phenomena and emergent properties of the dynamical processes. This strategy, going from microscopic interaction to emergent collective phenomena, has its roots in statistical physics methodology and population dynamics, and is currently viewed as a general paradigm to bridge the gap between the local and the large-scale properties of complex systems. It is important to stress, however, that the following material is a greatly abbreviated presentation of a huge field of research and by necessity just scratches the surface of the statistical theory of dynamical processes. Interested readers who want to dive into the mathematical and formal subtleties of the subject should refer to classic textbooks such as those by Ma (1985), Chandler (1987), Huang (1987), and Balescu (1997).

4.1 A microscopic approach to dynamical phenomena

Two main modeling schemes are adopted when dealing with dynamical processes on networks. In the first one, we identify each node of the network with a single individual or element of the system. In the second, we consider dynamical entities such as people, information packets, energy or matter flowing through a network whose nodes identify locations where the dynamical entities transit. In both cases, however, the dynamical description of the system can be achieved by introducing for each node i the notion of a corresponding variable σ_i characterizing its dynamical state. If each node represents a single individual, the variable σ_i may describe a particular attribute of the individual. A typical example that we will study in detail in Chapter 9 is the spread of an epidemic where the variable σ_i indicates if the individual is healthy or infected by a given disease. In the case of dynamical entities moving in a network substrate, the state variable σ_i generally depends on the entities present at that node. Without losing any generality, we can enumerate all possible states $\sigma_i = 1, 2, \ldots, \kappa$ for each node, and the knowledge of the state variable of all nodes in the network therefore defines the microscopic state (microstate) of the whole system. In other words, we can denote a particular configuration of the network at time t by the set $\sigma(t) = (\sigma_1(t), \sigma_2(t), \ldots, \sigma_N(t))$, where the index $i = 1, \ldots, N$ runs over all the nodes of the network of size N.

The dynamical evolution of the system is simply given by the dynamics of the configuration $\sigma(t)$ in the phase space of the system, defined by all possible configurations σ. The dynamical process is described by the transitions $\sigma^a \to \sigma^b$, where the superscripts a and b identify two different configurations of the system. In general, it is impossible to follow the microscopic dynamics of large-scale systems because of the large number of variables and the stochastic nature of most phenomena. For this reason, the basic dynamical description of the system relies on the master equation (ME) approach that we have briefly introduced in the previous chapter in the context of the dynamical evolution of networks. The ME approach focuses on the study of the probability $P(\sigma, t)$ of finding the system at time t in a given configuration σ. This probability has to be normalized, $\sum_\sigma P(\sigma, t) = 1$, and provides a probabilistic description of the system which encodes most relevant information. The so-called ME consists of the evolution equation for $P(\sigma, t)$, which reads (in the continuous time approximation)

$$\partial_t P(\sigma, t) = \sum_{\sigma'} \left[P(\sigma', t) W(\sigma' \to \sigma) - P(\sigma, t) W(\sigma \to \sigma') \right], \qquad (4.1)$$

where the sum runs over all possible configurations σ', and the terms $W(\sigma' \to \sigma)$ represent the transition rates from one configuration to the other owing to the microscopic dynamics of the system. The ME is thus made of two terms representing the gain and loss contributions for the probability distribution of the system to be in a given configuration σ. It is important to stress that the terms $W(\sigma' \to \sigma)$ are rates and carry the unit $[\text{time}]^{-1}$ as the ME is a differential equation representing the variation per unit time. The transition rates, in principle, depend on the whole configurations $\sigma = (\sigma_1, \sigma_2, \ldots, \sigma_N)$ and $\sigma' = (\sigma_1', \sigma_2', \ldots, \sigma_N')$, but in many cases of interest they can be simplified by considering that the change of state of a node i is determined only by the local interaction with the nodes directly connected to it. In the case where the local dynamics have the same parameters for all nodes, the transition rates can be simplified and read

$$W(\sigma' \to \sigma) = \prod_i w(\sigma_i' \to \sigma_i | \sigma_j). \tag{4.2}$$

The terms on the right-hand side represent the transition rates for the node i from the state σ_i' to the state σ_i, conditional to the state σ_j on the set of nodes j directly connected to it, i.e. of the neighbors $j \in \mathcal{V}(i)$ of i. The previous form decomposes the transition rates for the system into products of single node transition rates that are, in many cases, easier to handle. The network structure enters the dynamics at this point since the transition rate for a given node depends on its neighborhood structure. The network topology will thus have a direct and strong influence on the dynamics, as we will see in the various chapters of this book.

In principle the formal solution of the ME allows the calculation of the expectation values of all quantities of interest in the system. Given any function of the state of the system $A(\sigma)$ it is indeed possible to compute its average value at time t as

$$\langle A(t) \rangle = \sum_\sigma A(\sigma) P(\sigma, t), \tag{4.3}$$

where $\langle \cdots \rangle$ represents the phase space average of the quantity of interest. It is clear that the ME provides only statistical information. Even in the case where the probability distribution in the initial state is a delta function in the phase space with unit probability on a single configuration, $P(\sigma, t = 0) = \delta_{\sigma, \sigma^a}$, the stochastic evolution of the system will spread the probability on different configurations. In this perspective, the phase space average $\langle \cdots \rangle$ can be thought of as an average over different stochastic realizations of the evolution of the same system starting with identical initial conditions. In this context, a case of particular interest is given by systems with a well-defined asymptotic limit $\lim_{t \to \infty} P(\sigma, t) = P_\infty(\sigma)$. The system is then said to be in a stationary state, where the average over the stationary distribution is representative of the system after a typical transient time.

4.2 Equilibrium and non-equilibrium systems

While it is impossible to obtain a solution of the ME in most cases, the ergodic hypothesis and the maximization of entropy axiom allow an explicit form to be obtained for the stationary distribution $P_\infty(\sigma) = P_{eq}(\sigma)$ of equilibrium physical systems. Equilibrium statistical mechanics assumes that an isolated system max-imizes its entropy and reaches a uniform stationary equilibrium distribution with the same probability of being in any of the fixed energy accessible configurations. In addition, the ergodic hypothesis states that, for isolated systems, the average over the time evolution of any quantity of interest is the same as the average over the stationary equilibrium distribution. In general, however, physical systems are not isolated, but coupled to the external environment which can be considered as a heat bath fixing the equilibrium temperature of the system. By considering that the maximization of entropy applies to the system coupled to the external heat bath and that the energy is globally conserved, it is not difficult to show that the stationary distribution is no longer uniform and that the equilibrium distribution is given by the Boltzmann–Gibbs distribution

$$P_{eq}(\sigma) = \frac{\exp\left(-H(\sigma)/k_B T\right)}{Z}, \tag{4.4}$$

where T is the temperature, k_B is the Boltzmann factor that provides the correct dimensional units, and $H(\sigma)$ is the system's Hamiltonian which expresses the energy associated to each configuration of the system. The partition function Z is the normalization factor obtained by the condition $\sum_\sigma P_{eq}(\sigma) = 1$ and reads $Z = \sum_\sigma \exp\left(-H(\sigma)/k_B T\right)$. In other words, in the case of equilibrium physical systems there is no need to solve the complicated ME, and the stationary properties of the system may be obtained by knowing the system's Hamiltonian. As we will see in Chapter 5, the knowledge of the equilibrium distribution does not imply that the system's behavior is trivial. The properties of the system are indeed defined by the trade-off between the probability associated with a certain configuration at a certain energy and the number of configurations (the entropic contribution) close to that energy. The temperature, T, acts as a parameter modulating these two con-tributions, and the system changes its properties, often abruptly as in the case of phase transition, when the value of T varies. At a mathematical level the changes in the system's properties are associated with the singularities of $P_{eq}(\sigma)$ whose calculation corresponds to an exact solution for the partition function Z, which is generally not possible.

The equilibrium distribution defines the solution of the ME and it is always possible to find transition rates that drive the system towards the equilibrium distribution by imposing the so-called *detailed balance condition* on the ME

$$P_{eq}(\sigma) W(\sigma \rightarrow \sigma') = P_{eq}(\sigma') W(\sigma' \rightarrow \sigma). \tag{4.5}$$

This relation states that the net probability current between pairs of configurations is zero when $P = P_{eq}$. We will see in the next sections that this fact will allow us to construct the basic dynamical algorithms needed to simulate the microscopic dynamics of equilibrium systems.

The detailed balance is a strong condition that implies that each pair of terms in the ME has a null contribution. This is not the case for systems out of equilibrium, for which the microscopic processes violate the detailed balance and the currents between microstates do not balance. This might simply be the case of a system that has not yet reached the equilibrium stationary state and is still in a transient state. However, a wide range of different systems can be found constantly out of the detailed balance condition. In general, most systems are not isolated but are subject to external currents or external driving forces, the addition of energy and particles, or the presence of dissipation, and are therefore out of equilibrium. This is generally due to the fact that even in the stationary state the transition to a certain subset of configurations may be favored. Furthermore, a large class of non-equilibrium systems is confronted with the presence of absorbing states. These are configurations that can only be reached but not left. In this case we always have a non-zero probability current for some configurations so that the temporal evolution cannot be described by an equilibrium distribution. This is indeed what happens for a large number of spreading and reaction–diffusion systems used to model epidemic processes, as we will see in the following chapters.

It must be clear, however, that the lack of detailed balance does not imply the absence of a stationary state. Indeed, while the detailed balance (4.5) is a sufficient condition to achieve $\partial_t P(\sigma, t) = 0$, it is not necessary and the same result may be obtained through more complicated cancellations among terms of the ME, which can still lead to a stationary state. A typical example is provided in Figure 4.1, where we consider an equilibrium and a non-equilibrium system which can exist in only four different configurations A, B, C, and D. In the equilibrium case we

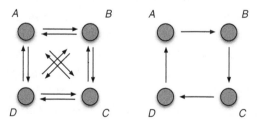

Fig. 4.1. Illustration of the detailed balance and non-equilibrium steady states in the case of different transition rates. The figure sketches two systems with four different microstates A, B, C, and D. On the left, the system obeying the detailed balance has equal transition rates among all pairs of states. On the right, transition rates allow only a specific directionality leading to a non-null probability current. In both cases, however, the system reaches a stationary state distribution function in the phase space with $P(A) = P(B) = P(C) = P(D) = 1/4$.

have, for all pairs of configurations X and Y, that $W(X \to Y) = W(Y \to X) = 1$, while in the non-equilibrium case the transition rates are $W(A \to B) = W(B \to C) = W(C \to D) = W(D \to A) = 1$ (the other rates being zero). Although the non-equilibrium system does not satisfy the detailed balance condition, both dynamics have the stationary state $P(A) = P(B) = P(C) = P(D) = 1/4$, as can easily be seen from an explicit computation of the ME. In the following we shall see that most of the considered dynamical processes relevant in real-world applications are indeed non-equilibrium phenomena for which it is not possible to provide equilibrium thermodynamic formulations.

4.3 Approximate solutions of the Master Equation

The complete solution of the master equation is rarely achievable even for very simple dynamical processes. For this reason we present suitable approximation schemes that can provide at least an approximate solution for the system behavior.

A first step in the simplification of the ME is the consideration of appropriate projections focusing on specific quantities of interest. For instance, we can inspect quantities such as

$$N_x(t) = \sum_{\sigma} \sum_{i} \delta_{\sigma_i, x} P(\sigma, t), \qquad (4.6)$$

where $\delta_{\sigma_i, x}$ is the Kronecker delta symbol, which is 1 if the node i is in the state x and 0 otherwise. The function $N_x(t)$ is simply the average number of nodes in the state x at time t, and we omit for the sake of clarity in the notation the average symbol $\langle \cdots \rangle$, which is, however, implicit in the definition of $N_x(t)$. As we are dealing with average quantities the obtained equations will be deterministic and will not account for the fluctuations inherent to the stochastic process.[1]

The deterministic projection, however, is not sufficient and further approximations are required to obtain a solvable and closed form of the ME. A typical approximation scheme is the so-called homogeneous assumption (HA) or mean-field (MF) theory. This scheme assumes that the system is homogeneous and that no relevant correlation exists between the microstate variables. In other words, all elements i of the system have the same properties and the interaction they feel is an average interaction due to the full system. This is generally encoded in a mass action or mean-field interaction with the other elements of the system. In more mathematical terms, the homogeneous assumption neglects all correlations and considers that the probability for a given element i to be in a given

---

[1] From the master equation it is possible to obtain stochastic differential equations accounting for the noise present in the system by using the Langevin formulation. Langevin equations are generally solvable in only a few cases.

state $\sigma_i = x$ is a quantity p_x independent of i. Furthermore, the probability of any system configuration is the product measure (no correlations) of single node probabilities:

$$P(\sigma) = \prod_i p_{\sigma_i}. \qquad (4.7)$$

The ME deterministic projection and the mean-field approximation allow writing sets of equations such as

$$\partial_t N_x(t) = F_x(N_1, N_2, \ldots, N_\kappa), \qquad (4.8)$$

where $x = 1, 2, \ldots, \kappa$ index the possible states of each individual node. These equations concern average values and are deterministic. The explicit form of the functions F_x depends on the specific interactions among the nodes, the transition rates, and the number of allowed microstates. In the rest of the book we will use the above approximation schemes as a general methodology in order to obtain an approximate analytic description of many systems. In constructing the MF or HA equations we will use general considerations on effective interactions and mass-action laws rather than a precise derivation from the basic ME. This allows a more rapid, versatile and intuitive derivation of the evolution equations of many systems, but it is important to keep in mind that the MF techniques are based on a general and formal approximation scheme. For this reason, we present here an instance of the MF approximation scheme in the case of a simple system and we show in detail how the basic deterministic MF equations are derived. This derivation is not necessary to understand the analytical schemes presented in the rest of the book, but for the sake of completeness we believe that it is important to detail the approximation scheme in at least one case.

Let us consider a very simple system in which each node can be in only two states $\sigma_i = A$ and $\sigma_i = B$. The dynamics of the system are simply described by a reaction process of the type $A + B \rightarrow 2B$. More precisely, the transition from A to B is irreversible and occurs with rate β each time a node in the state A is connected to at least one node in the state B. This simple reaction scheme defines the transition rates $w(A \rightarrow A|\sigma_j = A) = w(B \rightarrow B|\sigma_j = A) = w(B \rightarrow B|\sigma_j = B) = 1$, $w(A \rightarrow B|\sigma_j = B) = \beta$, where σ_j represents the state of the nodes j connected to i and we considered unitary rates $\sum_{\sigma'} w(\sigma \rightarrow \sigma'|\sigma_j) = 1$. In order to define deterministic equations we use the quantities

$$N_A(t) = \sum_\sigma \sum_i \delta_{\sigma_i, A} P(\sigma, t), \qquad (4.9)$$

and

$$N_B(t) = \sum_\sigma \sum_i \delta_{\sigma_i,B} P(\sigma,t),\qquad(4.10)$$

which are the average number of nodes in the state A or B at time t, respectively. By plugging the above projection in the ME we obtain the equation for the average number of nodes in the state B:

$$\partial_t N_B(t) = \sum_\sigma \sum_i \delta_{\sigma_i,B} \partial_t P(\sigma,t)$$

$$= \sum_i \sum_{\sigma'} \sum_\sigma \left[\delta_{\sigma_i,B} \prod_k w(\sigma'_k \to \sigma_k|\sigma'_j) P(\sigma',t)+ \right.$$

$$\left. - \delta_{\sigma_i,B} \prod_k w(\sigma_k \to \sigma'_k|\sigma_j) P(\sigma,t) \right],\qquad(4.11)$$

which can be readily simplified by noting that the term on the right-hand side can be rewritten using the normalization conditions

$$\sum_{\sigma'} \prod_k w(\sigma_k \to \sigma'_k|\sigma_j) = 1,\qquad(4.12)$$

$$\sum_\sigma \delta_{\sigma_i,B} \prod_k w(\sigma'_k \to \sigma_k|\sigma'_j) = w(\sigma'_i \to \sigma_i = B|\sigma'_j),\qquad(4.13)$$

finally yielding

$$\partial_t N_B(t) = \sum_i \sum_{\sigma'} \left[w(\sigma'_i \to \sigma_i = B|\sigma'_j) P(\sigma',t) \right] - N_B(t).\qquad(4.14)$$

So far we have not yet used the MF approximation, which is introduced by stating that the probability for each node to be in the state A or B is $p_A = N_A/N$ and $p_B = N_B/N$, respectively. This implies that all nodes have the same probabilities independent of the states of the other nodes. In addition, neglecting correlations allows us to write $P(\sigma',t) = \prod_i p_{\sigma'_i}$. By using this approximation we have

$$\sum_{\sigma'} w(\sigma'_i \to \sigma_i = B|\sigma'_j) P(\sigma',t)$$

$$= \sum_{\sigma'_j} \left[w(\sigma'_i = A \to \sigma_i = B|\sigma'_j) p_A \prod_{j\in\mathcal{V}(i)} p_{\sigma'_j}+ \right.$$

$$\left. + w(\sigma'_i = B \to \sigma_i = B|\sigma'_j) p_B \prod_{j\in\mathcal{V}(i)} p_{\sigma'_j} \right],\qquad(4.15)$$

where the sum is now restricted to the nodes j connected to the node i. This expression can be further simplified by noticing that $w(\sigma_i' = B \to \sigma_i = B|\sigma_j') = 1$ whatever the configuration of j and that $w(\sigma_i' = A \to \sigma_i = B|\sigma_j') = \beta$ if at least one of the connected nodes j is in the state B. This will happen with probability $1 - (1 - p_B)^k$, where k is the number of neighbors of i, which we assume to be the same for all nodes of the network. By substituting the previous expressions in the deterministic equation (4.14) one obtains

$$\partial_t N_B(t) = \sum_i \left(\beta p_A (1 - (1 - p_B)^k) + p_B \right) - N_B(t), \qquad (4.16)$$

and by using the expression for p_B and p_A and summing over all nodes i we obtain

$$\partial_t N_B(t) = \beta N_A (1 - (1 - N_B/N)^k). \qquad (4.17)$$

A final simplification can be obtained in the limit $N_B/N << 1$ that yields the dynamical equation

$$\partial_t N_B(t) = \beta k \frac{N_A N_B}{N}. \qquad (4.18)$$

This deterministic equation is easily solved analytically or by numerical integration. The equation for N_A in this case follows trivially from the conservation rule $N_A = N - N_B$. Equation (4.18) is the MF expression for the basic $A + B \to 2B$ process which also describes a wide range of epidemic spreading phenomena without recovery (the so-called SI model, see Chapter 9). The above equation is analogous to a mass action equation in which the per capita force of *transition* is given by the density of particles B times the number of contacts k per unit time. This also shows that in the case of a small probability p_B the MF approximation is equivalent to the homogeneous assumption using mass action laws. The deterministic MF equations can also be used to provide phenomenological Langevin equations in which Poissonian noise is added to the reaction terms. Phenomenological Langevin equations are, however, valid only under a set of precise conditions which in general are only partially satisfied. Finally, it is worth stressing that we have considered here a completely homogeneous network in which all nodes have the same degree k. We will see in the forthcoming chapters how to generalize similar equations to the case of heterogeneous networks.

4.4 Agent-based modeling and numerical simulations

In more complicated models, even the deterministic approach might not lead to solvable equations. In addition, this framework is intrinsically considering a coarse-grained perspective that does not take into account individual heterogeneity or other possible fluctuations. Numerical integration on the computer of the obtained equations therefore does not provide a complete picture of the system. In this situation, microscopic computer models, often defined as agent-based models (ABM), can be applied.[2] In these approaches each individual node is assumed to be in one of several possible states. At each time step, the model-specific update procedure that depends on the microscopic dynamics is applied to each node, which as a result changes its state depending on the state of neighboring nodes or other dynamical rules. Notably, the model's stochasticity may be introduced using Monte Carlo simulations in which rates and probabilities are mimicked in the computer with the use of random number generators. The microscopic perspective of this approach is evident in the fact that one can follow the dynamics of each individual element. In addition, the defining dynamics occur at the level of the microscopic interactions among elements, and the statistical regularities and macroscopic properties of the system are studied by looking at aggregate or average quantities. In principle, this kind of approach recreates the system within the computer, providing access to the microscopic dynamics of the system that is in general hindered by the mathematical complexity and the large number of degrees of freedom inherent to large-scale systems. The computer is therefore used as an in-silico laboratory to study complex realities not accessible mathematically or experimentally.

This way of approaching large-scale systems has a long tradition in physics. The physical laws ruling the interactions of molecules and atoms have been used in computer models of fluid dynamics, condensed matter, statistical physics and so on to provide realistic microscopic numerical simulations of systems ranging from material science and engineering to meteorology. Moreover, this tradition is one of the reasons why so many physicists enter other disciplines to exploit numerical and microscopic approaches in contexts outside physics.

In statistical physics the tradition of microscopic numerical simulations has been triggered by the advent of Monte Carlo methods in equilibrium systems. This technique has been introduced to solve the sampling problems in the numerical evaluation of physical observables. In order to provide an average of any physical quantities we must sample the quantity of interest over all the possible configurations σ that the system can assume. Even in the case of a simple system with only two states A and B this corresponds to 2^N configurations, where N is the number of elements in the system. Exhaustive enumeration is feasible only for modest sizes $N \lesssim 100$ and sampling techniques have to be used for large system sizes. On the other hand, each configuration contributes with a factor $P(\sigma)$ and most

---

[2] In some cases the name individual-based model is a more appropriate characterization in social and economical systems where each agent can be thought of as an individual performing a specific task or pursuing a set of defined tasks.

configurations are virtually inaccessible, so that sampling must take into account this heterogeneity. In the case of equilibrium systems this problem can be tackled by taking advantage of the detailed balance condition stating that

$$\frac{W(\sigma \to \sigma')}{W(\sigma' \to \sigma)} = \frac{\exp(-H(\sigma')/k_B T)}{\exp(-H(\sigma)/k_B T)}. \tag{4.19}$$

Any transition rate or probability at the microscopic level satisfying this relation ensures the convergence to the correct equilibrium distribution and can then be used to generate a microscopic dynamics to sample the configuration space. This relation does not, however, specify the transition rates uniquely as it involves ratios, and a rescaling factor has to be used for both terms. A common choice is the so-called Metropolis algorithm in which the transition probability from a configuration σ to σ' is defined as

$$W(\sigma \to \sigma') = \exp\left(-\frac{H(\sigma') - H(\sigma)}{k_B T}\right) \qquad \text{if } H(\sigma') > H(\sigma)$$

$$= 1 \qquad\qquad\qquad\qquad \text{if } H(\sigma') \le H(\sigma). \tag{4.20}$$

The Metropolis algorithm can be used to define a microscopic dynamics in which single elements are randomly chosen to change their state. The relative change is therefore accepted or rejected with the above rates (or probability in the case of finite time steps). The physics behind this is very simple. Every microscopic rearrangement that moves the system into a lower energy state is accepted with unit rate, while a move that increases the energy is accepted with a rate progressively smaller (exponentially decaying) according to the Boltzmann factor. It is trivial to see that the above expression satisfies by definition the detailed balance condition. The method can therefore be used in the computer to generate a series of configurations that, in the large time limit, properly samples the phase space. In this framework, the computer generates a Markov chain of states that can be used to correctly sample the phase space and evaluate the average of statistical observables.[3]

While in physical systems energetic considerations provide an intuitive basis to the algorithm, it is clear that in any system obeying the detailed balance condition, an appropriate microscopic dynamics can be defined by starting from the equilibrium distribution. Similar Monte Carlo methods and Markov chains are, for example, used in the modeling of the exponential random graph family presented in Chapter 3. It is important to remark that the dynamics used correctly samples only the equilibrium state and the approach to equilibrium can therefore depend upon the details of implementation, such as the choice of whether to update more

---

[3] Such a chain is Markovian since every state depends only on the previous one, and not on the full history of the system.

than one element at a time in choosing the new configuration, and the choice of time scale used. The initial starting configuration can also have an impact on the transient to the equilibrium distribution. We refer the interested reader to the classic books of Newman and Barkema (1999), Binder and Heermann (2002), Krauth (2006) for a thorough survey of Monte Carlo methods.

The success of Monte Carlo methods has paved the way for the use of microscopic computer simulations for the study of large-scale systems and their macroscopic properties. Unfortunately, the lack of detailed balance conditions in the case of non-equilibrium systems requires a different strategy. Indeed, the only way to ensure the study of the correct dynamical evolution and stationary state is the implementation of the actual microscopic dynamics ruling the evolution of the elements of the system. This implies a shift of focus from the static equilibrium perspective to dynamical models where the microscopic interaction is at the center of the system's description. This approach has been exploited in a wide range of non-equilibrium systems and is now customarily used to check the reliability and consistency of analytical results obtained through approximate methods. The microscopic approach has been historically used in other areas including theoretical epidemiology, population ecology, and game theory, often with a different name, such as mechanistic approach or individual-based modeling.

The next chapters will provide numerous examples of the application of such approaches, used in parallel with and as a complementary tool to approximate analytical approaches. For completeness of the present chapter, let us consider the reaction process $A + B \rightarrow 2B$ described in the previous section. To this end, we consider an Erdős–Rényi network of N sites, with average degree $\langle k \rangle = 10$. We start from a configuration in which all sites are in state A except for one in state B. The simulation proceeds by updating at each time step the state of all nodes in state A that are in contact with a node in state B. The reaction rate is set to $\beta = 10^{-2}$. It is straightforward to follow the evolution of the number of nodes in state B over time, as shown in Figure 4.2. The different symbols correspond to different stochastic realizations of the dynamics, while the solid line is an average over realizations of the dynamics and of the underlying network. Such simulations give easy access to quantities averaged either for a single run over the nodes or over many realizations of the dynamical process. The extent to which a single realization of the process can deviate from the average behavior can also be inferred from such simulations.

One of the advantages of agent-based numerical simulations is that they allow extremely detailed information to be obtained and, in fact, provide a way to monitor the single state of each agent at any time. Figure 4.3 illustrates such a possibility for the same reaction process $A + B \rightarrow 2B$ on a Watts–Strogatz network of $N = 150$

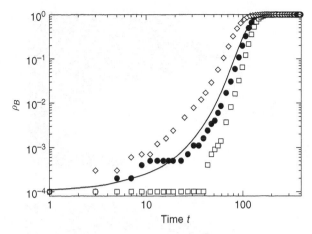

Fig. 4.2. Numerical simulations of the reaction process $A + B \rightarrow 2B$ on an Erdős–Rényi network of size $N = 10^4$ and average degree $\langle k \rangle = 10$, for a reaction rate $\beta = 10^{-2}$. The symbols show the evolution of the density $\rho_B = N_B/N$ of sites in state B for various runs, starting from a single node in state B at time 0, while the full line corresponds to the average of ρ_B over 20 realizations of the network, with 20 runs on each network realization. Especially at short times, the behavior of a single run can deviate substantially from the average evolution.

nodes (see Chapter 2).[4] The network is shown on the left of the figure as a vertical line with shortcuts. Note that periodic boundary conditions are used, i.e. the node at position $N = 150$ and the node at position 1 are considered to be neighbors. Time is shown as the horizontal axis, and the state of each node is coded as white if in state A and black if in state B. The initial state is given by one single node i_0 (here at position 108) in the B state. The first steps correspond to a propagation of the B state to the neighbors of i_0, i.e. to a coarsening phenomenon. When nodes connected to shortcuts change state, they allow new coarsening processes to be nucleated from seeds which are distant on the line. For comparison, we also show in Figure 4.3 the case in which no shortcuts are present on the one-dimensional line: a clear spatial pattern emerges and a single coarsening process is then observed as the B state is passed from each site to its neighbors along the line.

In recent times ABM approaches have become fashionable in a wide range of scientific areas such as economics, urbanscience, ecology, management, and organizational sciences. In some cases, ABMs have been put forward by enthusiasts as alternative approaches destined to revolutionize entire scientific areas and have therefore generated a reaction of mixed skepticism and mistrust from

---

[4] For the sake of clarity, we use a small network; storing the complete information on the dynamical states of all nodes soon becomes impossible as N grows, but ABMs, in any case, allow control over the precise evolution of any subset of nodes that may have particular relevance in the characterization and understanding of the process dynamics.

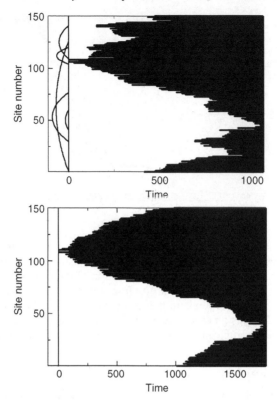

Fig. 4.3. Top: Evolution in time of one realization of the reaction process $A + B \rightarrow 2B$ on a small-world network of $N = 150$ sites (rewiring probability $p = 0.02$, average degree $\langle k \rangle = 4$, reaction rate $\beta = 10^{-2}$), with periodic boundary conditions (node 150 and node 1 are neighbors). Each site is shown in white when in state A, and in black when in state B. The dynamics starts with the single site 108 in state B. It first proceeds by coarsening around the initial site, and the long-range links depicted on the left of the figure allow new infection seeds to be nucleated at distant positions: for example, when site 109 changes to state B, it allows the process to reach site 141 directly, before the intermediate sites have changed state. On the bottom figure, we show the same process taking place on a one-dimensional line with no shortcuts. Note that the change of state of node 1 and its neighbors at time $t \approx 1000$ is due to the periodic boundary conditions for which nodes 150 and 1 are neighbors.

many scientists. This is mostly due to works that use a large number of unjustified assumptions, unphysical and/or non-measurable parameters in the definition of the agent-based model. In such cases, the models used are, in general, not falsifiable and propose microscopic rules that can be hardly, if at all, experimentally observed in the real systems.

Indeed, microscopic modeling approaches have the important advantage of fully incorporating stochastic effects and simulating very complicated dynamical processes. On the other hand, while extremely powerful, ABMs are often not transparent, since it quickly becomes difficult to discriminate the impact of any given modeling assumption or parameter as the number of parameters grows. ABMs in general have very few analytical tools by which they can be studied, and often no backward sensitivity analysis can be performed because of the large number of parameters and dynamical rules incorporated. This calls for a trade-off between the level of details included in the dynamics and the possibility of some understanding of the basic properties of the system beyond the simple computer simulations just described.

5

Phase transitions on complex networks

Statistical mechanics has long studied how microscopic interaction rules between the elements of a system at equilibrium are translated into its macroscopic properties. In particular, many efforts have been devoted to the understanding of the phase transition phenomenon: as an external parameter (for example the temperature) is varied, a change occurs in the macroscopic behavior of the system under study. For example, a liquid can be transformed into a solid or a gas when pressure or temperature are changed. Another important example of phase transition is by the appearance in various metallic materials of a macroscopic magnetization below a critical temperature. Such spontaneous manifestations of order have been widely studied and constitute an important paradigm of the emergence of global cooperative behavior from purely local rules.

In this chapter, we first recall some generalities and definitions concerning phase transitions and the emergence of cooperative phenomena. For this purpose we introduce the paradigmatic Ising model, which is a cornerstone of the statistical physics approach to complex systems and, as we will see in other chapters, is at the basis of several models used in contexts far from physics such as social sciences or epidemics. After a brief survey of the main properties of the Ising model, we show how the usual scenarios for the emergence of a global behavior are affected by the fact that the interactions between the microscopic elements define a complex network.

5.1 Phase transitions and the Ising model

Phase transitions refer to the change between various states of matter, such as the solid, liquid, and gaseous states. The transition from one state to the other depends upon the value of a set of external parameters such as temperature, pressure, and density characterizing the thermodynamic state of the system. A classic example is provided by liquid water at atmospheric pressure which becomes solid ice at

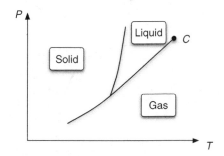

Fig. 5.1. Qualitative phase diagram of a material which, depending on the imposed pressure P and temperature T, can be in the solid, liquid or gaseous states. Note that the line separating liquid from gas ends at the critical point C.

precisely $0\,°\text{C}$ and boils at $100\,°\text{C}$, entering the gaseous state. For each value of the pressure P, the transition occurs at a specific temperature. In Figure 5.1 we draw schematically the corresponding *phase diagram* of the system which shows the coexistence curve between the two phases and for which values of the control parameter(s) each phase is obtained. Generally, phase transitions involve latent heat: the system absorbs or releases energy when changing state, and a discontinuity appears in the parameter describing the state of the system in the two phases, such as the density in the case of liquid–gas transitions. At some special points of the phase diagram, however, the transition is continuous and the system is characterized by a change of state with no latent heat.

The paradigmatic example of continuous phase transitions is provided by the ferromagnetic phase transition in materials such as iron, nickel, or cobalt, where the magnetization increases continuously from zero as the temperature is lowered below the Curie temperature. From a microscopic point of view, the spin of the electrons of each atom in the material, combined with its orbital angular momentum, results in a magnetic dipole moment and creates a magnetic field. We can then depict each atom as carrying a permanent dipole often called simply "spin". In paramagnetic materials, the magnetization is zero in the absence of any external magnetic field h, while the spins align in the direction of any non-zero value of h. In contrast, for ferromagnetic materials, neighboring spins tend to align in the same direction even for $h = 0$, as this reduces the configuration energy of neighboring atoms.[1] Below the transition temperature, this creates large domains of atoms with aligned spins, endowing the material with a spontaneous magnetization M. As the temperature increases the thermal fluctuations counteract the alignment, inducing spins to flip because of the increasing

---

[1] It is important to stress that the lower energy for paired spins is a purely quantum mechanical effect deriving from the spin-statistics theorem.

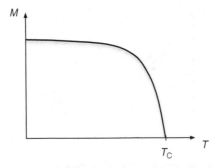

Fig. 5.2. Qualitative evolution of the macroscopic spontaneous magnetization M of a ferromagnet as a function of the temperature T. The magnetization M distinguishes between a ferromagnetic phase at $T < T_c$ and a paramagnetic phase at $T > T_c$.

energy available. When the temperature hits a critical value T_c (called the Curie temperature) the thermal fluctuations destroy the magnetic ordering and the system can no longer keep the spontaneous magnetization, settling in a paramagnetic phase. This change is continuous as more and more spins tend to escape the aligned state when temperature increases, as sketched in Figure 5.2. At the critical point there is a balance of spins in any direction and there is no net magnetization. The striking difference between the existence of a preferred alignment direction determined by the magnetization, and the disordered state with no favored direction, is a typical example of a *spontaneous symmetry breaking*. The magnetization defines the *order parameter*, which quantifies the degree of order in the system.

In order to have a microscopic description of the ferromagnetic phase transition, the most important ingredient is the interaction between spins, which tends to align them. So it seems natural, instead of considering the whole complexity of the atoms, to replace each one of them by a spin. These spins are placed on the sites of the crystalline lattice which describes the arrangement of atoms in the material. In most cases regular lattices (linear, square, or cubic) are considered. Moreover, the interaction between spins is short-ranged, so that, as a first approximation, only couplings between neighboring spins are retained. A further simplification resides in replacing the spin vectors σ_i by numbers σ_i which can take only two values, $+1$ or -1. This is equivalent to the idea that spins can take only two possible orientations (up for $\sigma_i = +1$, down for $\sigma_i = -1$).[2] In this simplified description of

---

[2] More realistic models such as the Heisenberg model or the *XY* model consider σ_i as matrices if quantum effects are taken into account or, more simply, three-dimensional or two-dimensional vectors. In particular, if the phase transition occurs at large enough temperature, the quantum effects can be completely neglected.

the ferromagnetic material, the Ising model is defined by the Hamiltonian which associates to each configuration of spins the energy

$$H = -\sum_{i \neq j} J_{ij}\sigma_i\sigma_j, \qquad (5.1)$$

where J_{ij} represents the energy reduction if spins are aligned and is either $J > 0$ if a coupling exists between i and j (i.e. if i and j are neighbors on the lattice), or 0 otherwise. Thanks to its simplicity, the Ising model has been extensively studied either when the elements (spins) are located on the sites of a D-dimensional regular lattice, or when they are all connected to each other – $J_{ij} = 1$ for all (i, j). The first case corresponds to situations close to real magnetic materials; in the second case, on the other hand, each spin is under the equal influence (or "mean field") of all the others, which allows fluctuations to be neglected and analytical solutions or approximations to be obtained.

The above Hamiltonian tells us that at $T = 0$ the system tends to the minimum energy that is achieved when all the spins are aligned (all spins in the $+1$ state or in the -1 state). To see what is happening to the system when the temperature is increasing it is possible to simulate the model on the computer by using Monte Carlo methods. These mimic the temperature-induced fluctuations by flipping spins according to transition probabilities that bring the system to the Boltzmann equilibrium (see Chapter 4 and Yeomans [1992]). The spin system is put in equilibrium with a thermostat (the external world) which imposes a certain temperature T quantifying the amount of thermal agitation which favors disordered configurations. The most important question then concerns the existence of a phase transition, i.e. of values of the control parameter compatible with either ordered or disordered behaviors. The order parameter is given by the average magnetization $M = \langle \sum_i \sigma_i \rangle / N$, where N is the number of spins in the system. For a one-dimensional chain of Ising spins, it can be shown that, owing to the short range of the interactions, the ordered configuration is obtained only when the temperature is rigorously 0 and no phase transition at a finite temperature is observed (Huang, 1987). The situation is radically different in two dimensions. Figure 5.3 shows two typical snapshots of $N = 100^2$ Ising spins at equilibrium on a square lattice, at two different temperatures. At low enough temperature, a symmetry breaking between the two possible states $+1$ and -1 takes place, while at high temperature the system is in disordered configurations with a globally vanishing magnetization. In the thermodynamic limit of an infinite N, the competition between thermal agitation and energy minimization leads, as previously described, to a phase transition at a critical value of the control parameter $T = T_c$, between a high-temperature paramagnetic phase $(T > T_c)$ with $M = 0$ and a low-temperature ferromagnetic ordered phase with a spontaneous magnetization

$T < T_C$ $T > T_C$

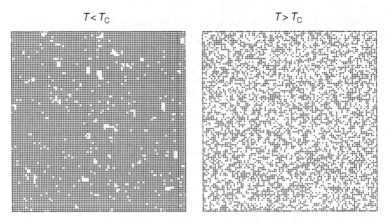

Fig. 5.3. Typical snapshots of a two-dimensional Ising system, below and above the critical temperature. Here the Ising spins are situated on the sites of a square lattice of linear size $L = 100$. Black dots represent spins $+1$, while empty spaces are left where spins take the value -1. At low temperature, a spontaneous symmetry breaking (here in favor of the $+1$ phase) is observed.

$M \neq 0$ (Figure 5.2). This was shown rigorously by Onsager in 1944 (Onsager, 1944), who completely characterized the corresponding *critical behavior* giving the evolution of the quantities of interest close to the transition.

It is important to note that, while the Ising model was originally defined in the purely physical context of magnetism, the concept of phase transitions is of course much wider and as mentioned earlier has far-reaching applications. In fact, the Ising model itself lies at the basis of many models of social behavior and opinion dynamics, as will be seen in Chapter 10. Thanks to the simplicity of its definition, it has also become the paradigm of phase transitions and of collective phenomena.

5.2 Equilibrium statistical physics of critical phenomena

For an analytical insight into phase transitions we have to rely on the equilibrium statistical physics framework, which considers that the system can be in any possible microscopic configuration. The Hamiltonian then associates an energy $H(\sigma)$ to each configuration σ. As discussed in Chapter 4, at equilibrium the probability for the system to be in each configuration is given by its Boltzmann weight

$$P(\sigma) = \frac{1}{Z} \exp[-\beta H(\sigma)] \tag{5.2}$$

where $\beta = 1/(k_B T)$ is the inverse temperature (k_B is the Boltzmann constant), and $Z = \sum_\sigma \exp(-\beta H(\sigma))$ is the partition function ensuring normalization. At low temperature, β is large and therefore low-energy configurations are favored. At high temperature, the Boltzmann probability distribution (5.2) becomes more and more uniform. Aligned spin configurations are therefore statistically favored in the low-temperature regime. Above the critical temperature, disordered configurations take over and the spontaneous magnetization is zero. The various properties of a model, defined by its Hamiltonian, are obtained through the calculation of (5.2). For example, given an observable O whose value depends on the configuration of the system, the average observed value of O at temperature T is given by

$$\langle O \rangle = \sum_\sigma O(\sigma) P(\sigma) = \frac{1}{Z} \sum_\sigma O(\sigma) \exp[-\beta H(\sigma)]. \tag{5.3}$$

The analytical solution of the partition function and the exact calculation for the order parameter are possible only in very special cases. In particular, at the transition point, the partition function is singular and the study of the critical properties has been one of the great challenges of statistical physics, culminating in the development of the renormalization group method (see, for instance, Ma, 2000).

The importance of critical phase transitions lies in the emergence at the critical point of cooperative phenomena and *critical behavior*. Indeed, close to the transition point, for T close to T_c, the thermodynamic functions have a singular behavior that can be understood by considering what is happening at the microscopic scale. By definition, the correlation between two spins is $\langle \sigma_i \sigma_j \rangle - \langle \sigma_i \rangle \langle \sigma_j \rangle$, and the correlation function $G(r)$ is the average of such correlations for pairs of spins situated at distance r. The correlation function measures the fluctuations of the spins away from their mean values, and vanishes for $r \to \infty$ at both low and high temperature where spins are either all aligned or fluctuate independently, respectively. In particular, at high temperature the correlation function decays with a correlation length ξ that can be considered as an estimate of the typical size of domains of parallel spins. As $T \to T_c$, long-range order develops in the system and the correlation length diverges: $\xi \to \infty$. More precisely, ξ increases and diverges as $|(T - T_c)/T_c|^{-\nu}$ and exactly at T_c no characteristic length is preferred: domains of all sizes can be found, corresponding to the phenomenon of *scale invariance* at criticality. Owing to the scale invariance at T_c, the ratio $G(r_1)/G(r_2)$ is necessarily a function only of r_1/r_2, say $\phi(r_2/r_1)$. Such identity, which can be rewritten as $G(r/s) = \phi(s)G(r)$, has for consequences $G(r/(s_1 s_2)) = \phi(s_1 s_2)G(r) = \phi(s_1)\phi(s_2)G(r)$, which implies that ϕ is a power law and we have therefore at the critical point

$$G(r) \sim r^{-\lambda}, \tag{5.4}$$

where λ is an exponent to be determined. For temperatures close to but different from T_c, one assumes that only one characteristic scale is present, namely ξ. The crucial *scaling assumption* then consists of writing that $G(r)$ *rescaled* by its value at the critical point is in fact only a function of r/ξ:

$$G(r, T) = r^{-\lambda} g(r/\xi(T)). \tag{5.5}$$

The quantity $g(r/\xi)$ tends to the constant $g(0)$ when ξ tends to infinity, recovering the power-law behavior $G(r) \sim r^{-\lambda}$. When the system is not exactly at the critical point, this behavior is modified in a way that depends only on $\xi \simeq |(T - T_c)/T_c|^{-\nu}$, i.e. on the distance of the control parameter from the critical point.

The system is therefore controlled by the characteristic length divergence and it is possible to express the singular behavior of any thermodynamic function $\langle O \rangle$ (assuming its limit exists) as a function of the adimensional deviation from the critical point $t = |(T - T_c)/T_c|$ as

$$\langle O \rangle = A|t|^{\mu}(1 + B|t|^{\mu_1} + \cdots), \tag{5.6}$$

where μ is the critical exponent characterizing the scaling of the associated thermodynamic quantity. The typical example is the average magnetization, which is found to follow the law

$$M \approx (T_c - T)^{\beta}, \tag{5.7}$$

where the value of β depends on the Euclidean dimension of the lattice. Power-law behaviors and critical exponents are also observed for other quantities, such as the magnetic susceptibility or response function $\chi \sim t^{-\gamma}$ which gives the variation dM/dh of the magnetization with respect to a small applied magnetic field h that tends to align the spins.

The interest in critical exponents is also related to the *universality* displayed by many models which, despite having different definitions, share the same sets of critical exponents. The scale invariance allows us to understand intuitively that, close to the transition, which is a large-scale cooperative phenomenon, the overall macroscopic behavior of the system should not depend on its very detailed features. For T close to T_c, the critical behavior will be determined by the symmetries of the microscopic degrees of freedom and the embedding Euclidean space, but, for example, not by the shape of the corresponding regular lattice. In general, critical exponents depend on the dimensionality of the system and the symmetry of the order parameter and not on the detailed form of the microscopic interactions. This is a major conceptual point that helps in understanding why the critical behavior of a fluid system can be the same as that of a ferromagnetic material, despite the difference in the physical interactions at play.

At this point, let us insist on the fact that universality does not mean that the systems under study are equivalent. Universality concerns only the occurrence of a common behavior in the critical region, near a phase transition phenomenon. The universal quantities are exponents, scaling behaviors, and scaling and correlation functions which describe the large-scale characteristics. At the local microscopic level, each system is still described by its own properties and details. The term universality thus refers to the possible similarities in the large-scale behavior of otherwise very different systems, and challenges us in searching for similarities in the underlying physical mechanisms producing these features.

5.2.1 Mean-field theory of phase transitions

The analytical calculation of the critical exponents and the singular behavior in critical phenomena has represented one of the toughest challenges tackled in statistical physics and culminated in the development of the renormalization group analysis (see, for instance, Ma, 2000). On the other hand, it is possible to grasp the basic physics of critical phenomena without getting into the complications of the renormalization group technique by using the most widely used approximation method: the mean-field theory. In mean field, one considers that each spin is under the equal influence of all the others. The contribution of a given spin σ_i to the total energy of the system, $-\sigma_i \sum_j J_{ij}\sigma_j$, is thus replaced by $-J\sigma_i \sum_j \langle\sigma_j\rangle$ where the sum runs over the neighbors of i, i.e. by $-J\langle k\rangle M\sigma_i$ where $\langle k\rangle$ is the average number of neighbors of each spin and M is the average magnetization. The self-consistency of the approximation imposes $M = \langle\sigma_i\rangle$, so that, by applying the usual statistical equilibrium mechanics to the spin σ_i at equilibrium at temperature T, one obtains[3]

$$
\begin{aligned}
M &= \frac{1}{Z} \sum_{\sigma_i = \pm 1} \sigma_i \exp\left(\frac{J\langle k\rangle M}{T}\sigma_i\right) \\
&= \frac{\exp\left(\frac{J\langle k\rangle}{T}M\right) - \exp\left(-\frac{J\langle k\rangle}{T}M\right)}{\exp\left(\frac{J\langle k\rangle}{T}M\right) + \exp\left(-\frac{J\langle k\rangle}{T}M\right)} \\
&= \tanh\left(\frac{J\langle k\rangle}{T}M\right).
\end{aligned}
\tag{5.8}
$$

This self-consistent equation has the form $M = F(M)$, where F has some important properties. First of all, one notes that $F(0) = 0$ which means that $M = 0$ is always a solution of this equation. Moreover, $\lim_{M\to\infty} F(M) = 1$, F increases strictly (the first derivative $F'(M)$ is positive) and is concave (the second derivative $F''(M)$ is negative), for $M > 0$. This allows the use of the graphical method sketched in Figure 5.4 to find the non-zero solution of the self-consistent equation.

---

[3] In the following for the sake of notation simplicity we will consider that the Boltzmann constant is unitary ($k_B = 1$). The complete expression can be recovered by simple dimensional arguments.

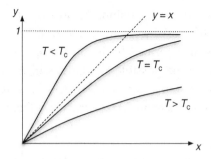

Fig. 5.4. Schematic representation of Equation (5.8). The dashed line represents $y = x$, continuous lines $y = F(x)$ for various values of T. A solution $M > 0$ exists if and only if the slope of the curve $y = F(x)$ in $x = 0$ is larger than 1.

The self-consistent solution for the magnetization is indeed given by the intersection of the curve $y = F(x)$ with the straight line $y = x$, which exists only if the derivative of F at $x = 0$ is larger than 1. This condition translates as

$$\frac{d}{dM}\left[\tanh\left(\frac{J\langle k\rangle}{T}M\right)\right]\Bigg|_{M=0} > 1. \tag{5.9}$$

This condition is satisfied if and only if T is smaller than a critical temperature T_c such that

$$\frac{J\langle k\rangle}{T_c} = 1. \tag{5.10}$$

Equation (5.8) is symmetric with respect to the change $M \to -M$, since the original Hamiltonian is itself symmetric if all spins are changed into their opposite ($\sigma_i \to -\sigma_i$) and therefore, as T decreases below $T_c = J\langle k\rangle$, two opposite non-zero solutions appear. It can be shown that these non-zero solutions are stable while the solution $M = 0$ is unstable for $T < T_c$. By expanding the self-consistent equation around the critical point T_c it is also possible to derive the behavior of the order parameter, leading to $M \approx (T_c - T)^{1/2}$, i.e. $\beta = 1/2$. Analogously the value of all critical exponents can be computed in the mean-field approximation.

While extremely simple, the mean-field theory is by nature only an approximation to the real solution. The main issue is that by definition the large fluctuations present at the critical point are not included in the theory, which is based on a homogeneous description of the system. Therefore, the mean-field results can only be valid when the fluctuations are not important. At the critical point, fluctuations are relevant and it is possible to show that the mean-field approximation is valid only above the so-called upper critical Euclidean dimension which is $D_c = 4$ for the Ising model. While this is a problem in most systems, we know that networks can be generally mapped into graphs that can be considered by definition as infinite-dimensional objects. In this perspective, mean-field techniques are particularly useful in calculating the behavior of critical phenomena in complex networks, as we will see in various chapters of this book.

5.3 The Ising model in complex networks

The Ising model (5.1) considers an ensemble of variables that interact pairwise. The couplings J_{ij} can therefore be seen as the presence ($J_{ij} > 0$) or the absence ($J_{ij} = 0$) of a link between i and j, defining the topology of the *interaction* network. The recent discovery of the relevance of network representations in many areas of science has stimulated the study of models in which the spins interacting via Equation (5.1) are located on the nodes of a network whose topology can be more complex than the previously described cases. A first motivation for such studies stems from the fundamental interest in understanding how the topology of the interactions change the nature of the phase transition. Moreover, the Ising variables can represent not only magnetic spins, but also, for example, opinions in the field of social sciences (see Chapter 10). In this case, complex networks are the natural framework for representing social interactions, and understanding the emergence of collective behavior of variables defined on a network becomes a natural question. In this chapter, we will show how the different characteristics of complex networks determine various behaviors for the phase transition, depending on the structure of the network.

5.3.1 Small-world networks

A first property of complex networks that may influence the emergence of collective phenomena is the small-world behavior. In this case, the distances between the nodes of the network are small and the diameter scales as the logarithm of the network size N. This strikingly contrasts with finite-dimensional lattices where for a lattice of N sites in D dimensions, the average distance between nodes is of order $N^{1/D} \gg \log N$. For a mean-field topology, on the contrary, the distance between two nodes is by definition 1. Small-world networks therefore appear as an intermediate situation between these two well-studied cases. The case of the Ising model on the small-world networks of Watts and Strogatz (1998) is particularly interesting in this respect: their construction allows continuous interpolation from a one-dimensional lattice topology, for which no phase transition is observed, to an almost random graph, for which a finite T_c exists on random graphs, with a mean-field behavior for the transition (Kanter and Sompolinsky, 1987).

Let us consider a system of N Ising spins $\sigma_i = \pm 1$, $i = 1, \dots, N$, with Hamiltonian

$$H(\sigma) = -J \sum_{i=1}^{N} \sigma_i \sum_{j=1}^{m} \sigma_{\mu(i,j)} \qquad (5.11)$$

where periodic boundary conditions are assumed (i.e. $\sigma_{N+j} = \sigma_j$), and where the independently and identically distributed numbers $\mu(i, j)$ are drawn from the probability distribution

$$P(\mu(i, j)) = (1 - p)\delta_{\mu(i,j),i+j} + \frac{p}{N} \sum_{l=1}^{N} \delta_{\mu(i,j),l}. \qquad (5.12)$$

For $p = 0$, this corresponds to a pure one-dimensional Ising model where every site is connected to its $2m$ nearest neighbors by ferromagnetic bonds of strength 1. At finite p, this corresponds to spins sitting on a Watts–Strogatz network in which each link of the one-dimensional chain has been rewired with probability p, yielding typically pmN long-range connections. For $p = 1$, the one-dimensional structure is completely replaced by randomly rewired links.

Barrat and Weigt (2000) have studied this model using the replica formalism. The obtained equations allow unveiling the behavior of the system. At high temperature, a stable paramagnetic phase is obtained for any p. The stability of this phase can be investigated at small p by means of a first order perturbation in p, i.e. an expansion around the topology of the one-dimensional model. The interesting observation is that the expansion contains powers of $p\xi_0$, where $\xi_0 \sim \exp[Jm(m+1)/T]$ is the correlation length of the one-dimensional system and implies that the first order approximation breaks down when $p\xi_0$ becomes of order 1. This condition is met either by increasing p or by decreasing T. At fixed temperature, there is therefore a crossover from a weakly perturbed one-dimensional system for $p \ll p_{cr}(T)$, with

$$p_{cr}(T) \propto \exp[-Jm(m+1)/T], \qquad (5.13)$$

to a regime dominated by the network disorder for larger p. Conversely, at fixed and finite p, lowering the temperature until $T < T_{cr}$ with

$$T_{cr} \propto -Jm(m+1)/\log(p) \qquad (5.14)$$

brings the system into a regime in which the network disorder dominates the system behavior, i.e. a mean-field regime in which a phase transition is expected. Different analytical treatments can also be applied. Viana Lopes *et al.* (2004) use combinatorial methods to show that the transition temperature in a modified Watts–Strogatz model (with deterministic shortcuts) behaves as $1/\ln p$, and that the critical behavior is of mean-field type. Another approach is provided by Nikoletopoulos et al. (2004), who solve the problem using the replica formalism within a replica symmetric approximation.

It is worth explaining the physical meaning of the crossover between the one-dimensional and the mean-field behavior, since it is general and does not depend on the precise model defined on the small-world network. Similar arguments can in fact be used in the study of different dynamical models defined on small-world networks, such as the Voter model (Castellano, Vilone and Vespignani, 2003) or the Naming Game (Dall'Asta *et al.*, 2006b). In the pure one-dimensional model ($p = 0$), domains of correlated spins have the typical size ξ_0 which, for the Ising model, corresponds to the typical size of a domain of spins having the same value. The average number of long-range links arriving in such a domain is, by definition of the Watts–Strogatz model, of order $p\xi_0$. It is then intuitively clear that the behavior of the system will be very different for $p\xi_0 \ll 1$ and $p\xi_0 \gg 1$. In the first case, the interactions will be dominated by the one-dimensional structure, and in the second case, the numerous long-range connections will lead to a mean-field-like regime. This argument can be rephrased in terms of a competition between the size ξ_0 of correlated regions and the average distance $N^* \sim 1/p$ between two long-range links. At fixed network disorder p, and as the temperature T is lowered, ξ_0 increases: this implies that a change in the system behavior is expected at the crossover temperature T_{cr} such that $p\xi_0(T_{cr}) \sim 1$.

Gitterman (2000) studies a slightly different construction of the network, obtained by addition of long-range links to a one-dimensional network. While a finite p as N goes to infinity yields a finite *density* of long-range links, this study considers a finite *number* of these links (which would correspond in the original formulation to rewiring probabilities $p = \mathcal{O}(1/N)$). In this case, a phase transition appears as soon as p is larger than a certain f_c/N, with f_c finite, i.e. as soon as a sufficiently large number of long-range links is present in the network.

Various Monte Carlo simulations of Ising spins on Watts–Strogatz networks have confirmed the above picture and the existence of a phase transition of mean-field character for any finite p (Barrat and Weigt, 2000; Pękalski, 2001; Hong, Kim and Choi, 2002b). Numerical studies have also considered the Ising model on Watts–Strogatz networks constructed starting from a D-dimensional network. For $D = 2$ or 3, a phase transition is present even in the ordered case (at $p = 0$). For any finite rewiring probability p, the character of this transition changes to mean-field, as indicated by the measure of the critical exponents (Herrero, 2002; Svenson and Johnston, 2002).

Finally, another paradigmatic model for the study of phase transitions – the *XY* model – has also been studied on small-world networks. In this model, spins have a *continuous* degree of freedom and are two-dimensional vectors of fixed length, determined by their angle θ_i with respect to a reference direction. Their interaction is given by their scalar product, thus yielding the Hamiltonian

$$H = -\sum_{i\neq j} J_{ij} \cos(\theta_i - \theta_j).\tag{5.15}$$

In this case, the order parameter is given by the average value of $\sum \exp(i\theta_j)/N$ (where in this expression i does not design a lattice site but the imaginary square root of -1: $i^2 = -1$). Numerical simulations of this model on small-world networks show that a phase transition with mean-field exponents is obtained even for very small values of p, suggesting a finite transition temperature for any finite p (Kim *et al.*, 2001; Medvedyeva *et al.*, 2003).

Interestingly, in all these cases, the physical argument taking into account the competition between the correlation length ξ_0 of the ordered system (at $p = 0$) and the average distance N^* between long-range connections also applies. Indeed, for the *XY* system, $\xi_0 \sim 1/T$ so that $\xi_0 \sim N^*$ implies $T_c \sim p$ at small p. For Watts–Strogatz networks constructed from 2- or 3- dimensional lattices, $\xi_0 \sim |T - T_c^0|^{-\nu}$, and $N^* \sim p^{-1/D}$ so that the shift in the critical temperature is given by $T_c - T_c^0 \sim p^{1/(D\nu)}$, where T_c^0 is the critical temperature for $p = 0$.

5.3.2 Networks with generic degree distributions

Another crucial ingredient of many complex networks, as explained in Chapter 2, lies in their heterogeneous connectivity structure signalled by a broad degree distribution. While Watts–Strogatz networks have bounded connectivity fluctuations, one can expect a heterogeneous structure for the underlying network to have dramatic consequences on the possible phase transitions of spin models defined on such networks.

Following numerical simulations of Aleksiejuk, Holyst and Stauffer (2002), this problem has been addressed analytically almost simultaneously by three different methods, by Bianconi (2002), Leone *et al.* (2002) and Dorogovtsev, Goltsev and Mendes (2002). These studies consider random graphs taken from an ensemble defined by a given degree distribution $P(k)$, i.e. without correlations between the degrees of neighboring sites. In this case, it is possible to write a simple mean-field approach, valid for generic networks (Leone *et al.*, 2002). As explained in Section 5.2.1, in the mean-field approximation each spin σ_i is under the influence of the average of its neighbors and the mean-field Hamiltonian reads as

$$H = -J\langle\sigma\rangle \sum_i k_i \sigma_i,\tag{5.16}$$

where k_i is the degree of node i. The various spins are therefore effectively decoupled. The average $\langle \sigma_i \rangle$ is then computed and has to be self-consistently equal to $\langle \sigma \rangle$. Solving the self-consistency equation allows its value to be obtained.

For spins located on the nodes of generic networks, the global average $M = \langle \sum_i \sigma_i \rangle / N$ mixes nodes of potentially very different degrees, so that it is more convenient to define the *average magnetization of the class of nodes with degree k*,

$$\langle \sigma \rangle_k = \frac{1}{N_k} \sum_{i/k_i=k} \langle \sigma_i \rangle, \tag{5.17}$$

where N_k is the number of nodes having degree k. We will see in other chapters that this approach, which consists of dividing the nodes according to their degree classes, is very effective and helps in obtaining useful analytical approximations.

We define u as the average magnetization *seen by a node on its nearest neighbors*. It is important to note that this definition is not equivalent to that of the average magnetization M. As detailed in Chapter 1, the probability that any edge is pointing to a nearest neighbor with degree k is $kP(k)/\langle k \rangle$, and therefore the average magnetization seen on any given neighbor is given by

$$u = \sum_k \frac{kP(k)}{\langle k \rangle} \langle \sigma \rangle_k. \tag{5.18}$$

This expression considers the proper average over all possible degrees of the magnetization $\langle \sigma \rangle_k$ of the neighbors and is different from the average net magnetization $M = \sum_k P(k) \langle \sigma \rangle_k$. Nevertheless, the magnetization will be non-zero, indicating a ferromagnetic phase, if and only if $u \neq 0$. The mean-field approach then amounts to writing the Hamiltonian as

$$H = -J \sum_k \sum_{i/k_i=k} \sigma_i ku \tag{5.19}$$

i.e. to consider that a spin σ_i on a node of degree k "feels" a local field ku. The mean-field equation in each class thus reads

$$\langle \sigma \rangle_k = \tanh(\beta J k u). \tag{5.20}$$

The combination of (5.18) and (5.20) yields the following closed equation for u:

$$u = \sum_k \frac{kP(k)}{\langle k \rangle} \tanh(\beta J k u). \tag{5.21}$$

Similar to (5.8), this equation is of the form $u = F(u)$, with $F(0) = 0$ and F increasing and concave, so that a non-zero solution is obtained if and only if the derivative of F in 0 is greater than 1. This condition reads

$$\sum_k \frac{k P(k)}{\langle k \rangle} \beta J k > 1, \tag{5.22}$$

which means that the critical temperature is

$$T_c = J \frac{\langle k^2 \rangle}{\langle k \rangle}. \tag{5.23}$$

This result, valid for any network, relies on a crude mean-field approximation which can in fact be refined. A detailed replica calculation (Leone *et al.*, 2002) yields the result

$$\frac{1}{T_c} = \frac{1}{2J} \ln \left(\frac{\langle k^2 \rangle}{\langle k^2 \rangle - 2\langle k \rangle} \right), \tag{5.24}$$

and also allows an analysis of the critical behavior. In particular, the above expressions indicate that, in heavy-tailed networks where the degree fluctuations $\langle k^2 \rangle$ diverge in the thermodynamic limit $N \to \infty$ and $\langle k^2 \rangle \gg \langle k \rangle$, the transition temperature T_c is in fact infinite. Let us consider the case of scale-free networks, which display power-law distributions $P(k) \sim k^{-\gamma}$. The second moment $\langle k^2 \rangle$ and thus the critical temperature T_c are finite if and only if $\gamma > 3$. In fact, the transition is of mean-field character if $\gamma \geq 5$, while non-trivial exponents are found for $3 < \gamma < 5$. The dependence of T_c with γ is shown in Figure 5.5: as γ decreases towards 3, T_c diverges, and only the ferromagnetic phase survives for $\gamma < 3$ (Leone *et al.*, 2002). The effect of the heterogeneous network topology on the physics of phase transition is also revealed in the way the system approaches the thermodynamic limit. Phase transitions are rigorously defined only in the thermodynamic limit of an infinite number of spins. At finite but large N, however, a qualitative change of behavior can be observed as the temperature is lowered, with a stronger tendency of the system to display ordered configurations at $T < T_c^{\text{eff}}(N)$. In the usual case of a finite transition temperature T_c, such a change of behavior appears at a temperature $T_c^{\text{eff}}(N)$ which tends to T_c as N grows. In the case of (5.23), on the other hand, the change of behavior appears at higher temperature as the system size increases. In the thermodynamic limit, $T_c^{\text{eff}} \to \infty$ and the transition disappears, because the hubs polarize all the network and the only possible phase is then an ordered, ferromagnetic phase.

Another simple way of obtaining the previous results (5.24) has been proposed by Dorogovtsev *et al.* (2002), for uncorrelated networks: the mere fact that the distribution of the degree of a *neighbor* of a vertex is $k P(k)/\langle k \rangle$ implies that the neighbors of a randomly chosen vertex have an average degree $\langle k^2 \rangle/\langle k \rangle$. For Ising spins

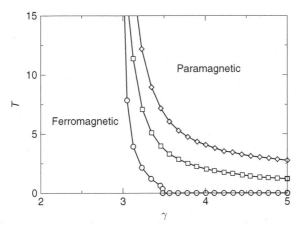

Fig. 5.5. Phase diagram of the Ising model on scale-free graphs with a power law degree distribution $P(k) = ck^{-\gamma}, m \leq k < \infty$. The ferromagnetic transition lines depend on the value of m, and are shown for $m = 1$ (circles), $m = 2$ (squares), and $m = 3$ (diamonds). For $m = 1$ and $\gamma > \gamma^* \simeq 3.47875$ the graph does not have a giant component so that $T_c = 0$. Data from Leone *et al.* (2002).

sitting on the nodes of a Cayley tree with coordination number q, the critical temperature is known and equal to $T_c = 2J / \ln[q/(q-2)]$ (Baxter, 1982). Assuming a local tree-like structure for the generic network allows this formula to be used with the substitution $q = \langle k^2 \rangle / \langle k \rangle$, recovering the previous result (5.24). More involved calculations (Dorogovtsev *et al.*, 2002) using recurrence relations show that this relation is in fact exact, still under the assumption of local tree-like structure and of the absence of correlations between the degrees of neighboring nodes.

It is also possible to extend the results for the Ising model to different statistical mechanics spin models. For example, the Potts model is a generalization of the Ising model to spins allowed to take q different values. The energy of a configuration of spins is then given by

$$H = -\sum_{i \neq j} J_{ij} \delta_{\sigma_i, \sigma_j}, \tag{5.25}$$

with couplings J_{ij} equal to $J > 0$ or 0, as in the ferromagnetic Ising case (which corresponds to $q = 2$). While the phase transition is of second order for the Ising model, it becomes of first order for $q \geq 3$, with a hysteresis region in temperature where both ferromagnetic and paramagnetic behaviors are possible. For this different type of transition, Dorogovtsev, Goltsev and Mendes (2004) show that the lower bound of the hysteresis region is given by

$$\frac{J}{T_c} = \ln \left(\frac{\langle k^2 \rangle + (q-2)\langle k \rangle}{\langle k^2 \rangle - 2\langle k \rangle} \right). \tag{5.26}$$

This temperature T_c thus diverges if $\langle k^2 \rangle$ diverges. In the case of scale-free networks with $P(k) \sim k^{-\gamma}$, this means that the first order phase transition survives if and only if $\gamma > 3$. The phenomenology is similar to what happens with the Ising model.

Equations (5.24) and (5.26) help one to grasp what makes phase transitions on heterogeneous networks remarkable. The appearance of the heterogeneity parameter $\kappa = \langle k^2 \rangle / \langle k \rangle$ shows that the topological characteristics of the interaction network strongly influence the behavior of the phase transitions. In particular, the critical temperature of an Ising model defined on a network with arbitrary degree distribution $P(k)$ is finite if and only if the second moment $\langle k^2 \rangle$ of the distribution is finite. In the opposite case, the large degrees of the hubs prevent any thermal fluctuation from destroying the long-range order: only a ferromagnetic phase is allowed. In general, if κ is large but finite, which can occur, for example, in networks for which the broad degree distribution is bounded by some structural cut-off, the phase transition occurs at a temperature that is much larger than in the case of homogeneous networks or finite-dimensional lattices.

5.4 Dynamics of ordering processes

Up to now in this chapter, we have reviewed the equilibrium properties of the Ising model, and shown how these properties are modified when the spins are located on the nodes of a network instead of regular lattices.

Physical systems are, however, not always at equilibrium. In particular, when the temperature is abruptly changed from T_1 to T_2, the system, if it was at equilibrium at T_1, is not at equilibrium at T_2 and has to adapt to the new temperature. This adaptation takes a certain time which corresponds to the approach to the new equilibrium state at T_2. If both T_1 and T_2 correspond to the same equilibrium phase, either disordered or ordered, this adaptation or "transient" is fast since the properties of the system are not strongly changed. If on the other hand this temperature change crosses a phase transition, with $T_1 > T_c$ and $T_2 < T_c$ ("quench" to the ordered phase), the system has to evolve from disordered configurations towards the ordered, low-temperature ones. This *ordering process* has been largely studied for spin models, as a framework for the understanding of the ordering dynamics of magnets, in particular for spins located on regular lattices (see the review by Bray [1994]).

As put forward in Chapter 4, a convenient way to study dynamical processes is to perform numerical Monte Carlo simulations of the model under scrutiny (Krauth, 2006). One then has to specify the probability of transition from one configuration to another. In order to describe the relaxation to equilibrium, these transition probabilities need to satisfy the detailed balance condition. The most popular choices are the Metropolis algorithm (see Chapter 4) and the so-called Glauber dynamics, in which the transition probability from one configuration σ to another σ' reads

$$W(\sigma \to \sigma') = \frac{1}{1 + \exp\left([H(\sigma') - H(\sigma)]/k_B T\right)}. \qquad (5.27)$$

It is straightforward to check that this choice respects the detailed balance condition (4.5) $W(\sigma \to \sigma')/W(\sigma' \to \sigma) = \exp([H(\sigma) - H(\sigma')]/k_B T)$ and that the shape of the transition rates as a function of the energy difference is similar to the Metropolis algorithm. More precisely, the Glauber and Metropolis rates become close to each other at small values of the temperature T. In particular, at exactly $T = 0$, the only difference between Metropolis and Glauber transition probabilities occurs for $H(\sigma') = H(\sigma)$, i.e. if the proposed change of configuration does not modify the energy of the system. The probability of accepting this change is then 1 for Metropolis and 1/2 for Glauber dynamics. According to the concept of universality, the main features of the dynamics should not depend on this particular choice.

On regular lattices, spins forming a finite-dimensional lattice in an initially disordered (random) configuration tend to order when they are quenched below the critical temperature. This process, known as *coarsening*, is illustrated in Figure 5.6 for a two-dimensional system of Ising spins quenched at zero temperature (Bray, 1994). Starting from a disordered configuration, domains of aligned spins start to form and grow. The typical size of such domains grows with time: small domains are absorbed by larger ones, and the interfaces between the domains tend to become smoother and to decrease their curvature. For any finite system, at large enough times one domain overcomes the other and all the spins become aligned.

Interestingly, if quenched at strictly zero temperature, the system may also remain trapped in states that are not fully ordered, but consist for instance in

Fig. 5.6. Evolution of the Ising model with Glauber dynamics at zero temperature for spins located on the sites of a two-dimensional square lattice of linear size $L = 200$. Black dots represent spins $+1$, while empty spaces are left where spins take the value -1. From left to right the system, starting from a disordered configuration, is represented at times $t = 1, 4, 16, 64$.

two dimensions of alternating stripes of $+$ and $-$ spins (Spirin, Krapivsky and Redner, 2001; 2002). In larger dimensions, the probability of reaching a fully ordered configuration decreases and vanishes in the thermodynamic limit. The system is likely to end up in particular configurations in which domains of opposite spins coexist, each spin being aligned with its local field, and therefore unable to evolve, except for some spins called "blinkers" which have an equal number of $+$ and $-$ neighbors and therefore can flip without any energy change.

As a natural complement to the study of the equilibrium properties of spin models on complex networks, it is interesting to investigate how the approach to equilibrium occurs in these cases, and how the ordering process unfolds.[4] Various efforts have therefore been devoted to the understanding of the prototypical dynamics of the Ising model quenched at zero temperature on complex networks. The main issues lie in the ability of the system to reach the perfectly ordered state corresponding to equilibrium, and the time needed for such ordering. Boyer and Miramontes (2003) have considered the Glauber dynamics at $T = 0$ of Ising spins located on the nodes of a small-world network obtained by adding, with probability p, to each site of a two-dimensional regular lattice, a link connected to a randomly chosen other site. Because of these shortcuts, the distances between nodes are smaller for $p > 0$ than for $p = 0$ and one could expect that this would favor the ordering process. On the contrary, it turns out that the shortcuts lead to an effective "pinning" of the interfaces because some sites have larger degree than others. At short times, domains of aligned spins form and grow according to the usual coarsening phenomenon. The coarsening, however, stops when the domains reach a typical size which depends on the disorder p, and the system remains frozen with coexisting domains of up and down spins.

For Ising spins on completely random networks, the situation bears some similarities. As noted by Svenson (2001), the Glauber dynamics at zero temperature does not always lead the system to full ferromagnetic ordering. In fact, Häggström (2002) has shown analytically that the dynamics always fails to reach the global energy minimum (ordered state) in the thermodynamic limit $N \to \infty$. Numerical investigations reveal that the system remains trapped in a set of configurations with two highly intertwined domains of opposite spins, having roughly the same size (Castellano *et al.*, 2005). The system is not frozen, and a certain number of spins keep flipping back and forth: these spins have an equal number of up and down neighbors, so that their state does not influence the energy of the system. In this stationary active state, the system wanders forever in an iso-energy set of configurations. For heterogeneous graphs, the global behavior of the Glauber dynamics at

---

[4] Note that the study of the dynamics at exactly $T = T_c$ also allows the investigation of the critical properties of a model, see e.g. Medvedyeva *et al.* (2003); Zhu and Zhu (2003).

$T = 0$ is similar to the one exhibited on random graphs (Castellano *et al.*, 2005): with a certain probability, the system reaches a state composed of two domains of opposite magnetization with a large number of interconnections, and a certain number of spins which flip without changing the energy. When the system size grows the probability of reaching this disordered stationary state tends to increase, making full ordering less likely.

A mean-field theory of the zero-temperature Glauber dynamics has been put forward by Zhou and Lipowsky (2005), and further analyzed by Castellano and Pastor-Satorras (2006b). By dividing the nodes of a network into degree classes, it is possible to write evolution equations for the probability of a node of degree k to carry a spin $+$. For random uncorrelated networks, one can then obtain an estimate of the ordering time as a function of the network size N. For a network with degree distribution $P(k) \sim k^{-\gamma}$, this approach predicts that the ordering time grows logarithmically with N if $\gamma < 5/2$, and tends to a constant as $N \to \infty$ for $\gamma > 5/2$. These predictions are, however, invalidated by numerical simulations, which point out that the value $5/2$ does not play any particular role. For all the values of γ investigated, the probability for the system to reach full order goes to 0 in the thermodynamic limit. Moreover, the ordering time for the runs that actually order grows as a power law of N (Castellano and Pastor-Satorras, 2006b). The failure of the mean-field analysis can be understood through a careful analysis of the simulations, which shows the breakdown of the basic assumption that the local field seen by a spin is independent of its value. On the contrary, the probability for a $+$ spin to have more $+$ than $-$ neighbors is very large (Castellano and Pastor-Satorras, 2006b), and strong correlations are present. The spins arrange themselves in two giant components of opposite signs which remain of similar size for a long time, competing through an extensive boundary, as shown in Figure 5.7.

Interestingly, although the heterogeneity of the network plays a crucial role in the equilibrium properties of the Ising model, it does not lead to strong differences in the behavior of the ordering Glauber dynamics: homogeneous and heterogeneous topologies lead to qualitatively similar behaviors.

5.5 Phenomenological theory of phase transitions

In the previous sections, we have considered a particular model, namely the Ising model, which represents the most paradigmatic microscopic model allowing for the emergence of collective phenomena through a phase transition. In this framework, the Hamiltonian (5.1) of the system is defined in terms of the microscopic variables, in this case the spins, which interact according to a certain topology. The order parameter is also defined as a function of the microscopic degrees of freedom, and the effects of the interaction topology are studied through

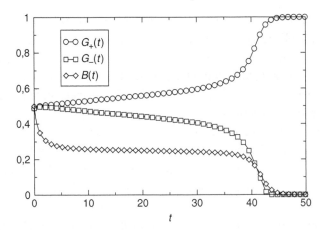

Fig. 5.7. Glauber dynamics at zero temperature for Ising spins located on the nodes of a random network with degree distribution $P(k) \sim k^{-\gamma}$. The figure shows the temporal evolution for one particular run (which finally reaches ordering) of the size, normalized by the total network size N, of the giant components of $+$ and $-$ spins, G_+ and G_-. The evolution of the boundary B, defined as the fraction of edges joining spins of opposite signs, is also shown. Here $N = 10^5$, $\gamma = 4$. Data from Castellano and Pastor-Satorras (2006b).

the solution of the model. It is possible to follow a different path to the understanding of phase transitions which is given by the phenomenological theory of Landau for phase transitions. In this approach, the starting point is the existence of the order parameter x, whose expression in terms of microscopic variables is not specified. The free energy of the system is written as an expansion in powers of x close to the phase transition and the properties of this transition are studied as a function of the expansion parameters. This expansion is justified by the fact that close to the transition the order parameter is supposed to be small as the phase transition separates an ordered phase with $x > 0$ at low temperature from a disordered one with $x = 0$ at high temperature. The thermodynamically stable state is obtained for the minimum of the free energy and this condition allows the calculation of the corresponding value of the order parameter as a function of the temperature (Huang, 1987).

The phenomenological approach has been adapted to the case of a model defined on a network with arbitrary $P(k)$ by Goltsev, Dorogovtsev and Mendes (2003). To take into account the possible heterogeneities of the network, the thermodynamic potential is written as

$$\Phi = -hx + \sum_k P(k)\phi(T, x, kx), \qquad (5.28)$$

where h is the field conjugated to x. The term $\phi(T, x, kx)$ is the contribution of vertices with degree k: at a mean-field level, one can consider that a site with degree k receives from its neighbors an effective field kx. The expansion of Φ will therefore be in x *and* kx, and one can already anticipate that the moments of $P(k)$ will appear in the theory. One also assumes that $\phi(T, x, y)$ can be expanded as

$$\phi(T, x, y) = \sum_{l,m} \phi_{l,m}(T) x^l y^m. \tag{5.29}$$

In the framework of Landau's theory, the potential Φ is moreover directly written as an expansion in powers of the order parameter:

$$\Phi = -hx + f_2 x^2 + f_3 x^3 + f_4 x^4 + \cdots, \tag{5.30}$$

where higher order terms are irrelevant and $f_2 = a(T - T_c)$ linearly vanishes at T_c, with $a > 0$. In the absence of an external field ($h = 0$), the minimum of Φ corresponds to the disordered phase $x = 0$ for $T > T_c$ and to an ordered phase with $x > 0$ for $T < T_c$. In the ordered phase, the value of the order parameter is determined by the condition that Φ is minimum, i.e.

$$\frac{d\Phi}{dx} = 0, \quad \frac{d^2\Phi}{dx^2} > 0. \tag{5.31}$$

The stability of the ordered phase near the transition thus implies $f_3 > 0$ or, if $f_3 = 0$, $f_4 > 0$.

The comparison of the two expansions of Φ allows writing, by identifying the coefficients of x^2,

$$f_2(T) = \phi_{20}(T) + \phi_{11}(T)\langle k \rangle + \phi_{02}(T)\langle k^2 \rangle, \tag{5.32}$$

which shows, since $f_2 \propto T - T_c$, that the critical temperature T_c depends on $\langle k^2 \rangle$. In particular, for $\phi_{20} = 0$, T_c is a function of the ratio $\langle k^2 \rangle / \langle k \rangle$, as it has indeed appeared in the previous subsections.

A careful analysis of the analytical properties of Φ at $x = 0$ can be carried out through its successive derivatives

$$\Phi^{(n)}(0) = n! \sum_{m=0}^{n} \phi_{n-m,m}(T)\langle k^m \rangle, \tag{5.33}$$

whose possible divergences are determined by the moments of $P(k)$: if $\langle k^m \rangle$ diverges for m larger than a certain p, then a singularity occurs in the pth derivative of Φ. The appearance of such a singularity allows us to obtain deviations from the standard mean-field results. Goltsev *et al.* (2003) use for definiteness a scale-free distribution $P(k) \sim k^{-\gamma}$, and explore the various possibilities emerging from the

Table 5.1 Various possible critical behavior on random networks with a degree distribution $P(k) \sim k^{-\gamma}$, for the order parameter x and the response function χ_x.

		x	χ_x
$f_3 = 0,$ $f_4 > 0$	$\gamma > 5$	$\tau^{1/2}$	
	$\gamma = 5$	$\tau^{1/2}/(\ln \tau^{-1})^{1/2}$	
	$3 < \gamma < 5$	$\tau^{1/(\gamma-3)}$	τ^{-1}
$f_3 > 0$	$\gamma > 4$	τ	
	$\gamma = 4$	$\tau/(\ln \tau^{-1})$	
	$3 < \gamma < 4$	$\tau^{1/(\gamma-3)}$	
arbitrary f_3 and f_4	$\gamma = 3$	e^{-cT}	T^{-1}
	$2 < \gamma < 3$	$T^{-1/(3-\gamma)}$	T^{-2}

The critical behavior depends on the coefficients f_3 and f_4, and on the exponent γ. Here $\tau \equiv 1 - T/T_c$, and c is a constant which is determined by the complete form of $P(k)$. If $f_3 < 0$, or if $f_3 = 0$ and $f_4 < 0$, at $\gamma > 3$, the system undergoes a first-order phase transition. From Goltsev *et al.* (2003).

variation of γ and the value of f_3 and f_4. In particular, the critical behavior of the order parameter x and of the susceptibility $\chi_x = dx/dh$ are summarized in Table 5.1. The main finding is that the usual mean-field behavior is recovered only if $\gamma > 5$ in the case $f_3 = 0$, $f_4 > 0$ or $\gamma > 4$ for $f_3 > 0$. In the other cases, the divergence of moments of $P(k)$ leads to the appearance of a singularity in the expansion of Φ, and thus to an anomalous critical behavior. For $\gamma \le 3$, the critical temperature is infinite in the thermodynamic limit, so that x and χ_x are not critical, but may be computed as a function of temperature (see Table 5.1).

The above results show that the effect of network connectivity fluctuations on the emergence of collective phenomena from microscopic rules can be generally described using the standard techniques of statistical physics. The randomness and small-world properties of the considered networks generically result in a mean-field character of the transition. For heterogeneous networks, however, a richer scenario emerges, which can be analyzed through mean-field approximate approaches.

In heterogeneous random networks, the local tree-like structure allows the system to be treated through mean-field methods. The vertices with large connectivity, however, induce strong correlations in their neighborhoods. Such hubs are much more numerous in heterogeneous networks than in Erdős–Rényi homogeneous random graphs. The possible divergence of moments of $P(k)$ determines the critical behavior, and the heterogeneity parameter, defined by the ratio of the two first moments of the degree distribution, appears as a crucial quantity which determines the existence of a phase transition at a finite value of the control parameter. If the degree fluctuations diverge, the hubs polarize the system at any finite temperature, so that only the ordered phase can exist and the high-temperature disordered phase disappears.

6

Resilience and robustness of networks

The large-scale collapses of the electric distribution grid or internet outages have shown that the stability of complex networks is a problem of crucial importance. The protection of critical infrastructures or the elaboration of efficient reaction strategies all rely on the identification of crucial (or weak) elements of the network and on the understanding of the progressive damage caused by successive removals or failures of nodes. In this context, it has been observed that complex network models usually display great stability even if they are confronted by a large number of repeated small failures, while at the same time major damage can be triggered, unpredictably, by apparently small shocks. This is also consistent with empirical experience in which unexpected, small perturbations may sometimes trigger large systemic failures.

In this chapter, we deal with the impact of the removal or failure of nodes in complex networks by studying their percolation behavior. Percolation models and the associated critical phenomena have been extensively studied in statistical physics and, even though they are not endowed with a high level of realism, they can be thought of as the zeroth order approximation to a wide range of damaging processes. In particular we focus on heterogeneous networks that typically display both a large robustness to random failures and a high vulnerability to targeted attacks on the most crucial nodes. These two aspects elucidate how the topology of a network influences its stability and integrity, and highlight the role of hubs and connectivity fluctuations.

6.1 Damaging networks

The simplest assessment of networks' behavior in the case of progressive levels of damage can be obtained by studying the effect of removal of nodes on the network structure. While this approach focuses on the bare topological impact of damage and neglects all the issues related to the detailed technical aspects of the system's elements and architecture, it provides an initial insight into the potential effects of failures in large-scale networks. In this spirit, Albert, Jeong and Barabási (2000) have made a comparative study of the topological effects of removal of nodes in various graph representations of real-world networked systems with either homogeneous or heterogeneous topologies. Albert *et al.* (2000) consider the damage created by removing a certain fraction of nodes or links, either in a random manner (to model a random failure), or in a "targeted" way to mimic intentional damage as illustrated in Figure 6.1. Targeted attacks are carried out by removing nodes preferentially according to their "centrality" (either measured by the degree or the betweenness centrality) and can describe intentional attacks supposed to maximize the damage by focusing on important hubs. The evidence put forward by Albert *et al.* (2000) is that heterogeneous and homogeneous topologies react very differently to increasing levels of damage.

In order to characterize the phenomenology associated to the damage process we need to define a quantitative measure of network damage. To this end, we consider a network in which a certain fraction f of the nodes has been removed and each time a node is removed, all links going from this node to other nodes of the network are deleted as well. Such damage may break an initially connected graph into

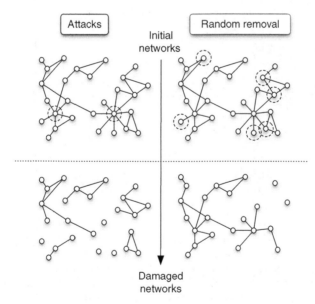

Fig. 6.1. Schematic comparison of random and targeted removal. On the left, we show the effects of targeted attacks on the two nodes with largest degrees. The resulting network is made of small disconnected components. In the case of random removal of six nodes (right column), the damage is much less significant as there is still a path between most of the nodes of the inital network.

various disconnected pieces. The simplest quantitative measure of damage is therefore given by the relative size of the largest connected component of the remaining network, S_f/S_0, where $S_0 = N$ is the original size of the network. The network will indeed keep its ability to perform its tasks as long as a "giant cluster" of size comparable to the original one (S_0) exists. When $S_f \ll S_0$, the network has been broken into many small disconnected parts and is no longer functional. The study of the evolution of S_f/S_0 as a function of the fraction of removed nodes f therefore characterizes the network response to damage, as shown in Figure 6.2 for both random and targeted removal of nodes. At low levels of damage, the resilience in the random removal process is essentially determined by local details, such as the minimum degree. At large levels of damage, homogeneous graphs exhibit a definite level of damage over which S_f/S_0 abruptly drops to zero, signaling the total fragmentation of the network. Heterogeneous networks on the contrary appear to lack a definite threshold and display higher tolerance to large random damage. It is possible to achieve a basic understanding of this evidence by recalling that the heavy-tailed form of the degree distribution of heterogeneous networks implies that the vast majority of vertices have a very small degree, while a few hubs collect a

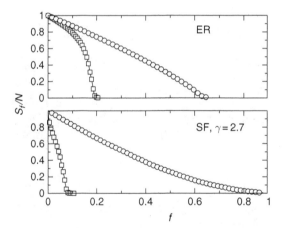

Fig. 6.2. Size S_f of the giant component (largest connected component) divided by the initial size $S_0 = N$ of the network, vs. the fraction f of removed nodes, for random removal (circles) or targeted attacks (squares). Two networks are considered: a homogeneous Erdős–Rényi (ER) network of $N = 10^5$ nodes and average degree $\langle k \rangle \sim 3$, and a generalized random scale-free (SF) network with same size, and average degree and distribution $P(k) \sim k^{-\gamma}$, $\gamma = 2.7$. In the heterogeneous (SF) network suffering random removals, the giant component decrease does not exhibit a definite threshold as in the case of the Erdős–Rényi network. The damage caused by targeted attacks, on the contrary, is much larger than in the case of the Erdős–Rényi network.

very large number of edges, providing the necessary connectivity to the whole network. When removing vertices at random, chances are that the largest fraction of deleted elements will have a very small degree. Their deletion will imply, in turn, that only a limited number of adjacent edges are eliminated. Therefore, the overall damage exerted on the network's global connectivity properties will be limited, even for very large values of f.

The picture is very different in the case of targeted attacks. The first vertex which is removed has the highest degree. Then the second highest degree vertex is removed, and so on until the total number of removed vertices represents a fraction f of the total number of vertices forming the original network. In Figure 6.2, the topological resilience to targeted attacks of heterogeneous graphs is compared with that of an Erdős–Rényi graph with the same average degree. In this case the emerging scenario appears in sharp contrast with the one found for the random removal damage. The heterogeneous graphs strikingly appear much more vulnerable than the Erdős–Rényi random graph. Obviously, they are more vulnerable than a regular lattice for which a degree-based targeted attack cannot be properly defined (all vertices have the same degree), which can thus be considered to have the same resilience as in the random removal case. In other words, in heavy-tailed networks, the removal of a small fraction of the largest degree nodes has a disproportionate effect that drastically reduces the size of the largest connected component of the network. The different behavior of heterogeneous and homogeneous networks is not unexpected and can be understood in terms of their degree distributions. The heavy-tailed nature of the heterogeneous networks makes the hubs extremely important to keep the graph connected. Their removal leads right away to network collapse. On the other hand, homogeneous graphs such as Erdős–Rényi graphs have exponentially decaying degree distributions in which the probability of finding a hub is exponentially small. Statistically, the vertices' degree is almost constant around their average values, and a targeted attack, while still performing better than a random removal, does not reach the extreme efficiency achieved in scale-free networks.

It is worth remarking that other possible quantitative measures of topological damage can be defined in networks. For example, the increase in the average path length between pairs of nodes indicates how communication becomes less efficient. When the removal of nodes leads to the disruption of the original network into disconnected clusters, the average shortest path length of the whole network in tact diverges (the distance between two non-connected nodes being infinite). An interesting way to avoid this problem consists in using the average inverse geodesic length (also called efficiency; Latora and Marchiori [2001]) defined as

$$\frac{1}{N(N-1)} \sum_{i \neq j} \frac{1}{\ell_{ij}}, \tag{6.1}$$

where ℓ_{ij} is the shortest path length between the nodes i and j. This quantity is finite even for disconnected graphs because each pair of disconnected nodes is assumed to have $\ell_{ij} = \infty$ and adds a null contribution to the efficiency expression. In weighted networks, other measures have to be defined in order to take into account that some nodes or links are more important than others: the integrity of a network is quantified by the largest sum of the weights in a connected component, giving the amount of connected traffic that the damaged network is still able to handle (see Section 6.5.2).

It is important to stress that different definitions of the damage do not change the general picture of the different behavior of heterogeneous and homogeneous networks under attack. In summary, while homogeneous networks suffer damage similarly in the different strategies, heterogeneous networks appear resistant to random failure and quite fragile with respect to targeted attacks. While not completely unexpected, the result is surprising in that the two kinds of damages do not just differ quantitatively but exhibit very distinct functional behavior in contrast to what happens in homogeneous networks. In particular, heterogeneous networks appear to lack a definite threshold in the case of random failures. This intuition and the handwaving considerations presented here find a solid theoretical ground in the analytical solution of percolation models on random graphs.

6.2 Percolation phenomena as critical phase transitions

Percolation theory provides the natural theoretical framework to understand the topological effect of removing nodes (Bunde and Havlin, 1991; Stauffer and Aharony, 1992). Simply stated, the percolation problem considers an arbitrary network topology in which each node is occupied with probability p and links are present only between occupied nodes. As the probability p increases, connected components – called clusters in this context – emerge. Such a set-up defines the site percolation problem that studies the properties of the clusters, and in particular their sizes, as a function of the occupation probability p. It is indeed intuitively clear that if p is small, only small clusters can be formed, while at large p the structure of the original network will be almost preserved (if $p = 1$ the network is completely recovered). In this context the removal of a fraction f of randomly selected nodes is just equivalent to considering that each site of the network is occupied with probability $p = 1 - f$.

For the sake of simplicity let us first consider a two-dimensional lattice such as the one depicted in Figure 6.3. At low occupation probability, only isolated nodes and small clusters of occupied sites are obtained. At large enough p, in contrast, one observes larger clusters and in particular a *percolating* cluster which spans the lattice by a connected path of occupied sites. Similarly, the *bond percolation* problem considers that the edges of the lattice can be either occupied or empty with

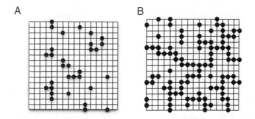

Fig. 6.3. Site percolation problem in two dimensions. Each lattice site is occupied (filled circles) with probability p. A, At small p, only small clusters of occupied sites are formed. B, At large p, a *percolating cluster* of occupied sites connects opposite boundaries.

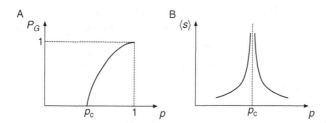

Fig. 6.4. A, Schematic plot of the probability P_G for a node to belong to the infinite percolating cluster as a function of the occupation probability p. For $p < p_c$, P_G is equal to zero in the thermodynamic limit, while it takes strictly positive values for $p > p_c$. B, Evolution of the average size of the finite clusters close to the transition.

probability p and $1 - p$ respectively, and is described by the same phenomenology of site percolation by increasing the percolation probability. As the size of the lattice goes to infinity, the change between the two possible behaviors sketched in Figure 6.3 defines a phase transition (see Chapter 5) at a critical value p_c. For $p < p_c$, all clusters have a finite size, while for $p > p_c$ a *giant* cluster appears and contains a finite fraction of all lattice sites, and thus becomes of infinite size in the thermodynamic limit corresponding to a system of infinite size. This allows us to generalize the concept of percolation transition to any kind of network.

In the percolation problem, the order parameter is defined as the probability P_G for a node to belong to the infinite percolating cluster, and evolves from 0 for $p \leq p_c$ to a finite value at $p > p_c$, as sketched in Figure 6.4. A basic quantity describing the system structure is the number $N_s(p)$ of clusters of size s in a network of size N, or as customarily used the *cluster number distribution* $n_s(p) = N_s(p)/N$, which is the number of clusters of size s per lattice vertex, at the percolation probability p. The probability for any node i to belong to a cluster of size s is simply given by $sn_s(p)$, where the fact that the vertex can be any one of the s cluster's elements has been considered. The probability p of a site to be occupied can then be rewritten as the sum of these probabilities over all possible sizes. If $p < p_c$, this translates into

$$p = \sum_s s n_s(p). \tag{6.2}$$

Above the critical point p_c, the infinite cluster appears and each node has a finite probability $P_G > 0$ to be part of it. As any occupied vertex belongs either to the infinite cluster or to a cluster of finite size, we can write for $p > p_c$ that

$$p = P_G + {\sum_s}' s n_s(p), \tag{6.3}$$

where ${\sum_s}'$ indicates that the giant cluster is excluded from the summation.

The function $n_s(p)$ also allows writing the conditional probability that an occupied vertex belongs to a cluster of size s as $s n_s(p)/{\sum_s}' s n_s(p)$, where the infinite cluster has been excluded to avoid divergences. The average size $\langle s \rangle$ of the cluster to which any occupied vertex belongs is therefore given by

$$\langle s \rangle = \frac{{\sum_s}' s^2 n_s(p)}{{\sum_s}' s n_s(p)}, \tag{6.4}$$

where again the divergence due to the infinite clusters is not taken into account by considering the constrained sums ${\sum_s}'$. At $p < p_c$ the average cluster size $\langle s \rangle$ is finite. For increasing values of p the average size increases. At p_c the appearance of the giant cluster corresponds to the divergence of the average size that above p_c is completely contained in the infinite connected cluster excluded by the sum. It then results that $\langle s \rangle$ is a singular function at p_c, as shown in Figure 6.4. This divergence contains much physical information about the system and is the fingerprint of a critical phase transition. Indeed, the singularity at p_c corresponds to the lack of a characteristic length for the clusters which is at the origin of the scaling behavior typical of phase transitions (see Chapter 5) and yields the following scaling form for $n_s(p)$, close to the transition

$$n_s(p) = \begin{cases} s^{-\tau} f_+(s/s_c), & \text{if } p \geq p_c \\ s^{-\tau} f_-(s/s_c), & \text{if } p \leq p_c \end{cases} \quad \text{with } s_c = |p_c - p|^{-1/\sigma}, \tag{6.5}$$

where $f_+(x)$ and $f_-(x)$ are two scaling functions which converge to the same finite constant when $x \to 0$ ($f_+(0) = f_-(0)$), and have a fast decay (e.g. exponential) at large x. Here τ and σ are exponents whose values depend on the dimensionality and other properties of the system. The quantity $s_c(p)$, equivalent to the characteristic length in equilibrium phase transitions, plays the role of the size cut-off: only connected clusters with a size smaller or comparable to $s_c(p)$ are present and define the physical properties of the system for any given value of p.

Using the scaling form for $n_s(p)$ and replacing the sum by an integral in the numerator of Equation (6.4.), one obtains for the average size of the finite clusters close to p_c the following relation

$$\langle s \rangle \sim \frac{1}{p} \int s^2 s^{-\tau} f(s/s_c) ds \sim s_c^{3-\tau} \int x^{2-\tau} f(x) dx \sim |p_c - p|^{(\tau-3)/\sigma}, \quad (6.6)$$

which predicts that the average size of finite clusters scales as a power law. Analogously, the probability for a node to be in the giant cluster P_G can be estimated close to p_c, by noting that $p_c = \sum_s s n_s(p_c)$ (since at $p = p_c$, $P_G = 0$), and rewriting Equation (6.3) for $p \approx p_c$ as

$$P_G \approx \sum_s s[n_s(p_c) - n_s(p)] \sim \int ds\, s^{1-\tau} [f(0) - f(s/s_c)]$$
$$\sim s_c^{2-\tau} \sim (p - p_c)^{(\tau-2)/\sigma}. \quad (6.7)$$

Note that in order for the sum $\sum_s s n_s(p_c)$ to converge, one needs $\tau > 2$, so that P_G goes to 0 as $p \to p_c$, as initially assumed.

From the scaling form for $n_s(p)$, it follows that both P_G and $\langle s \rangle$ obey the power-law scaling forms close to p_c

$$\langle s \rangle \sim |p_c - p|^{-\gamma}, \quad (6.8)$$
$$P_G \sim (p - p_c)^{\beta}, \quad (6.9)$$

which define the critical exponents $\gamma = (3 - \tau)/\sigma$ and $\beta = (\tau - 2)/\sigma$. The last two equations are the classical scaling relations customarily found in critical phenomena.

Percolation theory has been extensively validated for different kinds of lattices in both analytical and computer simulations studies (Stauffer and Aharony, 1992). The power-law behavior and singular functions found at the percolation threshold are a typical example of critical phase transition, where the onset of a macroscopically ordered phase (for instance the presence of a global connected structure) is anticipated by large fluctuations in the statistical properties of the system (Binney *et al.*, 1992). When these fluctuations become of the order of the system size itself, at the critical point, the macroscopic order arises and the system enters the new phase. In addition, as with equilibrium critical phenomena, percolation exhibits the universality of critical behavior and exponents. Indeed, the exact value of the critical exponents does not depend on the fine details of the percolation model. In general, they just depend on the system's dimensionality and symmetries of the order parameter. Thus, while the exact value of p_c depends on the lattice geometry, the critical exponents do not.

6.3 Percolation in complex networks

As described in the previous section, percolation phenomena are generally considered on regular lattices embedded in a D-dimensional space. In random graphs with N vertices, however, there is no embedding space and any vertex can in principle be connected to any other vertex. This is equivalent to working in a space which is $(N - 1)$-dimensional where any vertex has $N - 1$ possible neighbors. Depending on the graph connectivity properties, we may or may not observe a giant connected component made of a finite fraction of nodes.[1] In the $N \to \infty$ limit often considered in random graph theory, the problem is therefore analogous to infinite-dimensional edge percolation. In network language, the critical point thus translates unambiguously into the existence of a given threshold condition that marks the separation between a regime in which the network is fragmented into a myriad of small subgraphs and a regime in which there is a giant component containing a macroscopic fraction of the network's vertices.

The study of the percolation transition as a function of the connectivity properties of generalized random graphs finds a convenient formulation in the generating functions technique (see Callaway *et al.* [2000]; Newman [2003c] and Appendix 2). We report here a non-rigorous argument which provides the condition for a giant cluster to arise in graphs that have a local tree structure with no cycles. We focus for simplicity on undirected graphs, while the case of directed graphs is described in Appendix 3. We consider an uncorrelated network with degree distribution $P(k)$ and we denote by q the probability that a randomly chosen edge does not lead to a vertex connected to a giant cluster. This probability can be self-consistently computed if cycles are neglected. It can indeed be written as the average over all possible degrees k of the products of two probabilities: (i) the probability $kP(k)/\langle k \rangle$ that the randomly picked edge leads to a vertex of degree k (see Chapter 1); (ii) the probability q^{k-1} that none of the remaining $k - 1$ edges lead to a vertex connected to a giant cluster. This translates into the self-consistent equation

$$q = \sum_k \frac{kP(k)}{\langle k \rangle} q^{k-1}. \tag{6.10}$$

The probability P_G for a given site to belong to a giant cluster can also be easily written: $1 - P_G$ is the probability of *not* belonging to this cluster, which is, for a

---

[1] For instance, the onset of the giant component of the Erdős–Rényi random graph is analogous to an edge percolation problem in which the $N - 1$ edges of each vertex can each be occupied with probability p.

node of degree k, the probability q^k that none of its edges lead to the giant cluster. We then obtain

$$P_G = 1 - \sum_k P(k)q^k. \qquad (6.11)$$

The value $q = 1$, which is always a solution of Equation (6.10), corresponds to $P_G = 0$, i.e. to the absence of any percolating cluster. On the other hand, the percolation transition is determined by the appearance of another solution to Equation (6.10). The necessary condition for such a possibility can be determined in a simple graphical way as sketched in Figure 6.5. Indeed, we can rewrite Equation (6.10) as $q = F(q)$ with $F(q) = \sum_k kP(k)q^{k-1}/\langle k \rangle$. The function F has the following properties: $F(0) = P(1)/\langle k \rangle$, $F(1) = 1$, $F'(q) > 0$ and $F''(q) > 0$ for $0 < q < 1$, which implies that F is monotonously growing and convex. Figure 6.5 shows schematically the curve $y = F(q)$ together with the straight line $y = q$. It is clear that an intersection between these two lines exists at $q < 1$ if and only if the slope of F at $q = 1$ is larger than the slope of $y = q$. Mathematically this leads to the condition for the presence of a giant component

$$\frac{d}{dq}\left(\sum_k \frac{kP(k)}{\langle k \rangle} q^{k-1} \right)\bigg|_{q=1} > 1, \qquad (6.12)$$

which can be rewritten as

$$\frac{\langle k^2 \rangle}{\langle k \rangle} > 2. \qquad (6.13)$$

The percolation condition (6.13), in which the heterogeneity parameter $\kappa = \langle k^2 \rangle / \langle k \rangle$ appears, is exact if cycles are statistically irrelevant, which is the case for random uncorrelated graphs in the $N \to \infty$ limit close to the transition point, and was first derived on more rigorous grounds by Molloy and Reed (1995). More precisely, it is also possible to show that the percolation threshold marks the critical point of a phase transition, separating the phase (for $\langle k^2 \rangle / \langle k \rangle < 2$) in which all the

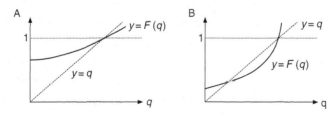

Fig. 6.5. Schematic graphical representation of Equation (6.10). A, The curves $y = F(q)$ and $y = q$ intersect only at $q = 1$. B, A second intersection with $q < 1$ is also obtained if and only if the slope of the function $F(q)$ in $q = 1$ is larger than 1.

components are trees and the size of the largest component scales at most as $\ln N$, from the phase (for $\langle k^2 \rangle / \langle k \rangle > 2$) in which there exists a giant component of size proportional to N, while the sizes of the other components scale at most as $\ln N$. At the point $\langle k^2 \rangle / \langle k \rangle = 2$, the largest cluster scaling is anomalous and follows the behavior $N^{2/3}$.

In this framework, the critical parameter is determined by the connectivity properties of the network. For instance, in the case of a Poisson random graph where $\langle k^2 \rangle = \langle k \rangle^2 + \langle k \rangle$, the condition of Equation (6.13) takes the form $\langle k \rangle_c = 1$. This means that any Poisson random graph with average degree larger than 1 exhibits a giant component and $\langle k \rangle$ can be considered as the critical parameter. In the case of the Erdős–Rényi model, in which edges are drawn with probability p, the average degree is given by the expression $\langle k \rangle = pN$ and the critical connection probability $p_c = 1/N$ is recovered.

6.4 Damage and resilience in networks

In the percolation framework, the macroscopic behavior of networks in the case of random removal of vertices or edges finds a natural characterization in terms of an *inverse* percolation process in a connected random graph. In this context, the lattice in which percolation takes place is the graph under consideration. In the intact graph, with $f = 0$, all the vertices are occupied. The deletion of a fraction f of vertices corresponds to a random graph in which the vertices are occupied with probability $p = 1 - f$. For small f, we are in the region of p close to 1, in which the infinite cluster (identified as the giant component) is present. The threshold for the destruction of the giant component, $f_c = 1 - p_c$, can be thus computed from the percolation threshold at which the infinite cluster first emerges. In this case the phase transition corresponds to the separation of a region of damages where there still exists a connected network of appreciable size from a region in which the system is fragmented into small clusters. The order parameter P_G is a function of $f = 1 - p$ and can be defined as $P_G = S_f / S_0$, where as in the previous sections S_f is the size of the largest component after a damage f and $S_0 = N$ is the size of the original network.[2] The strategy for calculating the damage threshold thus consists in finding the damage density f_c at which the surviving network fulfills the percolation condition $\langle k^2 \rangle_f / \langle k \rangle_f = 2$, where $\langle k^2 \rangle_f$ and $\langle k \rangle_f$ refer to the degree distribution moments of the damaged graph.

---

[2] It is worth remarking, however, that only in the infinite size limit does the relation $P_G = \lim_{S_0 \to \infty} S_f / S_0$ unambiguously define the transition point. In finite systems, the transition is smoother and the order parameter never attains a null value above the threshold, relaxing to its minimum value $P_G = 1/S_0$.

Starting from an undamaged network with degree distribution $P_0(k)$, average degree $\langle k \rangle_0$, and second moment $\langle k^2 \rangle_0$, it is possible to compute the corresponding $P_f(k)$, $\langle k \rangle_f$ and $\langle k^2 \rangle_f$ after the random removal of a fraction f of nodes (Cohen *et al.*, 2000). For each surviving node of initial degree k_0, the random removal amounts to the deletion of a certain number of its neighbors (each one independently with probability f), and the node has remaining degree k with probability

$$\binom{k_0}{k}(1-f)^k f^{k_0-k}. \tag{6.14}$$

The resulting degree distribution $P_f(k)$ is obtained by summing expression (6.14) over all possible values of k_0, each one weighted by its probability of appearance $P_0(k_0)$ in the initial network

$$P_f(k) = \sum_{k_0 \geq k} P_0(k_0)\binom{k_0}{k}(1-f)^k f^{k_0-k}. \tag{6.15}$$

One finally obtains $\langle k \rangle_f = (1-f)\langle k \rangle_0$ and $\langle k^2 \rangle_f = (1-f)^2 \langle k^2 \rangle_0 + f(1-f)\langle k \rangle_0$. The critical value f_c is such that for $f > f_c$, no giant component can be found in the network. According to the Molloy and Reed criterion (6.13) obtained in the previous section, this occurs if and only if $\langle k^2 \rangle_f < 2\langle k \rangle_f$, which can be rewritten as

$$f > 1 - \frac{\langle k \rangle_0}{\langle k^2 \rangle_0 - \langle k \rangle_0}. \tag{6.16}$$

The critical value f_c is thus given by

$$f_c = 1 - \frac{\langle k \rangle_0}{\langle k^2 \rangle_0 - \langle k \rangle_0}$$
$$= 1 - \frac{1}{\kappa - 1}, \tag{6.17}$$

where $\kappa = \langle k^2 \rangle_0 / \langle k \rangle_0$ quantifies the heterogeneity of the initial network. This formula allows us to understand the different resilient behaviors exhibited by homogeneous and heterogeneous topologies. If the degree fluctuations are bounded, $\langle k^2 \rangle_0$ is finite and f_c is strictly less than 1. As more nodes are removed, the size of the largest component thus decreases and reaches 0 when a certain fraction (strictly less than 1) of the nodes has disappeared. On the other hand, heavy-tailed networks display very large fluctuations and $\langle k^2 \rangle_0$ diverges in the thermodynamic limit which leads to $f_c = 1$. This means that a giant component is present *for any fraction of removed sites strictly less than 1*. Random failures of an arbitrary fraction of the nodes therefore do not affect the functionality of the network. This extreme robustness is due to the presence of hubs which, although in small proportion, hold the network together and are unlikely to be removed in the random process as small degree nodes are the huge majority.

The critical threshold derived above corresponds to the limit of infinite size. However, all real networks have a finite number of nodes N, so that the fluctuations $\langle k^2 \rangle_0$ do not strictly diverge and f_c will not be strictly 1. Let us consider for concreteness the case of a scale-free network of size N, with degree distribution $P(k) = ck^{-\gamma}$ for $k = m, m+1, \ldots, k_c(N)$, where m is the minimum degree and c a normalization constant ($\gamma > 1$ in order to ensure convergence of $\int^\infty P(k)$). The cut-off $k_c(N)$ is the maximal degree observed in a system of size N and is given by the fact that typically only one node will have degree $k_c(N)$ or larger,[3] which can be written as

$$N \int_{k_c(N)}^{\infty} P(k)dk = 1, \tag{6.18}$$

yielding $k_c(N) = mN^{1/(\gamma-1)}$. The ratio $\kappa = \langle k^2 \rangle / \langle k \rangle$ can then easily be computed as

$$\kappa = \frac{2-\gamma}{3-\gamma} \cdot \frac{k_c(N)^{3-\gamma} - m^{3-\gamma}}{k_c(N)^{2-\gamma} - m^{2-\gamma}}, \tag{6.19}$$

leading to the different possible behaviors according to the value of γ:

- if $\gamma > 3$, $\kappa \sim (\gamma-2)m/(\gamma-3)$ is finite, so that f_c is strictly less than 1: a percolation threshold exists, just as for ordinary random graphs.
- if $\gamma < 3$, κ diverges as N goes to infinity ($\kappa \sim N^{(3-\gamma)/(\gamma-1)}$ if $2 < \gamma < 3$, and $\kappa \sim N^{1/(\gamma-1)}$ if $\gamma < 2$). The percolation threshold f_c therefore becomes closer and closer to 1 as N increases. Although, for any finite network, the size of the largest component goes to values close to 0 at a value of f_c smaller than 1, $1 - f_c$ is effectively small and the robustness increases as the initial network size increases.

In particular, for $2 < \gamma < 3$, which is encountered in many real networks, the threshold is given by

$$f_c \approx 1 - \frac{3-\gamma}{\gamma-2}m^{2-\gamma}k_c(N)^{\gamma-3}, \tag{6.20}$$

and is close to 1 even at relatively small network sizes. For example, for $N = 1000$, $m = 1$, $\gamma = 2.5$, one obtains $k_c = 100$ and $f_c \approx 0.9$.

The cut-off in the degree distribution can also arise because of some physical constraints. The physical Internet is an example of this, since the routers cannot be linked to an arbitrarily large number of other routers. In this case, the degree fluctuations do not diverge, even in the infinite size limit. A typical example of such degree distribution can be written as $P(k) = ck^{-\gamma} \exp(-k/k_c)$ ($2 < \gamma < 3$).

---

[3] See Appendix 1 for a detailed derivation of this result.

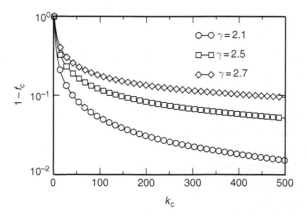

Fig. 6.6. Threshold for the random removal of vertices, for graphs with degree distribution of the form $P(k) = ck^{-\gamma}\exp(-k/k_c)$. The extension of the fragmented region $1 - f_c$, is plotted as a function of the cut-off k_c for various exponents γ. For $\gamma \leq 3$ f_c tends to 1 as k_c increases. Networks with lower exponents have a larger resilience to damage at given cut-off.

It is then possible to compute κ and f_c (Callaway *et al.*, 2000), obtaining that f_c tends to 1 as k_c increases, and this more and more rapidly as γ decreases, as shown in Figure 6.6.

Finally, we note that the theory presented so far refers to *uncorrelated* networks: the probabilities that two neighboring vertices have degree k and k', respectively, are independent. In fact, most real networks bear correlations, which are described by the conditional probability $P(k|k')$ that a node has degree k given that it is the neighbor of a node of degree k' (see Chapter 1). The behavior of the percolation transition and of the damage threshold then depend on the form of the correlations (Newman, 2002a; Moreno and Vázquez, 2003). It can in fact be shown that $f_c = 1 - 1/\Lambda$, where Λ is the largest eigenvalue of the *correlation matrix* having elements $C_{k'k} = (k - 1)P(k|k')$. The resilience properties of a correlated network thus depend on the possible divergence of this eigenvalue. Generically, it turns out that it does diverge for heavy-tailed networks with divergent second moment (Boguñá, Pastor-Satorras and Vespignani, 2003b). For concreteness, one can state that, for finite networks in which f_c is always lower than 1, assortative networks in which high degree vertices tend to stick together are more resilient than disassortative networks.

6.5 Targeted attacks on large degree nodes

So far we have considered only random percolation models suited at best to simulate the occurrence of random failure. We have also discussed the behavior of networks in the case of targeted attacks and it is interesting to explain this phenomenology in the framework of percolation phenomena. In order to do this we have to consider the percolation transition of the removal of nodes according to deterministic strategies.

Let us first consider that nodes are knocked out in the order defined by their degree rank; that is, the nodes are removed in descending order of degree centrality. In this case the analytical calculation has been worked out explicitly in generalized uncorrelated random networks with arbitrary degree distribution $P(k)$ (Callaway *et al.*, 2000; Cohen *et al.*, 2001). The removal of a fraction f of the highest degree nodes corresponds to the removal of all nodes of degree larger than a certain value $k_c(f)$, implicitly defined by the equation

$$f = \sum_{k=k_c(f)+1}^{\infty} P(k). \tag{6.21}$$

Moreover, such a removal will modify the degree distribution of the remaining nodes, owing to the deletion of the edges between removed and remaining nodes. The probability that one of the neighbors of any given node is removed equals the probability that the corresponding link points to a node of degree larger than $k_c(f)$, i.e.

$$r(f) = \sum_{k=k_c(f)+1}^{\infty} \frac{kP(k)}{\langle k \rangle}. \tag{6.22}$$

The surviving graph is thus equivalent to a network with cut-off $k_c(f)$, from which the neighbors of each node are randomly removed with probability $r(f)$. The Molloy and Reed criterion yields the following equation for the threshold value f_c at which the giant component disappears:

$$r(f_c) = 1 - \frac{1}{\kappa(f_c) - 1}, \tag{6.23}$$

where

$$\kappa(f_c) = \frac{\sum_k^{k_c(f_c)} k^2 P(k)}{\sum_k^{k_c(f_c)} kP(k)}. \tag{6.24}$$

Explicit values of the damage threshold can be calculated in scale-free graphs of minimum degree m and degree distribution $P(k) = ck^{-\gamma}$. By using the continuous degree approximation which replaces the sums by integrals in the various equations, we obtain

$$k_c(f) \approx mf^{1/(1-\gamma)}, \quad r(f) \approx f^{(2-\gamma)/(1-\gamma)}, \tag{6.25}$$

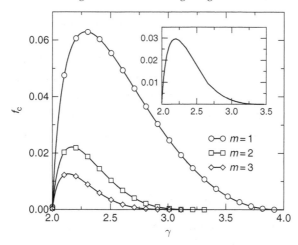

Fig. 6.7. Threshold f_c for the targeted removal of the most connected nodes, for scale-free graphs ($P(k) \sim k^{-\gamma}$) of minimum degree m, as a function of the exponent γ. The symbols correspond to an approximate computation treating the degrees as continuous variables (Cohen *et al.*, 2001), while the inset shows the results of the discrete formalism of Dorogovtsev and Mendes (2001), for $m = 1$.

and

$$\kappa(f_c) \approx \frac{2-\gamma}{3-\gamma} \times \frac{k_c^{3-\gamma} - m^{3-\gamma}}{k_c^{2-\gamma} - m^{2-\gamma}}. \tag{6.26}$$

Plugging these expressions into Equation (6.23) yields the implicit threshold condition

$$f_c^{(2-\gamma)/(1-\gamma)} = 2 + \frac{2-\gamma}{3-\gamma} m \left(f_c^{(3-\gamma)/(1-\gamma)} - 1 \right), \tag{6.27}$$

which can be solved numerically providing an estimate of f_c always of the order of a few percent for $\gamma \in [2, 3]$, as shown in Figure 6.7. Dorogovtsev and Mendes (2001) have in fact shown that a more rigorous approach can also be undertaken which treats the degrees as discrete variables, and that the obtained equation can be solved numerically. The results, shown in the inset of Figure 6.7, point to an even greater fragility since f_c is lower than in the continuous degree approximation. These results clearly show the fragility of heterogeneous networks where the removal of a very small fraction of the hubs is sufficient to shatter the network into small pieces and destroy its functionalities. The divergence of the degree fluctuations and the relative abundance of very high degree nodes appear again as the origin of the extreme fragility in front of targeted attacks of heavy-tailed networks.

The analytical results we have shown assume a perfect knowledge of the ranking of nodes in the network according to the degree centrality. Of course, in real-world large-scale networks it is very difficult to reach such a global knowledge of the connectivity pattern, and intermediate targeting strategies have been investigated (Gallos *et al.*, 2004; 2005) in which the probability to remove a node of degree k is proportional to k^α. The limits $\alpha = 0$ and $\alpha \to \infty$ then represent, respectively, random removal and deterministic attacks on the most connected nodes. As could be intuitively expected, it turns out that the critical point f_c at which the network becomes disconnected increases as α decreases: for a given fraction of nodes undergoing a failure, larger probabilities that these removed nodes are hubs lead to smaller remaining connected components.

6.5.1 Alternative ranking strategies

As already stated, the most intuitive topological measure of importance (and centrality) of a node is given by its degree. However, by considering solely the degree of a node we overlook that nodes with small degree may be crucial for connecting different regions of the network by acting as bridges. With this in mind, a detailed study of the effects of various attack strategies on heterogeneous real-world as well as on model networks has been realized by Holme *et al.* (2002). In particular, the effect of targeting nodes with largest degree has been compared with the removal of nodes with the largest betweenness centrality.

In many model networks, such as the Barabási–Albert model, betweenness centrality and degree are in fact strongly correlated, and in particular the most connected nodes typically coincide with the most central ones (see Chapter 1). The two types of attacks thus yield very similar effects. In many real-world networks, however, the situation is more complex: although centrality and degree are always correlated, large fluctuations are often observed and some nodes may have large betweenness centrality despite a relatively modest degree. Such features have been uncovered, for example in the worldwide airport network (see Chapter 2), and appear generically in networks where the tendency to form hubs is contrasted by geographical constraints (Barrat *et al.*, 2005).

Another important consideration is taken into account by Holme *et al.* (2002): each time a node is removed, the next on the list in order of decreasing degree or betweenness centrality may in fact change. This is particularly clear for the betweenness centrality since the removal of one node alters the routes of the shortest paths and thus may affect the betweenness of all nodes. As a result, four different attack strategies can be considered: (i) removing the nodes in decreasing order of degree according to the *initial* list of degrees, or (ii) the *initial* list of betweenness centralities, or according after each removal to the (iii) *recalculated* list of degrees (RD) or (iv) the *recalculated* list of betweenness centralities (RB).

Systematically, it turns out that the RB strategy is the most lethal for the network (see Figure 6.8 A for a comparison of RD and RB). This is quite understandable

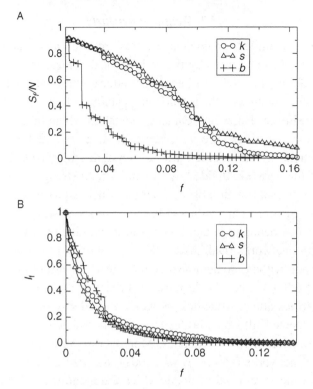

Fig. 6.8. Effect of different node removal strategies on the structure of the worldwide airport network. Nodes are removed in order of decreasing degree k, strength s or betweenness centrality b. A, Decrease of the size of the largest connected component. B, Decrease of the network's integrity I_f defined as the largest traffic or strength still carried by a connected component of the network, divided by the initial total strength of the undamaged network. Data from Dall'Asta *et al.* (2006c).

since at each step, the node through which the largest number of shortest paths goes is removed. For comparison, the result of the same attacks on homogeneous graphs is much less dramatic: the absence of hubs makes a network more resilient to targeted attacks since all nodes have approximately the same importance. Analogous results have been obtained in the case of the North American power grid which displays simultaneously an exponentially decaying degree distribution and strong heterogeneities in the betweenness centrality (Albert, Albert and Nakarado, 2004). Finally, another interesting point made by Holme *et al.* (2002) is that for a given average degree, large clustering values renders the network more vulnerable. This can be intuitively understood by considering that the edges creating triangles are locally redundant but are not used for holding together distant parts of the network.

6.5.2 *Weighted networks*

In real networks, complex topological features are often associated with a diversity of interactions as measured by the weights of the links. In order to study the vulnerability of such networks to intentional attacks, these attributes must therefore be considered along with the topological quantities. Recent works have indeed proposed centrality measures and damage indicators which take into account the links' weights or capacities. The *integrity* (Dall'Asta *et al.*, 2006c) of a damaged network is defined as the largest traffic or strength still carried by a connected component of the network, divided by the initial total strength of the undamaged network. Its study as a function of the fraction of nodes provides interesting complementary information on the vulnerability of complex networks. While the theory of inhomogeneous percolation can be used to generalize the Molloy–Reed criterion to weighted networks (Dall'Asta, 2005), we will employ the example of the worldwide airport network to demonstrate the effects of various kinds of malicious attacks. Figure 6.8 illustrates the decrease of the size of the largest connected component, for various attack strategies, as well as the decay of the integrity of the network. As expected, all strategies lead to a rapid breakdown of the network with a very small fraction of removed nodes. Moreover, in agreement with the results on unweighted networks (Holme *et al.*, 2002), the size of the giant component decreases faster upon removal of nodes which are identified as central according to global properties (i.e. betweenness), instead of local ones (i.e. degree, strength). Interestingly, when the attention shifts to the behavior of the integrity measure, one finds a different picture in which all the strategies achieve the same level of damage. Most importantly, its decrease is even faster and more pronounced than for topological quantities: for S_f/N still of the order of 80%, the integrity is typically smaller than 20%. This emphasizes how the purely topological measure of the size of the largest component does not convey all the information needed. In other words, the functionality of the network can be temporarily jeopardized in terms of traffic even if the physical structure is still globally well connected. This implies that weighted networks appear to be more fragile than thought by considering only topological properties. All targeted strategies are very effective in dramatically damaging the network, and reach the complete destruction at a very small threshold value of the fraction of removed nodes.

As discussed in the previous section, attacks based on initial centrality ranking lead to smaller *topological* damage compared with the case where the centrality measure (degree or betweenness) is recalculated after each node removal. However, when traffic integrity measures are studied, differences are negligible: a very fast decrease of the integrity is observed for all strategies, based either on initial or recalculated quantities.

In summary, the study of the vulnerability of weighted networks to various targeted attack strategies has shown that complex networks are more fragile than expected from the analysis of topological quantities when the traffic characteristics are taken into account. In particular, the network's integrity in terms of carried traffic is vanishing significantly before the network is topologically fragmented. Moreover, the integrity of the network is harmed in a very similar manner by attacks based on initial or recalculated centrality rankings. All these results warn about the extreme vulnerability of the traffic properties of weighted networks and signal the need to pay particular attention to weights and traffic in the design of protection strategies.

6.6 Damage in real-world networks

The percolation model is a very stylized model of networks' reactions to local damage and clearly does not capture most of the features contributing to the resilience and robustness of real-world networks. It would be a gross oversimplification to conclude that what is presented in this chapter may explain a real blackout or internet congestion.

First of all, we have presented a purely topological picture. As the network is progressively damaged by the removal of nodes, its topology is modified, and in particular some node characteristics depending on the whole structure, such as the betweenness centrality, may change strongly, as mentioned in Section 6.5. This naturally leads to the concept of cascading failures: the failure of a single node leads to a redistribution of traffic on the network which may trigger subsequent overloads and failure of the next most-loaded node. While avalanche and cascading phenomena will be analyzed in Chapter 11, another main issue of real-world networks is not at all considered. As we discussed in the early chapters of the book, the network models we use are generally unstructured. They are good at capturing large-scale statistical properties but do not take into account the inherent level of local engineering of most of the real-world infrastructures. In addition, engineering in most cases is aimed at reducing risk and damage spreading in actual failure occurrence.

On the other hand, it would be inconclusive to approach the problem of network robustness in a "deterministic" way focusing only on the technical engineering and the (ubiquitous) human error. In particular the stylized approaches presented here show that, on the large scale, the inherent complexity of our infrastructures has built-in critical response properties to local damages which may be responsible for large-scale failures. In other words, we need to capitalize conceptually on these simple models by introducing higher levels of realism that might tackle at once the engineering and the globally emerging properties in risk evaluation and prediction.

7

Synchronization phenomena in networks

Many natural systems can be described as a collection of oscillators coupled to each other via an interaction matrix. Systems of this type describe phenomena as diverse as earthquakes, ecosystems, neurons, cardiac pacemaker cells, or animal and insect behavior. Coupled oscillators may display synchronized behavior, i.e. follow a common dynamical evolution. Famous examples include the synchronization of circadian rhythms and night/day alternation, crickets that chirp in unison, or flashing fireflies. An exhaustive list of examples and a detailed exposition of the synchronization behavior of periodic systems can be found in the book by Blekhman (1988) and the more recent reviews by Pikovsky, Rosenblum and Kurths (2001), and Boccaletti *et al.* (2002).

Synchronization properties are also dependent on the coupling pattern among the oscillators which is conveniently represented as an interaction network characterizing each system. Networks therefore assume a major role in the study of synchronization phenomena and, in this chapter, we intend to provide an overview of results addressing the effect of their structure and complexity on the behavior of the most widely used classes of models.

7.1 General framework

The central question in the study of coupled oscillators concerns the emergence of coherent behavior in which the elements of the system follow the same dynamical pattern, i.e. are synchronized. The first studies were concerned with the synchronization of periodic systems such as clocks or flashing fireflies. More recently, much interest has also been devoted to the possible synchronization of chaotic systems (see Boccaletti *et al.* [2002] for a recent review), whose existence can seem paradoxical but occurs in systems as diverse as lasers (Otsuka *et al.*, 2000) or neural networks (Hansel and Sompolinsky, 1992), and is very relevant in many physiological processes (Glass, 2001). Chaotic systems are characterized by a very strong sensitivity to initial conditions, and two identical chaotic systems will become uncorrelated at large times even if they start from very similar (but not identical) states. Nevertheless, the coupling of such systems can lead them to follow the same chaotic trajectories in the available phase space.

In order to frame precisely how the network structure enters the problem of synchronization, let us consider a large number N of oscillators (units) in interaction. Each oscillator i is described by an internal degree of freedom $\phi_i(t)$ ($i = 1, \ldots, N$) which evolves in time both because of an internal dynamics and because of the coupling with the other units. A quite general form to describe the evolution of the system is given by the set of equations

$$\frac{d\phi_i}{dt} = f_i(\{\phi\}), \tag{7.1}$$

where $\{\phi\}$ represents the set of all ϕ_i's. Each unit, if isolated, can reach at large times either a stable fixed point, a limit cycle, or a chaotic strange attractor (Bergé, Pomeau and Vidal, 1984). The set of units and their evolution equations (7.1) can be seen as a network in which each node represents an oscillator and two nodes i and j are linked by a directed edge from j to i if the evolution equation of i depends on the state ϕ_j of oscillator j. While this definition leads to the emergence of a directed network, the case of symmetric interactions is most often considered: the evolution of ϕ_j depends on ϕ_i and vice versa, so that the resulting network is undirected.

When a large number of units is coupled through a complex network of interactions, various types of synchronization behaviors are (a priori) possible. The equality of all internal variables (which evolve in time) $\{\phi_i(t) = s(t), \forall i\}$ is called *complete synchronization* and is the most commonly studied synchronization phenomenon (Pecora and Carroll, 1990). *Phase synchronization* (Rosenblum, Pikovsky and Kurths, 1996) is a weaker form of synchronization which can be reached by oscillators described by a phase and an amplitude: it consists in a locking of the phases while the correlation between amplitudes is weak. *Generalized synchronization* of two dynamical units is an extension of the synchronization concept in which the output of one unit is constantly equal to a certain function of the output of the other unit (Rulkov *et al.*, 1995).[1] More complex phenomena such as intermittent bursts of non-synchronized behaviors in between periods of synchronized evolution can also be observed. Most studies have, however, focused on the case of complete synchronization (see Boccaletti *et al.* [2006] for a recent review), and we will limit our presentation to this case in the present chapter.

---

[1] The generalized synchronization is therefore an involved phenomenon which may not easily be detected by the simple observation of the evolution of the dynamical variables $\{\phi\}$.

In general, the synchronization behavior is the result of a combination of the connectivity pattern and the specific oscillators and interaction functions. The remaining sections of this chapter are therefore devoted to specific models or classes of models as identified by the properties of each oscillator and their interaction function.

7.2 Linearly coupled identical oscillators

Given a system of coupled oscillators, the issue of the existence and stability of a synchronized state can be tackled analytically by assuming a specific form for the oscillators' interactions. The first model that is natural to consider is a collection of N identical dynamical units, each of them endowed with an internal degree of freedom ϕ_i (for simplicity we consider scalar ϕ_i, but the discussion can easily be extended to the vectorial case). Each unit, if isolated, evolves according to an identical internal dynamics governed by a local ordinary differential equation of the form

$$\frac{d\phi_i}{dt} = F(\phi_i), \quad i = 1, \ldots, N. \tag{7.2}$$

When the units are interconnected through an interaction network, the previous equation is modified since each unit i interacts with its neighbors $j \in \mathcal{V}(i)$. A simple example corresponds to the *linear* coupling, for which each unit i is coupled to a linear superposition of the outputs of the neighboring units, and for which the evolution equations of the system take the form

$$\frac{d\phi_i}{dt} = F(\phi_i) + \sigma \sum_{j=1}^{N} C_{ij} H(\phi_j). \tag{7.3}$$

Here H is a fixed output function, σ represents the interaction strength, and C_{ij} is the *coupling matrix*. In this section, we restrict the discussion to identical oscillators and output functions (same functions F and H for all units). One can moreover assume that the coupling between two units depends only on the difference between their outputs. In this case, each node i evolves according to both its internal dynamics and the sum of the differences between its output function and the ones of its neighbors as

$$\frac{d\phi_i}{dt} = F(\phi_i) + \sigma \sum_{j \in \mathcal{V}(i)} \left[H(\phi_i) - H(\phi_j) \right]. \tag{7.4}$$

This corresponds to the coupling $C_{ij} = L_{ij}$ where L_{ij} is the Laplacian matrix of the interaction network. As recalled in Appendix 4, $L_{ij} = -1$ for $i \neq j$ if and only if i and j are connected by a link (otherwise $L_{ij} = 0$), and L_{ii} is the degree of node i, i.e. the number of units to which i is connected.

The functions F and H and the structure of the evolution equations are crucial for the subsequent analysis and the existence of a fully synchronized solution of the evolution equations. Let us denote by $s(t)$ the evolution of the uncoupled oscillators, according to Equation (7.2). It is then straightforward to see that the fully synchronized behavior $\{\phi_i(t) = s(t), \forall i\}$ is also a solution of (7.4). The stability of this synchronized state can be studied by using the *master stability function* approach that considers a small perturbation $\phi_i = s + \xi_i$ of the system, with $\xi_i \ll s$, close to the synchronized state.[2] The synchronized state is then *stable* if and only if the dynamical evolution drives the system back to a synchronized state by a steady decrease of the perturbation ξ_i. If the perturbation increases the synchronized state is unstable. In order to distinguish between these two cases, we have to determine the evolution equation of the perturbation. To this aim, we can expand the quantities appearing in the evolution equation of ϕ_i as $F(\phi_i) \approx F(s) + \xi_i F'(s)$ and $H(\phi_j) \approx H(s) + \xi_j H'(s)$, where F' and H' denote the derivatives of the functions F and H with respect to s. Since $s(t)$ is by definition solution of $ds/dt = F(s(t))$, the evolution equation for the perturbations $\xi_i(t)$ can be written as

$$\frac{d\xi_i}{dt} = F'(s)\xi_i + \sigma \sum_j \left[L_{ij} H'(s)\right] \xi_j. \tag{7.5}$$

The Laplacian matrix L is symmetric, and its N eigenvalues, λ_i $(i = 1, \ldots, N)$ are real and non-negative. The fact that $\sum_j L_{ij} = 0$ ensures that one of these eigenvalues is null, we can therefore order them as $0 = \lambda_1 \le \lambda_2 \le \cdots \le \lambda_N \equiv \lambda_{max}$. The system of equations can be decoupled by using the set of eigenvectors ζ_i which are an appropriate set of linear combinations of the perturbations ξ_j, obtaining

$$\frac{d\zeta_i}{dt} = \left[F'(s) + \sigma \lambda_i H'(s)\right] \zeta_i. \tag{7.6}$$

At short times, one can assume that s almost does not vary, and these decoupled equations are easily solved, with the solutions

$$\zeta_i(t) = \zeta_i^0 \exp\left\{[F'(s) + \sigma \lambda_i H'(s)]t\right\}, \tag{7.7}$$

where ζ_i^0 is the initially imposed perturbation. These equations show that the perturbation will either increase or decrease exponentially, depending on the signs of the quantities $\Lambda_i \equiv F'(s) + \sigma \lambda_i H'(s)$. The eigenvalue $\lambda_1 = 0$ gives $\Lambda_1 = F'(s)$ and is only related to the evolution of each single unit, which is either chaotic (if

---

[2] We present here the case in which the units obey continuous time evolutions, but the case of time-discrete maps can be analyzed along similar lines; see Jost and Joy (2001), Jalan and Amritkar (2002), Lind, Gallas and Herrmann (2004).

$F'(s) > 0$) or periodic. For the synchronized state to be stable, *all* the other components of the perturbation ζ_i have to decrease, which means that all the Λ_i, for $i = 2, \ldots, N$, need to be negative. This has to be true for all values of s taken by the system during its synchronized evolution. The master stability function is thus defined as (Heagy, Pecora and Carroll, 1995; Pecora and Carroll, 1998; Barahona and Pecora, 2002)

$$\Lambda(\alpha) = \max_s \left(F'(s) + \alpha H'(s) \right), \tag{7.8}$$

where the max is taken over the trajectory defined by $ds/dt = F(s(t))$,[3] and the stability condition of the synchronized state translates into the condition that all $\sigma\lambda_i$, for $i = 2, \ldots, N$, are located in the negative region of the master stability function, i.e. are such that $\Lambda(\sigma\lambda_i) \leq 0$. For $\alpha > 0$, various cases can be distinguished: if $\Lambda(\alpha)$ is always positive, the synchronized state is never stable. If $\Lambda(\alpha)$ decreases and becomes negative for all $\alpha > \alpha_c$, it is on the other hand enough to have a large enough coupling strength ($\sigma > \alpha_c/\lambda_2$) to ensure synchronization.

A more interesting situation, which corresponds to a large class of functions F and H (Barahona and Pecora, 2002) is sketched in Figure 7.1. In this case $\Lambda(\alpha)$ takes negative values for α between α_1 and α_2 and the necessary condition for stability of the synchronous state is then $\sigma\lambda_i \in [\alpha_1, \alpha_2]$ for $i = 2, \ldots, N$. This criterion is not straightforward: it seems indeed quite natural that the coupling σ should be large enough, which explains the lower bound $\sigma\lambda_2 \geq \alpha_1$, but on the other hand it cannot be increased indefinitely as it might cause destabilization of

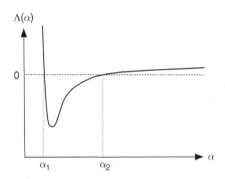

Fig. 7.1. Typical master stability function. The synchronized state is stable only if all the positive eigenvalues λ_i of the Laplacian are such that $\Lambda(\sigma\lambda_i) \leq 0$, which corresponds to $\sigma\lambda_i \in [\alpha_1, \alpha_2]$ for $i = 2, \ldots, N$. Adapted from Barahona and Pecora (2002).

---

[3] If the variables ϕ_i are d-dimensional vectors, then the ζ_i are also vectors, $F'(s)$ and $H'(s)$ are $d \times d$ matrices, and $\Lambda(\alpha)$ is the maximal eigenvalue of the matrix $F'(s) + \alpha H'(s)$.

the synchronous state when $\sigma \lambda_{max}$ becomes larger than α_2. The condition for synchronization is then satisfied if the coupling σ lies between α_1/λ_2 and α_2/λ_{max} and a compact way to express if the network is synchronizable (Barahona and Pecora, 2002) reads as

$$\frac{\lambda_{max}}{\lambda_2} < \frac{\alpha_2}{\alpha_1}, \tag{7.9}$$

where λ_2 and λ_{max} are the first non-zero and maximum eigenvalues of the Laplacian matrix, respectively. The condition (7.9) is particularly interesting since it separates the spectral properties of the network, which determine the *eigenratio* λ_{max}/λ_2, from the properties of the oscillators (F and H), which govern α_2 and α_1. Strikingly, for given functions F and H, some networks will have too large a value of λ_{max}/λ_2 and therefore cannot exhibit synchronized behavior, whatever the value of the coupling parameter σ.

The previous result clearly shows the role of the interaction pattern in synchronization phenomena and has led to the study of the general propension towards synchronization of different network topologies, *independently of the nature of the oscillators*. Networks with smaller eigenratio λ_{max}/λ_2 will indeed favor synchronization of whichever oscillators they couple. The synchronizability of different networks is therefore obtained by investigating the eigenvalues of the corresponding coupling matrices $\mathbf{C}$. Various studies have been devoted to this problem and in the following sections we will see how the eigenratio λ_{max}/λ_2 varies according to the properties of networks, and which kinds of networks can minimize it.

7.2.1 Small-world networks

As mentioned in the previous chapter, one of the motivations behind the formulation of the Watts–Strogatz model (Watts and Strogatz, 1998) was the investigation of the synchronization properties of real-world networks and how clustering might favor synchronizability (Gade and Hu, 2000; Lago-Fernández *et al.*, 2000). In the case of identical oscillators with linear coupling, Barahona and Pecora (2002) have provided a thorough comparison of the synchronizability of small-world systems. At first they analyzed the two limiting cases of a random Erdős–Rényi graph and a one-dimensional ordered ring. For a ring of N nodes, each coupled to its $2m$ nearest neighbors, the eigenvalues λ_{max} and λ_2 of the Laplacian matrix are given by $\lambda_{max} \sim (2m+1)(1+2/3\pi)$ for $m \gg 1$, and $\lambda_2 \sim 2\pi^2 m(m+1)(2m+1)/(3N^2)$ for $m \ll N$, so that the eigenratio

$$\frac{\lambda_{max}}{\lambda_2} \propto \frac{N^2}{m(m+1)} \tag{7.10}$$

increases rapidly with the number of oscillators at m fixed. For large N the synchronizability is less likely. This, however, improves by increasing m, and in the limit of a complete graph with $m = (N - 1)/2$, the system always exhibits a synchronizability interval.

The Erdős–Rényi random graphs with connection probability p (average degree pN) represent the opposite limiting case. In this case the eigenratio can be explicitly written as (Mohar, 1997)

$$\frac{\lambda_{max}^{ER}}{\lambda_2^{ER}} \simeq \frac{Np + \sqrt{2p(1-p)N \log N}}{Np - \sqrt{2p(1-p)N \log N}}. \tag{7.11}$$

The Erdős–Rényi random graph thus becomes synchronizable, in the sense that $\lambda_{max}^{ER}/\lambda_2^{ER}$ becomes finite, for p larger than $2 \log N/N$. This is close to the threshold value $\log N/N$ at which the Erdős–Rényi becomes "almost surely" connected. It is important to note that a random graph consisting of several disconnected components is by definition not synchronizable, since the various components are not able to communicate.

In order to interpolate between the ordered ring structure and the completely random Erdős–Rényi graph, Barahona and Pecora (2002) have altered the ordered ring by adding to each node additional edges randomly wired in the ring. In this way small-world networks with different values of the connectance $\mathcal{D} = 2E/(N(N-1))$ are obtained. The synchronizability of networks can be visually inspected in Figure 7.2 by comparing the behavior of the eigenratio

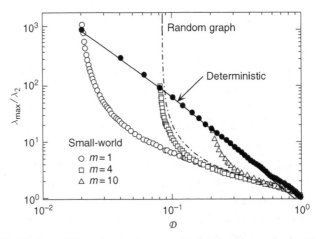

Fig. 7.2. Evolution of the eigenratio λ_{max}/λ_2 as the connectance $\mathcal{D}$ is varied, for deterministic ring structures (black filled circles), small-world networks with various m (open symbols), and for the Erdős–Rényi random graph (dot-dashed line). The continuous line corresponds to Equation (7.10). Data from Barahona and Pecora (2002) with $N = 100$.

λ_{max}/λ_2 as a function of the graph connectance in the two limiting cases and for the small-world graphs. In general, an increasing connectance corresponds to an improvement in synchronizability. The small-world topology, however, exhibits better synchronizability properties than those found in ordered rings. Figure 7.2 provides evidence that the small-world topology is a viable route to improve synchronizability generally by only acting on the system connectivity pattern.

7.2.2 Degree fluctuations: the paradox of heterogeneity

While a naive generalization of the results of Barahona and Pecora (2002) could lead to the conclusion that networks get easier to synchronize when their diameter gets smaller, Nishikawa *et al.* (2003) have shown that the situation is not so simple. Small network diameters can be obtained by considering networks with power-law degree distribution $P(k) \sim k^{-\gamma}$ with decreasing γ. The surprising result of such a procedure is that the eigenratio λ_{max}/λ_2 of the Laplacian's largest and smallest non-zero eigenvalues *increases* strongly when γ decreases. In other words, as the network becomes more heterogeneous and shrinks in diameter, it becomes less synchronizable. The intuitive explanation of this effect lies in the "overloading" of the hubs by too many different inputs which come from the many oscillators to which they are connected (Nishikawa *et al.*, 2003). In general the inputs cancel each other if they have different phases or frequencies and hubs are not easily synchronizable. This effect is not a particular feature of the scale-free heterogeneity and is recovered in any network characterized by large hubs as shown by Nishikawa *et al.* (2003) on a variant of the Watts–Strogatz model. This variant of the model considers the basic small-world construction with N initial nodes forming a one-dimensional ring, each node having $2m$ connections to its nearest neighbors. In order to introduce a given number of hubs, n_c "centers" are chosen at random, and a fixed number of shortcuts are added between the n_c centers and nodes chosen at random. The level of heterogeneity is therefore tuned by n_c: smaller n_c leads to a large degree of the centers as well as to a small diameter of the network. The results obtained by Nishikawa *et al.* (2003) clearly show that this construction leads to large eigenratios λ_{max}/λ_2 and decreased synchronizability.

The fact that networks with heavy-tailed degree distributions and heterogeneous connectivity patterns are more difficult to synchronize than homogeneous networks has been dubbed the *paradox of heterogeneity* (Motter, Zhou and Kurths, 2005a; 2005b; 2005c). Interestingly, this evidence has led Motter *et al.* (2005a) to explore alternative ways to improve synchronizability in heterogeneous networks by relaxing the assumption of symmetric couplings between oscillators.

In particular, Motter *et al.* (2005a) have proposed the following coupling structure

$$\frac{d\phi_i}{dt} = F(\phi_i) + \frac{\sigma}{k_i^\beta} \sum_j L_{ij} H(\phi_j), \tag{7.12}$$

where L_{ij} is, as before, the Laplacian matrix, k_i is the degree of node i and β is a tunable parameter. The obtained coupling matrix $C_{ij} = L_{ij} k_i^{-\beta}$ is no longer symmetric. It can, however, be written as $C = D^{-\beta} L$, where $D = \text{diag}\{k_1, \ldots, k_N\}$ is a diagonal matrix. The identity $\det(D^{-\beta} L - \lambda I) = \det(D^{-\beta/2} L D^{-\beta/2} - \lambda I)$, valid for any λ, shows that the eigenspectrum of C is the same as that of the symmetric matrix $D^{-\beta/2} L D^{-\beta/2}$, hence is real and non-negative.

The influence of β can be first understood through a mean-field approach (Motter *et al.*, 2005a). The evolution equation of ϕ_i, (7.12), can indeed be written as

$$\frac{d\phi_i}{dt} = F(\phi_i) + \frac{\sigma}{k_i^\beta} \left[k_i H(\phi_i) - \sum_{j \in \mathcal{V}(i)} H(\phi_j) \right]$$

$$= F(\phi_i) - \sigma k_i^{1-\beta} \left[\bar{H}_i - H(\phi_i) \right], \tag{7.13}$$

where $\bar{H}_i \equiv \sum_{j \in \mathcal{V}(i)} H(\phi_j)/k_i$ is the average "field" of i's neighbors. If the network is sufficiently random, and if the system is close to the synchronized state s, $\bar{H}_i$ can be approximated by $H(s)$, so that one obtains the set of *decoupled* equations

$$\frac{d\phi_i}{dt} = F(\phi_i) - \sigma k_i^{1-\beta} \left[H(s) - H(\phi_i) \right], \tag{7.14}$$

and the condition for all oscillators to be synchronized by the common mean-field forcing $H(s)$ is that $\alpha_1 < \sigma k_i^{1-\beta} < \alpha_2$ for all i. For $\beta \neq 1$, it is then clear that as soon as one node has a degree different from the others, this condition will be more difficult to meet. On the other hand, $\beta = 1$ allows the network characteristics to be removed from the synchronizability condition. In the context of the above mean-field approximation the eigenratio reads as

$$\frac{\lambda_{\max}}{\lambda_2} = \begin{cases} \left(\frac{k_{\max}}{k_{\min}} \right)^{1-\beta} & \text{if } \beta \leq 1 \\ \left(\frac{k_{\min}}{k_{\max}} \right)^{1-\beta} & \text{if } \beta \geq 1 \end{cases}, \tag{7.15}$$

where $k_{\max}$ and $k_{\min}$ are respectively the maximum and minimum degree in the network. The minimum value for the eigenratio is therefore obtained for $\beta = 1$. The numerical computations of eigenvalue spectra confirm the mean-field result, and show a global minimum of $\lambda_{\max}/\lambda_2$ at $\beta = 1$ (see Figure 7.3). This minimum (almost) does not depend on the degree distribution exponent γ for various models of scale-free networks. Thanks to the compensation of the large degree of the hubs

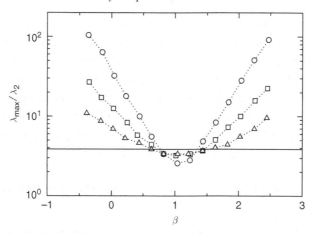

Fig. 7.3. Eigenratio λ_{max}/λ_2 as a function of β for random scale-free networks with $\gamma = 3$ (circles), $\gamma = 5$ (squares), $\gamma = 7$ (triangles), and $\gamma = \infty$ (line). Each symbol corresponds to an average over 50 different networks, with minimum degree 10 and size 1024. Data from Motter *et al.* (2005a).

by the $k_i^{-\beta}$ term, with $\beta = 1$, scale-free networks in fact become as synchronizable as random homogeneous networks. This effect is moreover confirmed by direct simulations of oscillators coupled with a scale-free interaction network (Motter *et al.*, 2005a; Motter *et al.*, 2005b).

The optimal value of $\beta = 1$ has a very intuitive explanation as each unit i is connected to exactly k_i other oscillators, so that the sum over $j \in \mathcal{V}(i)$ contains k_i terms. When $\beta < 1$, the oscillators with larger degree are more strongly coupled than the ones with smaller degree. When $\beta > 1$, the opposite situation happens. In both situations, some oscillators will be more coupled than others and the most or least coupled ones will limit the synchronizability (Motter *et al.*, 2005c). This consideration can be simply generalized to weighted networks by showing that the synchronizability decreases when the nodes' strengths (see Chapter 1) become more heterogeneous (Zhou, Motter and Kurths, 2006). The effect of clustering and assortativity can be investigated by using analogous considerations (Motter *et al.*, 2005a; 2005c). In particular, while disassortativity enhances the synchronizability (Sorrentino, di Bernardo and Garofalo, 2007), the eigenratio λ_{max}/λ_2 generally increases if the clustering coefficient increases. In other words, local cohesiveness does not favor global synchronization, as also found in networks characterized by communities identified by cohesive groups of nodes (Donetti, Hurtado and Muñoz, 2005; Park *et al.*, 2006).

As we have seen in previous chapters, the centrality of nodes is not just characterized by their degree. An important indicator is the betweenness centrality that identifies nodes having a key role in keeping the network connected, thus allowing the exchange of signals anywhere in the network. It is then natural to imagine that high betweenness nodes are the hot spots that are difficult to synchronize as they bridge parts of the networks which are otherwise disconnected and thus not synchronized. More precisely, Nishikawa *et al.* (2003) have shown that the increase of λ_{max}/λ_2 mirrors the one of the betweenness centrality (see also Hong *et al.* [2004]). In this context, Chavez *et al.* (2005) have pushed further the reasoning of Motter *et al.* (2005c) to achieve synchronizability, and proposed a rescaling of the couplings of the equations (7.3) that reads as

$$\frac{d\phi_i}{dt} = F(\phi_i) - \frac{\sigma}{\sum_{j \in \mathcal{V}(i)} b_{ij}^a} \sum_{j \in \mathcal{V}(i)} b_{ij}^a [H(\phi_i) - H(\phi_j)], \qquad (7.16)$$

where a is a parameter and b_{ij} is the betweenness centrality of the edge (i, j).[4] The case $a = 0$ corresponds to the optimal synchronizability of Equation (7.12), since in this instance $b_{ij}^a = 1$ and $\sum_{j \in \mathcal{V}(i)} b_{ij}^a = k_i$, while positive values of a give larger coupling strengths to the edges with large centrality. These edges are indeed crucial in holding the network together and ensuring communication (see Chapters 6 and 11). Numerical computation of eigenratios for various scale-free networks, as well as direct simulations of oscillators coupled according to (7.16), show that synchronizability is increased with respect to the case (7.12) for a close to 1. In fact, the procedure (7.16) is more efficient than the uniformization by the degree (7.12) because it weighs the links according to the global structure of network, rather than to local information.

7.2.3 Degree-related asymmetry

The couplings considered in the previous subsection are not symmetric, in contrast with the initial formulation (7.3). Asymmetric couplings between oscillators can in fact be studied with more generality. In most real-world networks indeed the interactions display some degree of asymmetry, due, for example, in social networks, to differences in age or social influence. In general, one can consider coupled oscillators evolving according to the set of equations (7.3) with an asymmetric coupling matrix $C_{ij} \neq C_{ji}$ (with obviously $C_{ij} = 0$ if i and j are not neighbors). The stability analysis of the synchronized state is then more involved, since the eigenvalues of $\mathbf{C}$ can be complex-valued. Let us simply mention that, if the eigenvalues $\lambda_j = \lambda_j^r + i\lambda_j^i$ ($i^2 = -1$) are ordered by increasing real parts λ_j^r ($0 = \lambda_1^r \leq \lambda_2^r \leq \cdots \leq \lambda_N^r \equiv \lambda_{max}^r$), the best propensity for synchronization is

---

[4] As in the case of (7.12), the coupling is not symmetric but the coupling matrix is the product of the symmetric matrix with elements $b_{ij}^a L_{ij}$ and of the diagonal matrix $\text{diag}\{1/\sum_j b_{1j}^a, \cdots, 1/\sum_j b_{Nj}^a\}$, so that its eigenspectrum is still real and non-negative.

obtained when the ratio of the largest to the smallest non-zero real parts, $\lambda^r_{max}/\lambda^r_2$, and the quantity $M = \max_j\{|\lambda^i_j|\}$ are simultaneously made as small as possible (Hwang *et al.* [2005] and Boccaletti *et al.* [[2006]] for details).[5]

It is of course possible to study arbitrarily complicated forms for the coupling between oscillators. In the context of networks, a natural classification of nodes is given by their degree. An interesting example of asymmetric couplings consists therefore in an asymmetry depending on the nodes' degrees (Hwang *et al.*, 2005). In other words, the value of the coupling C_{ij} depends on the relative values of the degrees of the nodes i and j. In particular, one can compare the two following opposite situations. In the first situation, the couplings are such that $C_{ij} > C_{ji}$ for $k_i < k_j$. The influence of j on i is then larger than the influence of i on j, if j has a larger degree than i. The large degree nodes are thus "driving" the nodes with smaller degree. In the second situation, the asymmetry is reversed ($C_{ij} < C_{ji}$ for $k_i < k_j$), and a node with small degree will have a stronger influence on a neighbor with larger degree. In each case, the most extreme asymmetry is obtained when, for each link (i, j), one of the couplings C_{ij} or C_{ji} is equal to 0. A parameter can be introduced to tune the system from one extreme situation to the other. Hwang *et al.* (2005) then obtain that the propensity for synchronization is maximal when hubs completely drive the small nodes. This case is obtained by taking $C_{ji} = 0$ for $k_i < k_j$ (with on the other hand $C_{ij} \neq 0$), and corresponds in fact to strictly unidirectional couplings on each link (if j has an influence on i, it does not in return feel the influence of i). Such results can be understood intuitively by considering a star network consisting of a single hub connected to many nodes of degree 1. If the coupling is directed from the hub to the other nodes, synchronization will clearly be much easier than in the opposite case, in which the hub would receive uncorrelated inputs from the other nodes. Interestingly, this result is robust and remains valid even if the interaction network does not display strong degree fluctuations, for example in an Erdős–Rényi network.

Investigation of the effect of mixing patterns and degree correlations more-over leads to the conclusion that synchronizability is enhanced in *disassortative* scale-free networks when smaller nodes drive larger degree ones, while the bet-ter propensity for synchronization when the hubs are driving the small nodes is slightly improved in assortative networks (Sorrentino *et al.*, 2006). Synchroniz-ability is therefore favored if hubs are strongly interconnected and drive the nodes with smaller degrees. In other words, a rich-club (Chapter 1) of large degree nodes will lead to a good synchronizability. Zhou and Kurths (2006) confirm this pic-ture by showing that, even when the system is not fully synchronized, it displays a

---

[5] See also Nishikawa and Motter (2006a; 2006b) for the extension of the master stability function framework to directed, *weighted*, and *non-diagonalizable* coupling matrices.

hierarchical synchronization in which the hubs are synchronized more closely and form the dynamical core of the network.

7.3 Non-linear coupling: firing and pulse

Let us now turn to some concrete examples of coupled oscillators and to their synchronization properties. In many biological systems the interaction between oscillators is not smooth but rather episodic and pulse-like. For example, neurons communicate by firing sudden impulses. Neurosciences are indeed particularly concerned with the understanding of collective dynamical phenomena such as the propagation of such coherent sensory signals through networks (Maass and Bishop, 1998; Koch, 1999). More generally, synchronization and rhythmic processes play an important role in physiology (Glass, 2001).

To shed light on how the interaction topology affects the collective behavior of coupled neurons, Lago-Fernández *et al.* (2000) have considered a system of non-identical neurons, each described by a set of coupled differential equations for the evolution of local currents and potentials. In this model, interaction between coupled neurons occurs through pulses, and various coupling topologies can be compared. On the one hand, regular topologies (grids) give rise to coherent oscillations but with a slow time scale for the system response and a relatively long transient. Completely random topologies on the other hand ensure a fast response of the system to external stimulation, but no coherent oscillations are observed. Interestingly, the Watts–Strogatz topology, with small distances between units and large clustering coefficient, displays both a fast system response and coherent global oscillations.

Such analysis can be extended to simpler and more general models of pulse driven oscillators. In particular, the integrate-and-fire model for synchronous firing of biological oscillators is based on Peskin's model (Peskin, 1975) of the cardiac pacemaker. This model describes a population of identical oscillators which communicate through sudden impulses. The firing of one oscillator is transmitted to its neighbors, possibly bringing them to the firing threshold. For simplicity, each oscillator is described by a phase $\theta_i(t) \in [0, 2\pi]$ which increases linearly in time. The evolution of the dynamical variable $\phi_i = \theta_i/2\pi$ is thus governed by the equation

$$\frac{d\phi_i}{dt} = 1, \tag{7.17}$$

until it reaches $\phi_i(t_i) = 1$. The oscillator's phase is then reset to 0 while it "fires" a signal to its neighbors $j \in \mathcal{V}(i)$, whose internal variables undergo a phase jump whose amplitude depends on their states. The global update rules when ϕ_i reaches 1 are thus given by

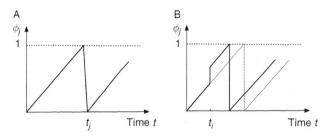

Fig. 7.4. A, Unperturbed evolution of a firing oscillator j according to (7.17) and (7.18). B, Evolution of ϕ_j when a neighbor i of j fires at t_i.

$$\phi_i(t_i) = 1 \Rightarrow \begin{cases} \phi_i(t_i^+) = 0 \\ \phi_j(t_i^+) = \phi_j(t_i) + \Delta\left[\phi_j(t_i)\right] \end{cases} \tag{7.18}$$

for all neighboring sites $j \in \mathcal{V}(i)$ of i, as illustrated in Figure 7.4. Here the function Δ, called the phase response curve, governs the interaction, and t_i^+ represents the time immediately following the firing impulse of node i. It is also possible to introduce a delay between the firing of the node i and the update of its neighbors. In this case we can consider that the firing signal takes a time τ to go from one oscillator to the other, yielding the update rule $\phi_j((t_i+\tau)^+) = \phi_j(t_i+\tau)+\Delta\left[\phi_j(t_i + \tau)\right]$. An interesting feature of the firing dynamics is the possibility of generating avalanches because if $\phi_j((t_i + \tau)^+)$ is larger than 1, the unit j fires in its turn and so on, triggering in some cases a chain reaction of firing events.

A generalization of the integrate-and-fire model (Mirollo and Strogatz, 1990) describes many different models of interacting threshold elements (Timme, Wolf and Geisel, 2002). In this generalization each oscillator i is characterized by a state variable function of the phase $x_i = U(\phi_i)$, which is assumed to increase monotonically toward the threshold 1. When x_i reaches the threshold, the oscillator fires and its state variable is set back to zero. The function $U(\phi)$ is assumed to be monotonic increasing ($\partial_\phi U > 0$), concave ($\partial_\phi^2 U < 0$), with $U(0) = 0$ and $U(1) = 1$. When the oscillator at node i fires, its neighbors $j \in \mathcal{V}(i)$ are updated according to the following rule

$$x_i(t_i) = 1 \Rightarrow x_j(t_i^+) = \min(1, x_j(t_i) + \varepsilon_{ji}), \tag{7.19}$$

where ε_{ji} gives the coupling strength between i and j. The interaction is excitatory if ε_{ji} is positive, and inhibitory if $\varepsilon_{ji} < 0$. In terms of the phase variable (see also Figure 7.5), the "pulse-coupled" model is therefore described by the set of equations

$$\phi_i(t_i) = 1 \Rightarrow \begin{cases} \phi_i(t_i^+) = 0 \\ \phi_j(t_i^+) = \min\{U^{-1}(U(\phi_j(t_i)) + \varepsilon_{ji}), 1\}. \end{cases} \tag{7.20}$$

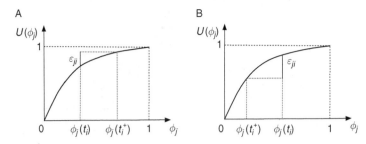

Fig. 7.5. Phase dynamics of oscillator j in response to the firing of a neighbor i of j at time t_i, according to (7.20), in the case of (A) an excitatory spike $\varepsilon_{ji} > 0$ and (B) an inhibitory spike $\varepsilon_{ji} < 0$.

In the case of zero delay ($\tau = 0$), Mirollo and Strogatz (1990) have shown that for arbitrary initial conditions a system of fully connected oscillators reaches a synchronized state in which all the oscillators are firing synchronously. The introduction of a positive delay $\tau > 0$ introduces a number of attractors increasing exponentially with the network size. Most of these attractors are periodic orbits that exhibit several groups of synchronized clusters of oscillators which fire and receive pulses alternatively. The above results refer to an operative and quantitative definition of the level of synchronization of the N interacting units expressed by the quantity

$$S = \frac{1}{N} \sum_{i=1}^{N} \left[1 - \phi_i(t_j^+) \right], \qquad (7.21)$$

measured at each firing event of a reference unit j (the choice of the reference unit being arbitrary). This function approaches 1 when the system is completely synchronized. The *synchronization time* t_s can be defined as the time needed for S to reach 1, starting from random phases ϕ_i (Guardiola *et al.*, 2000), as shown in Figure 7.6. Interestingly, oscillators located on the nodes of regular lattices reach synchronization in a shorter time than in the case in which they are coupled through a random network with similar average degree. Further investigation shows that small-world topologies such as those induced by the Watts–Strogatz construction yield intermediate results. More precisely, the synchronization time is an increasing function of the rewiring parameter of the Watts–Strogatz model.

Like what is observed in the synchronizability of networks (see Section 7.2.2), the decrease in synchronizability as the network disorder increases is due to the variations in the number of neighbors and the ensuing fluctuations in the amplitude of the signals received by each unit. The role of degree fluctuations is clearly illustrated (Guardiola *et al.*, 2000) by using, for each oscillator i of degree k_i, a rescaled

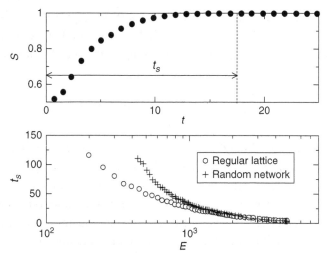

Fig. 7.6. Top: Evolution of the synchronization parameter given by Equation (7.21), and definition of the synchronization time t_s as the time needed to reach $S = 1$. In this case a population of 300 oscillators with random initial conditions is simulated. Bottom: Synchronization time t_s as a function of the number of edges E with a fixed number of oscillators $N = 100$ and interaction strength $\varepsilon = 0.01$ for the regular lattice and the Erdős–Rényi random graph. The regular lattice always performs better than the random graph although the difference quickly vanishes as E increases, i.e. as the network becomes more globally coupled. Data from Guardiola *et al.* (2000).

phase response $\Delta(\phi_i) \rightarrow \Delta(\phi_i)\langle k \rangle / k_i$ when a neighbor of i fires. This modification, in the same spirit as Equation (7.12) in Section 7.2.2, has the effect of making the couplings uniform and improving the synchronizability of networks. Some other results generally valid for arbitrary topologies can be obtained by studying particular coupling functions. For instance, it turns out that the stability of the synchronized state is ensured for inhibitory couplings ($\varepsilon_{ij} < 0$), whatever the network structure (Timme *et al.*, 2002). For arbitrary couplings on the other hand, firing times can be highly irregular (Brunel and Hakim, 1999; van Vreeswijk and Sompolinsky, 1996), and a coexistence of both regular and irregular dynamics may also be observed (Timme *et al.*, 2002). Finally, large heterogeneities in the coupling values tend to disrupt the synchronous state, which is replaced by aperiodic, asynchronous patterns (Denker *et al.*, 2004).

7.4 Non-identical oscillators: the Kuramoto model

Real coupled oscillators are rarely identical. In this spirit, a widely studied paradigm is defined by the Kuramoto model, which considers an ensemble of N

planar rotors, each one characterized by an angular phase ϕ_i and a natural frequency ω_i ($i = 1, \ldots, N$) (Kuramoto, 1984). These oscillators are coupled with a strength K, and their phases evolve according to the set of non-linearly coupled equations

$$\frac{d\phi_i}{dt} = \omega_i + K \sum_{j \in \mathcal{V}(i)} \sin(\phi_i - \phi_j), \qquad (7.22)$$

where $\mathcal{V}(i)$ is the set of neighbors of i (see Acebrón *et al.* [2005] for a recent review). The natural frequencies and the initial values of the phases are generally assigned randomly from an a priori given distribution. For the sake of simplicity, the frequency distributions generally considered are unimodal and symmetric around a mean value Ω. The level of synchronization achieved by the N coupled oscillators is quantified by the complex-valued order parameter

$$r(t)e^{i\psi(t)} = \frac{1}{N} \sum_{j=1}^{N} e^{i\phi_j(t)}, \qquad (7.23)$$

which can be seen as measuring the collective movement produced by the group of N oscillators in the complex plane. As shown schematically in Figure 7.7, each oscillator phase ϕ_j can be represented as a point moving around the circle of unit radius in the complex plane. Equation (7.23) then simply gives the position of the center of mass of these N points. The radius $r(t)$ measures the coherence, while the phase $\psi(t)$ is the average phase of the rotors. If the oscillators are not synchronized, with scattered phases, $r(t) \to 0$ in the so-called thermodynamic limit with $N \to \infty$. In contrast, a non-zero value of r denotes the emergence of a

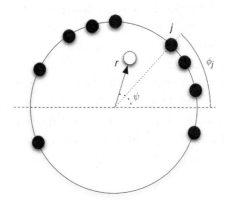

Fig. 7.7. Schematic presentation of the Kuramoto model: the phases ϕ_j correspond to points moving around the circle of unit radius in the complex plane. The center of mass of these points defines the order parameter (7.23), with radius r and phase ψ.

state in which a finite fraction of oscillators are synchronized. A value of r close to 1 corresponds to a collective movement of all the oscillators that have almost identical phases.

7.4.1 The mean-field Kuramoto model

A commonly studied version of the Kuramoto model considers a fully connected interaction network in which all the oscillators are coupled with the same strength $K = K^0/N$, with finite K^0 in order to ensure a regular limit $N \to \infty$ (Kuramoto, 1984). Effectively, each oscillator is affected by a non-linear coupling that is the average of the coupling with all other units, therefore defining a mean-field version of the model. The evolution equations (7.22) can be rewritten by introducing the order parameter (7.23). Multiplying both sides of (7.23) by $e^{-i\phi_\ell}$ and taking the imaginary part yields

$$r \sin(\psi - \phi_\ell) = \frac{1}{N} \sum_{j=1}^{N} \sin(\phi_j - \phi_\ell), \qquad (7.24)$$

which allows Equation (7.22) to be written in the form

$$\frac{d\phi_\ell}{dt} = \omega_\ell + K^0 r \sin(\phi_\ell - \psi). \qquad (7.25)$$

This rewriting of the oscillators equations highlights the mean-field character of the model, in which the interaction term is expressed as a coupling with the mean phase ψ, with intensity proportional to the coherence r (Strogatz, 2000b). The presence of a positive feedback mechanism is then clear: as more rotors become coherent, the effective coupling strength $K^0 r$ increases, which pushes still other rotors to join the coherent group. This mechanism arises in fact only if the coupling K increases above a critical value K_c. For $K^0 < K_c$, phases become uniformly distributed and $r(t)$ decays to 0 (or to fluctuating values of order $\mathcal{O}(N^{-1/2})$ at finite population size). When K^0 is increased above the critical value K_c, some elements tend to lock their phases and r becomes finite. Closer examination shows that the population of rotors is then divided into two groups: a certain number of oscillators oscillate at the common frequency Ω, while others (which have natural frequencies farther from the average) keep drifting apart. In this partially synchronized state, the group of synchronized nodes is composed by an increasing number of oscillators as K^0 grows, and r increases as schematically shown in Figure 7.8 (Strogatz, 2000b). More precisely, the transition between incoherent and coherent

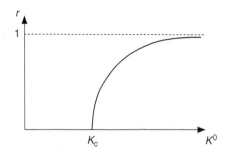

Fig. 7.8. Schematic mean-field result for the Kuramoto model. Below the critical coupling K_c there is no synchronization ($r = 0$). Above the threshold a finite fraction of the nodes are synchronized.

global dynamics is continuous: the order parameter behaves in the double limit of large sizes (thermodynamic limit $N \to \infty$) and times ($t \to \infty$) as

$$r = \begin{cases} 0 & K^0 < K_c \\ (K^0 - K_c)^\beta & K^0 \geq K_c \end{cases},$$
(7.26)

where the precise value of K_c depends on the distribution of natural frequencies ω_i, and $\beta = 1/2$.

7.4.2 The Kuramoto model on complex networks

Strogatz and Mirollo (1988) have shown that the probability of phase-locking for Kuramoto oscillators located at the vertices of regular lattices vanishes as the number N of oscillators diverges. On the other hand, numerical simulations (Watts, 1999) have pointed out the existence of synchronized states when a small amount of disorder is introduced into the lattice through shortcuts, as in the small-world model network construction (Watts and Strogatz, 1998). By performing a detailed numerical investigation, Hong, Choi and Kim (2002a) have shown that the order parameter r, averaged over time and realizations of the intrinsic frequencies, obeys the finite-size scaling form

$$r(N, K) = N^{-\beta/\nu} F\left[(K - K_c)N^{1/\nu}\right],$$
(7.27)

where F is a scaling function and the exponent ν describes the divergence of the typical correlation size $(K - K_c)^{-\nu}$. The form (7.27) implies that at $K = K_c$, the quantity $rN^{\beta/\nu}$ does not depend on the system size N. This method allows numerical determination of K_c as well as β and ν. For any finite rewiring probability p in the Watts–Strogatz model, a finite K_c is obtained, and the exponents β and ν are compatible with the mean-field values $\beta \approx 0.5$ and $\nu \approx 2.0$, valid for the fully connected graph (Hong, Choi and Kim, 2002a).

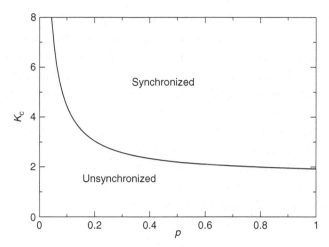

Fig. 7.9. Critical coupling K_c versus p for a small-world network of Kuramoto oscillators. Above the curve, the system is synchronized while below it does not show any extensive synchronization. Adapted from Hong, Choi and Kim (2002a).

As $p \to 0$, K_c diverges, as shown in Figure 7.9, in agreement with the results of Strogatz and Mirollo (1988). Moreover, the time needed to achieve synchronization decreases significantly when p increases. This global behavior is reminiscent of the case of the equilibrium Ising model evolving on small-world networks for which, as explained in Chapter 5, a phase transition of mean-field character appears at finite temperature as soon as p is finite.

Similar to other dynamical processes, the synchronization of Kuramoto oscillators may also be modified by large degree fluctuations in the coupling network. In order to analytically study the model in the case of oscillators with varying degree, it is possible to use an equivalence assumption for oscillators within the same degree class analogous to those used in other chapters of this book. By using this approximation, Ichinomiya (2004) finds that the critical coupling for uncorrelated random networks with arbitrary degree distribution is given by the expression

$$K_c = c\frac{\langle k \rangle}{\langle k^2 \rangle},\qquad(7.28)$$

where the constant c depends on the distribution of the individual frequencies ω (see also Restrepo, Ott and Hunt [2005a; 2005b]). The quantity $\kappa = \langle k^2 \rangle / \langle k \rangle$, which determines the heterogeneity level of the degree distribution, turns out once again to be the relevant parameter as for other processes such as equilibrium phase transitions (Chapter 5), percolation (resilience) phenomena (Chapter 6) or epidemic spread (Chapter 9). If the degree fluctuations are bounded, $\langle k^2 \rangle$ is finite as the size of the network approaches infinity, and K_c is finite. In the case of a heavy-tailed degree distribution with a diverging second moment, on the other hand, the synchronization threshold goes towards zero in the limit of infinite size network. Hubs drive the dynamics and lead to synchronization for any finite coupling between nodes, just as, in the case of Ising spins, the hubs polarize the network and lead to a ferromagnetic state at any finite temperature (see Chapter 5). The role of the hubs is highlighted by the calculation of the average time $\langle \tau \rangle$ for a node to re-synchronize after a perturbation, as a function of its degree. Moreno and Pacheco (2004) obtain numerically the behavior $\langle \tau \rangle \sim k^{-\mu}$ with μ close to 1: the larger the degree of a node, the more stable its synchronized behavior. Let us finally note that the mean-field approach also allows one to compute the critical behavior of the order parameter (Lee, 2005). For finite K_c, one obtains

$$r \sim \Delta^\beta, \tag{7.29}$$

where $\Delta \equiv (K - K_c)/K_c$ and, for a scale-free network with degree distribution $P(k) \sim k^{-\gamma}$, the exponent takes the value $\beta = 1/2$ if $\gamma > 5$ and $\beta = 1/(\gamma - 3)$ if $5 \geq \gamma > 3$.

Since the analytical results rely on a certain number of assumptions and on the degree classes equivalence approximation, it is worth mentioning that extensive numerical simulations have been performed for the evolution of Kuramoto oscillators on complex networks. An interesting result is reported by Moreno and Pacheco (2004) and Moreno, Vazquez-Prada and Pacheco (2004c) who find a non-zero threshold K_c for the onset of synchronization of oscillators interacting on a Barabási–Albert network. Such a discrepancy with the mean-field results could be due either to the inadequacy of the mean-field approximation, or to the fact that $\langle k^2 \rangle$ diverges very slowly – logarithmically – with the network size in Barabási–Albert networks, possibly leading to an apparent finite value of K_c even at very large network sizes.

7.5 Synchronization paths in complex networks

The Kuramoto model of coupled oscillators described in the previous section represents a very interesting framework to study the effect of network topology on the synchronization mechanisms. The study of the order parameter r defined in (7.23) can be enriched by the definition of the link order parameter (Gómez-Gardeñes, Moreno and Arenas, 2007b)

$$r_{\text{link}} \equiv \frac{1}{2E} \sum_j \sum_{\ell \in \mathcal{V}(j)} \lim_{\Delta t \to \infty} \frac{1}{\Delta t} \int_\tau^{\tau + \Delta t} e^{i[\phi_j(t) - \phi_\ell(t)]}, \tag{7.30}$$

where E is the total number of links in the network and τ is a time large enough to let the system relax from its initial conditions. The quantity r_{link} quantifies the fraction of links that are synchronized in the network, and therefore gives more detailed information than r. In a hypothetical situation in which nodes would be synchronized by pairs for example, r, as the sum of $N/2$ random phases, could be very small while r_{link} would take a non-zero value. Even more insights are obtained from the full matrix

$$D_{j\ell} \equiv x_{j\ell} \left| \lim_{\Delta t \to \infty} \frac{1}{\Delta t} \int_{\tau}^{\tau + \Delta t} e^{i[\phi_j(t) - \phi_\ell(t)]} \right|, \qquad (7.31)$$

where $x_{j\ell}$ is the adjacency matrix. A threshold T can then be used to distinguish pairs of synchronized nodes for $D_{j\ell} > T$ from unsynchronized neighbors for $D_{j\ell} < T$.

The combined study of r and r_{link} as a function of the coupling strength of Kuramoto oscillators located on the nodes of Erdős–Rényi and Barabási–Albert networks with same average degree reveals a very interesting picture (Gómez-Gardeñes, Moreno and Arenas, 2007b; 2007c). By varying the coupling strength it is first observed that the critical coupling is strongly decreased by the heterogeneity of the network. At large coupling however, the homogeneous topology leads to a stronger synchronization with larger r values. Strikingly, r_{link} takes non-zero values as soon as the coupling is non-zero, even in the non-synchronized phase with $r = 0$, illustrating the existence of local synchronization patterns even in the regime of global incoherence. The detailed study of the clusters of synchronized nodes, available through the measure of the matrix elements (7.31), shows that the largest cluster of synchronized oscillators contains almost half of the network nodes as the coupling approaches the critical value, and that many smaller clusters of synchronized nodes are present as well.

Figure 7.10 schematically displays the evolution of synchronized clusters as the coupling is increased. For homogeneous networks, a percolation type process is observed in which small synchronized clusters appear and coalesce at larger coupling strength in a sharp merging process towards the complete synchronization of the network. In contrast, the synchronization of a scale-free network is organized around its hubs, which first form a central synchronized core and then aggregate progressively the other small synchronized clusters as the coupling is strengthened. These results emphasize the role of the hubs in the emergence of cohesion in complex networks, as also noticed in other problems such as network resilience and rumor spreading (Chapters 6 and 10).

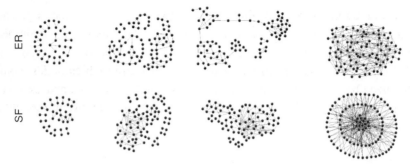

Fig. 7.10. Synchronized clusters for two different topologies: homogeneous and scale-free. Small networks of 100 nodes are displayed for the sake of visualization. From left to right, the coupling intensity is increasing. Depending on the topology, different routes to synchronization are observed. In the case of Erdős–Rényi networks (ER), a percolation-like mechanism occurs, while for scale-free networks (SF), hubs are aggregating more and more synchronized nodes. Figure courtesy of Y. Moreno, adapted from Gómez-Gardeñes, Moreno and Arenas (2007b).

7.6 Synchronization phenomena as a topology probing tool

The various studies on the synchronizability of complex networks have shed light on the basic mechanisms through which the topological features affect the capacity of oscillators to reach a synchronized state. Networks' topological characteristics are, however, not limited to their homogeneous or heterogeneous character. For instance, Park *et al.* (2006) have shown how community structures strongly influence the synchronizability of networks. Analogously, Oh *et al.* (2005) and Timme (2006) show for Kuramoto oscillators that rather different synchronization properties are obtained for different kinds of modular structure, even with similar degree distributions.

In other words, the detailed structure and connectivity pattern of networks are extremely relevant in determining the synchronizability of systems. This evidence suggests an inverse approach to the problem in which the synchronization behavior is used as a probe for the understanding of the community and modular structure of networks. The community detection problem in large-scale networks is in fact a very difficult task, which has generated a large body of work. Intuitively, and using a simple topological approach, a community in the network can be defined as a densely interconnected subgraph whose number of connections to nodes outside the subgraph is much smaller than the number of connections within the subgraph. The precise mathematical definition of a community is more elusive, and various formal or operative definitions have been put forward in the literature. At the same time, various algorithms have been proposed in order to find partitions of a network's subgraphs in order to capture the presence of communities according to various rigorous and less rigorous definitions (for recent reviews, see Newman [2004] and Danon *et al.* [2005]). In this context, the study of synchronization behavior and the eigenvalue spectra of networks represents an interesting alternative. In particular, it turns out that, during the *transient towards synchronization*, modules synchronize at different times and in a hierarchical way. In this way the approach to the synchronized regime is able to progressively identify the basic modules and community of the network. The study of the synchronization properties can then be approached by looking at the dynamical behavior of the system or by the analysis of the eigenvalue spectrum of the network (Donetti and Muñoz, 2004; Capocci *et al.*, 2005; Arenas, Díaz-Guilera and Pérez-Vicente, 2006a; 2006b). In both approaches a wealth of information on the detailed structure of the network can be obtained by simply looking at the basic characterization of the dynamical properties of coupled oscillators.

8

Walking and searching on networks

The navigation and exploration of complex networks are obviously affected by the underlying connectivity properties and represent a challenging scientific problem with a large number of practical applications. The most striking example is the World Wide Web (WWW), an immense data repository which acquires a practical interest and value only if it is possible to locate the particular information one is seeking. It is in this context and having in mind large-scale information and communication technologies (ICTs) that most of the models and studies are developed. Because information discovery and retrieval rely on the understanding of the properties of random walks and diffusion phenomena in complex networks, considerable activity has been devoted to the investigation of these basic processes.

In this chapter we will review the main strategies that can be used to explore and retrieve information from the vertices of a network. In particular, we want to expose how the topological properties of networks might affect search and retrieval strategies and how these strategies can effectively take advantage of the network's connectivity structure. We start from basic considerations on diffusion processes taking place on networks and naturally dive into more applied works that consider well-defined problems encountered in the ICT world.

8.1 Diffusion processes and random walks

The discovery process and the related algorithms in large-scale networks are clearly related to the properties of a random walk diffusing in a network of given topology. In other words, the simplest strategy to explore a network is to choose one node, to follow randomly one of the departing links to explore one of its neighbors, and to iterate this process until the required information is found or a satisfactory knowledge of the network connectivity pattern is acquired. It is clear that a random

160

walk lies underneath this simple strategy, and in order to improve the search and navigation a good understanding of this basic process is necessary.

As a starting point in the study of diffusion processes in complex networks, we first analyze the simple unbiased random walk in an undirected network with given connectivity pattern. The basic properties of this process can be understood as the random diffusion of W walkers (or particles) in a network made of N nodes. We can consider that each node i has an occupation number W_i counting the number of walkers present on it. The total number of walkers in the network is $W = \sum_i W_i$ and each walker diffuses along the edges with transition rates that depend on the node degree and other factors. In the case of a simple Markovian random walk, a walker located on a node i diffuses out of i along an edge (i, j) with a rate given by

$$d_{ij} = \frac{r}{k_i}, \tag{8.1}$$

where k_i is the degree of i. This relation simply defines a uniform rate of diffusion along any one of the edges departing from the node i, and corresponds to a total rate of escape $\sum_{j \in \mathcal{V}(i)} d_{ij}$ from any node equal to r.

The simplicity of random walk processes allows for the derivation of detailed analytical results. Before describing some of them, we take this opportunity to illustrate a general method of wide applicability in the context of heterogeneous networks. In Chapters 1 and 2, we have seen that it is convenient to group the nodes in degree classes for the statistical characterization of networks. The assumption of statistical equivalence of nodes with the same degree is in fact a very general tool that applies in the study of many dynamical processes, such as epidemics or opinion dynamics (Chapters 9 and 10). In this framework, we consider that nodes are only characterized by their degree and that their statistical properties are the same as long as their degree is the same. This assumption allows the use of degree block variables defined as

$$W_k = \frac{1}{N_k} \sum_{i \mid k_i = k} W_i, \tag{8.2}$$

where N_k is the number of nodes with degree k and the sum runs over all nodes i having degree k_l equal to k. The variable W_k represents the average number of walkers in nodes within the degree class k and conveniently allows the wide range of degrees present in the system to be taken into account. We now consider a network with degree distribution $P(k)$ in which the walkers move from a node with degree k to another with degree k' with a transition rate r/k. The dynamics of walkers is simply represented by a mean-field dynamical equation expressing the variation in time of the walkers $W_k(t)$ in each degree class. This can easily be written as:

$$\partial_t W_k(t) = -r W_k(t) + k \sum_{k'} P(k'|k) \frac{r}{k'} W_{k'}(t). \tag{8.3}$$

The first term on the right-hand side of this equation just considers that walkers move out of the node with rate r. The second term accounts for the walkers diffusing from the neighbors into the node of degree k. This term is proportional to the number of links k times the average number of walkers coming from each neighbor. This is equivalent to an average over all possible degrees k' of the conditional probability $P(k'|k)$ of having a neighbor of degree k' times the fraction of walkers coming from that node, given by $W_{k'}(t)r/k'$. In the following we consider the case of uncorrelated networks in which the conditional probability does not depend on the originating node and for which it is possible to use the relation $P(k'|k) = k'P(k')/\langle k \rangle$ (see Chapter 1). The dynamical rate equation (8.3) for the subpopulation densities reads then as

$$\partial_t W_k(t) = -r W_k(t) + \frac{k}{\langle k \rangle} \sum_{k'} P(k')r W_{k'}(t). \tag{8.4}$$

The stationary condition $\partial_t W_k(t) = 0$ does not depend upon the diffusion rate r which just fixes the time scale at which the equilibrium is reached and has the solution

$$W_k = \frac{k}{\langle k \rangle} \frac{W}{N}, \tag{8.5}$$

where we have used that $\sum_k P(k) W_k(t) = W/N$ represents the average number of walkers per node that is constant. The above expression readily gives the probability $p_k = W_k/W$ to find a single diffusing walker in a node of degree k as

$$p_k = \frac{k}{\langle k \rangle} \frac{1}{N}. \tag{8.6}$$

This last equation is the stationary solution for the visiting probability of a random walker in an uncorrelated network with arbitrary degree distribution. As expected, the larger the degree of the nodes, the larger the probability of being visited by the walker. In other words the above equation shows in a simple way the effect of the diffusion process that brings walkers, with high probability, into well-connected nodes, thus showing the impact of a network's topological fluctuations on the diffusion process.

As previously noted, the simplicity of the diffusion process allows for much more detailed analysis and results. In particular, it is possible to identify specific network structures where powerful analytical techniques can be exploited to find rigorous results on the random walk properties and other physical problems such as localization phenomena and phase transitions (see for instance the review by Burioni and Cassi [2005]). In the case of complex networks with arbitrary degree distribution, it is actually possible to write the master equation for the probability $p(i, t|i_0, 0)$ of a random walker visiting site i at time t (starting at i_0 at time $t = 0$) under the form

$$\partial_t p(i, t|i_0, 0) = -\left(\sum_j x_{ij}d_{ij}\right) p(i, t|i_0, 0) + \sum_j x_{ji}d_{ji}p(j, t|i_0, 0), \quad (8.7)$$

where x_{ij} is the adjacency matrix of the network. The first term on the right-hand side corresponds to the probability of moving out of node i along the edges connecting i to its neighbors, while the second term represents the flux of walkers arriving from the neighbors of i. With the choice of transition rates given by Equation (8.1), one can obtain explicitly the probability for a walker to be in node i in the stationary limit of large times (Noh and Rieger, 2004) as

$$p_i^\infty = \frac{k_i}{\langle k \rangle} \frac{1}{N}, \quad (8.8)$$

showing again the direct relation between the degree of a node and the probability of finding random walkers in it.

Another particularly important quantity for diffusion processes is the return probability $p_0(t)$ that a walker returns to its starting point after t steps, and is directly linked to the eigenvalue density $\rho(\lambda)$ (also called the spectral density) of the modified Laplacian operator associated with this process[1]

$$L'_{ij} = \delta_{ij} - \frac{x_{ij}}{k_j}. \quad (8.9)$$

As shown in Appendix 5, these two quantities are related by the equation

$$p_0(t) = \int_0^\infty d\lambda e^{-\lambda t} \rho(\lambda), \quad (8.10)$$

so that the behavior of $p_0(t)$ is connected to the spectral density, and in particular its long time limit is directly linked to the behavior of $\rho(\lambda)$ for $\lambda \to 0$. Notably, this problem is relevant not only to the long time behavior of random walks but also to many other processes including synchronization or signal propagation (Samukhin, Dorogovtsev and Mendes, 2008).

---

[1] We emphasize that this modified Laplacian is different from the Laplacian discussed in Chapter 7 and Appendix 4, each column being divided by k_j: $L_{ij} = k_j L'_{ij}$.

Before turning to the case of complex networks, let us first consider some simple examples. For a D-dimensional regular lattice, the low eigenvalue behavior of the spectral density is given by (Economou, 2006) $\rho_D(\lambda) \sim \lambda^{D/2-1}$, which, using Equation (8.10), leads to the well-known result

$$p_0(t) \sim t^{-D/2}. \tag{8.11}$$

For the case of a random Erdős–Rényi graph, Rodgers and Bray (1988) have demonstrated that for $\lambda \to 0$ the spectral density has the form $\rho_{ER}(\lambda) \sim e^{-c/\sqrt{\lambda}}$ which leads to the long time behavior of the form

$$p_0(t) \sim e^{-at^{1/3}} \tag{8.12}$$

where a and c are constants depending on the specific network. In the case of a Watts–Strogatz network for which shortcuts are added to the regular lattice, Monasson (1999) has shown that the eigenvalue density can be written as the product of ρ_D and ρ_{ER} leading to

$$\rho_{WS}(\lambda) \sim \lambda^{D/2-1} e^{-p/\sqrt{\lambda}}, \tag{8.13}$$

where p is the density of shortcuts. This form implies the following behavior for the return probability

$$p_0(t) - p_0(\infty) \sim \begin{cases} t^{-D/2} & t \ll t_1 \\ \exp(-(p^2 t)^{1/3}) & t_1 \ll t \end{cases}, \tag{8.14}$$

where $p_0(\infty) = 1/N$ and the crossover time is $t_1 \sim 1/p^2$. In the first regime, the diffusion is the same as on a D-dimensional lattice with a typical behavior scaling as $1/t^{D/2}$. Accordingly, the number of distinct nodes visited is $N_{cov} \sim t^{D/2}$ (Almaas, Kulkarni and Stroud, 2003). After a time of order t_1, the walkers start to feel the effect of the shortcuts and reach a regime typical of the Erdős–Rényi network with a stretched exponential behavior $\exp(-t^{1/3})$ and $N_{cov} \sim t$. At very long times, the number of distinct nodes saturates at the size of the network $N_{cov} \sim N$ (see Figure 8.1).

Interestingly, Samukhin *et al.* (2008) show that the spectral density for uncorrelated random networks with a local tree-like structure is rather insensitive to the degree distribution but depends mainly on the minimal degree in the network (in the infinite size limit). In particular, for uncorrelated random scale-free networks with minimum degree m equal to one or two, the spectral density is given by

$$\rho(\lambda) \sim \lambda^{-\xi} e^{-a/\sqrt{\lambda}}, \tag{8.15}$$

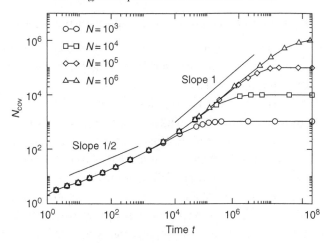

Fig. 8.1. Average number of visited distinct sites for a random walk on a (one-dimensional) Watts–Strogatz network with shortcut density $p = 0.01$. At short times, $N_{cov} \sim t^{1/2}$, and for longer times, $N_{cov} \sim t$. In the last regime, the coverage number saturates to the size N of the network. Data from Almaas *et al.* (2003).

where a is a constant, $\xi = 9/10$ for $m = 1$ and $\xi = 4/3$ for $m = 2$. The return probability on scale-free networks is thus different from the Watts–Strogatz case and turns out to be given by

$$p_0(t) \sim t^{\eta} e^{-bt^{1/3}}, \qquad (8.16)$$

with $\eta = -7/30$ for $m = 1$ and $\eta = 1/18$ for $m = 2$ (Samukhin *et al.*, 2008).

Many other quantities can be investigated both numerically and analytically, and we refer the interested reader to previous work (Gallos, 2004; Kozma, Hastings and Korniss, 2005; Sood, Redner and ben Avraham, 2005; Bollt and ben Avraham, 2005; Kozma, Hastings and Korniss, 2007). For example, the average time to visit or to return to a node is inversely proportional to its degree (Noh and Rieger, 2004; Baronchelli and Loreto, 2006). All these results confirm the importance of hubs in diffusion processes and show how scale-free topologies have an important impact on the dynamics, highlighting the relevance of topological heterogeneity. Finally, the diffusion process equations can also be generalized to the case of weighted networks and more complicated diffusion schemes (Wu *et al.*, 2007; Colizza and Vespignani, 2008), but in all cases the degree variability plays an important role and alters the stationary visiting or occupation probability, favoring high degree nodes. This has a noticeable impact on most of the algorithms aimed at navigating and searching large information networks, as we will show in the next sections.

8.2 Diffusion in directed networks and ranking algorithms

The basic considerations of the previous section are at the core of one of the most celebrated applications of the internet world: the PageRank algorithm. This algorithm has been the winning feature of the search engine "Google" and represented a revolution in the way we access information on the WWW.

In order to explore and index web pages, search engines rely on automatic programs called *web crawlers* that follow a list of links provided by a central server or follow recursively the links they find in the pages that they visit, according to a certain set of searching instructions. When a crawler finds a new web page in its search, it stores the data it contains and sends it to a central server. Afterwards, it follows the links present in the page to reach new websites. Web crawlings are repeated at periodic time intervals, to keep the index updated with new pages and links. The information retrieved by the crawlers is analyzed and used to create the index. The index stores information relative to the words present in the web pages found, such as their position and presentation, forming a database relating those words with the relevant hyperlinks to reach the pages in which they appear, plus the hyperlinks present in the pages themselves. The final element in a search engine is the user interface, a search software that accepts as an input words typed by the user, explores the index, and returns the web pages that contain the text introduced by the end user, and are considered as most relevant. In this process, important information is given by the *ranking* of the pages returned, i.e. the order in which they are presented after the query. Obviously, nobody is willing to visit dozens of uninteresting pages before discovering the one that contains the particular information that is sought. Therefore, the more relevant the first page returns are, the more successful and popular will the search engine be. The search engines available in the market make use of different ranking methods, based on several heuristics for the location and frequency of the words found in the index. Traditionally, these heuristics combine information about the position of the words in the page (the words in the HTML title or close to the top of the page are deemed more important than those near the bottom), the length of the pages and the meaning of the words they contain, the level of the directory in which the page is located, etc.

The PageRank algorithm is in this respect a major breakthrough based on the idea that a viable ranking depends on the topological structure of the network, and is provided by essentially simulating the random surfing process on the web graph. The most popular pages are simply those with the largest probability of being discovered if the web-surfer had an infinite time to explore the web. In other words "Google" defines the importance of each document by a combination of the probability that a random walker surfing the web will visit that document, and some heuristics based in the text disposition. The PageRank algorithm just gauges the importance of each web page i by the PageRank value $P_R(i)$ which is the probability that a random walker surfing the web graph will visit the page i. According to the considerations of the previous section it is clear that such a diffusion process will provide a large P_R to pages with a large degree, as the random walker has a much higher probability of visiting these pages. On the other hand, in the web graph we have to take into account the directed nature of the hyperlinks. For this reason, the PageRank algorithm (Brin and Page, 1998) is defined as follows

$$P_R(i) = \frac{q}{N} + (1 - q) \sum_j x_{ji} \frac{P_R(j)}{k_{\text{out},j}}, \qquad (8.17)$$

where x_{ij} is the adjacency matrix of the Web graph, $k_{\text{out},j}$ is the out-degree of vertex j, N is the total number of pages of the web graph and q is the so-called *damping factor*. In the context of web surfing, the damping q is a crude modeling of the probability that a random surfer gets bored, stops following links, and proceeds to visit a randomly selected web page. The set of equations (8.17) can be solved iteratively and the stationary P_R can be thought of as the stationary probability of a random walk process with additional random jumps, as modulated by q. If $q = 0$, the PageRank algorithm is a simple diffusion process that just accumulates on the nodes with null out-degree (since the web is directed, random walkers cannot get out of such nodes) or in particular sets of nodes with no links towards the rest of the network (such as the out-component of the network, see Chapter 1). If $q \neq 0$ (applications usually implement a small $q \simeq 0.15$) the stationary process gives the PageRank value of each node.

While we know from the previous section that the visiting probability in an undirected network increases with the degree of the node, in the case of the directed web graph and the PageRank algorithm this fact is not completely intuitive as the P_R visiting probability depends on the global structure of the web and on the hyperlinks between pages. In particular the structure of the equations appears to give some role to the out-degree as well. It is possible, however, to show with a mean-field calculation that in uncorrelated networks the average PageRank value is just related to the in-degree of the nodes. Analogously to what has been shown for regular diffusion, let us define the PageRank for statistically equivalent nodes of degree class $\mathbf{k} = (k_{\text{in}}, k_{\text{out}})$ as

$$P_R(k_{\text{in}}, k_{\text{out}}) = \frac{1}{N_{\mathbf{k}}} \sum_{i|k_{\text{in},i}=k_{\text{in}}; k_{\text{out},i}=k_{\text{out}}} P_R(i), \qquad (8.18)$$

where $N_{\mathbf{k}}$ is the number of nodes with in- and out-degree k_{in} and k_{out}, respectively. In this case Fortunato *et al.* (2005) have shown that in the stationary limit

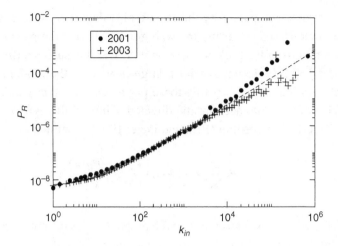

Fig. 8.2. Average PageRank as a function of in-degree for two samples of the Web graph obtained by two large crawls performed in 2001 and 2003 by the WebBase collaboration at Stanford (http:// dbpubs. stanford.edu:8091/~testbed/doc2/WebBase/). The 2001 crawl indexes $80, 571, 247$ pages and $752, 527, 660$ links and the 2003 crawl has $49, 296, 313$ pages and $1,185,396,953$ links. The PageRank values are obtained with a $q = 0.15$ and averaged over logarithmic bins of in-degree. The dashed line corresponds to a linear fit. Data courtesy of S. Fortunato.

$$P_R(k_{in}, k_{out}) = \frac{q}{N} + \frac{(1-q)}{N} \frac{k_{in}}{\langle k_{in} \rangle}, \qquad (8.19)$$

where $\langle k_{in} \rangle$ is the average in-degree in the web graph considered. In other words, the average PageRank of nodes just depends on the in-degree of the node. This expression is obtained in a statistical equivalence assumption and for uncorrelated networks, while the real web graph obviously does not satisfy these conditions. However, numerical inspection of the relationship between in-degree and P_R in real web graph samples shows that the linear behavior is a good approximation to the real behavior (see Figure 8.2 and Fortunato *et al.* [2006b]). The PageRank algorithm is therefore a measure of the popularity of nodes that is mostly due to the in-degree dependence of the diffusion process. This is, however, a mean-field result and important fluctuations appear: some nodes can have for example a large PageRank despite a modest in-degree, and one of the refinements of search engines consists in their ability to uncover such outliers. Overall, the striking point is that the effectiveness of the algorithm is due to the heterogeneous properties of the network which induce a discovery probability varying over orders of magnitude and truly discriminating the most popular pages. The self-organized complex structure of the web is therefore a key element in our capability of making use of it.

Interestingly, the use of PageRank has spilled out in other areas where the ranking and retrieval of information in large-scale directed networks is particularly relevant. A recent example is provided by the problem of measuring the impact of a scientific publication. The usual measure given by the simple number of citations can hide papers that appear as important only after some time, or overestimate some other papers because of temporary fads. Network tools prove useful in this context: scientific publications are considered as nodes, and each citation from a paper A to a paper B is drawn as a directed link from A to B. The number of citations of a paper is then given by its in-degree. Detailed analysis of the corresponding directed network for the *Physical Review* journals can be found in Redner (1998; 2005). The retrieval of scientific information often proceeds in this network in a way similar to what occurs on the web. Typically, a scientist starts from a given paper and then follows chains of citation links, discovering other papers. In this context, Chen *et al.* (2007) propose to use the PageRank of a paper, instead of its in-degree, to quantify its impact. The fundamental reasons why such a quantity is better suited are two-fold: (i) it takes into account the fact that being cited by papers that are themselves important is more important, and (ii) for PageRank, being cited by a paper with a small number of references (small out-degree) gives a larger contribution than being cited by a paper which contains hundreds of references (see Equation (8.17)). The analysis of the citation network of the *Physical Review* shows that the PageRank of a paper is correlated with its in-degree, as in the case of the World Wide Web. However, a number of outliers with moderate in-degree and large PageRank are found. For example, old papers having originated a fundamental idea may not be cited anymore because the idea has become folded into a field's common practice or literature, so that these old papers' in-degree remains moderate, but PageRank measures more accurately their impact over the years. Walker *et al.* (2007) explore these ideas further by defining the CiteRank index, which takes into account the fundamental differences between the web and a citation network: first, links between papers cannot be removed or updated; second, the time-arrow implies that papers age. The CiteRank algorithm therefore simulates the dynamics of scientists looking for information by starting from recent papers (the probability of starting from a random paper decays exponentially with its age) before following the directed links. In a different context – but still in the area of academic ranking – Schmidt and Chingos (2007) propose to rank academic doctoral programs based on their records of placing their graduates in faculty positions. In network terms, academic departments link to each other by hiring their faculty from each other (and from themselves) and the PageRank index can be computed on this directed network. Wissner-Gross (2006) also uses a weighting method similar to PageRank to generate customized reading lists based on the link structure of the online encyclopedia Wikipedia (http://wikipedia.org).

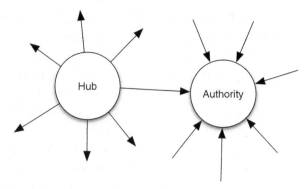

Fig. 8.3. Authorities and hubs in the web graph. Authorities are the most relevant web pages on a certain topic and hubs are pages containing a large number of links to authorities.

Finally, let us mention that alternative methods in the same spirit of PageRank have been proposed in order to improve the ranking of search engines by considering both the in- and the out-degree distribution of the web graph and other directed graphs such as citation and authorship networks (Kleinberg, 1998). This method relies on the distinction between *authorities* and *hubs* (see Figure 8.3). Authorities are web pages that can be considered the most relevant source of information about a given topic. Given the large amount of knowledge that these kinds of pages encode, it is natural to assume that they have a large number of incoming links. Hubs, on the other hand, are pages dealing with a given topic, which are not authorities themselves but which contain a large number of outgoing links pointing to related authorities. In this situation, the set of hubs and authorities on a topic form a bipartite clique, in which all hubs point to all authorities. Therefore, by focusing on the detection of bipartite cliques, it should be possible to identify which are those authorities and rank them in the highest position. Following this approach, Kleinberg (1998) has proposed the Hyperlink-Induced Topic Search (HITS) algorithm, which has been the seed for several variations and improvements (Marendy, 2001).

8.3 Searching strategies in complex networks

The problem of searching in complex networks is a very important and practical one which arises in many contexts. Complex networks can often be seen as large reservoirs of information, which can be passed from one part of the network to another thanks to the existence of paths between nodes (see Chapter 1). As described in the previous section, the most obvious example is given by the World Wide Web, which is nowadays one of the most important sources of information used on a daily basis. Peer-to-Peer (P2P) networks also allow for the retrieval of desired files, while the Internet is the support of such virtual networks and makes it possible to transmit messages in electronic form. The exploration of a network has therefore the typical goal of retrieving a particular item of information, of transmitting messages to a node (be it a computer or an individual), or of finding in a social network the right person to perform a given task.

In this respect, it is very important to consider the experiment elaborated by Milgram (1967) which has been largely celebrated as putting on solid ground the familiar concept of the small world (see also Chapter 2). Randomly selected individuals in the Midwest were asked to send a letter to a target person living in Boston. The participants knew the name and occupation of the target but were only allowed to pass the letter to a person they knew on a first-name basis. Strikingly, the letters that successfully reached the target had been passed only a small number of times, traveling on average through a chain of six intermediate hops. This experiment was duplicated and repeated in various environments, with similar results (Travers and Milgram, 1969; Korte and Milgram, 1970; Lundberg, 1975; Bochner, Buker and McLeod, 1976). Very recently, it was also performed and analyzed by Dodds, Muhamad and Watts (2003) in a more modern set-up: the participants were asked to use email instead of regular mail. Besides the small-world effect, another important and puzzling conclusion can be drawn from the outcome of the experiments of Milgram (1967) and Dodds *et al.* (2003), with more far-reaching implications: the existing short paths *can be found without global knowledge of the social network*. People receiving the message had to forward it to one of their neighbors in the social network, but were not given any information on the structure of the network (which indeed remains largely unknown). The standard algorithms for finding the shortest paths between pairs of nodes involve the exploration of the whole network. On the contrary, the participants in the experiments used only their local knowledge consisting of the identity and geographical location of their neighbors, so that the search for the final target was decentralized. Even if there is no guarantee that the true shortest paths were discovered during the experiment, it is quite striking that very short paths were found at all.

Obviously, the task of searching for a specific node or a specific piece of information will be more or less difficult depending on the amount of information available to each node. The most favorable case arises when the whole structure of nodes and edges is known to all vertices. The straightforward strategy is then to take advantage of this global knowledge to follow one of the shortest paths between the starting node and the target vertex. Clearly, such a case of extensive knowledge is seldom encountered and other strategies have to be used, as we will review in the following (see also Kleinberg [2006] for a recent review on this topic).

8.3.1 Search strategies

The task of searching can correspond to various situations: finding a particular node (possibly given some information on its geographical position, as in Milgram's experiment) in order to transmit a message, or finding particular information without knowing a priori in which node it is stored (as is the case in Peer-to-Peer networks). In all cases, search strategies or algorithms will typically consist of message-passing procedures: each node, starting from the initial node, will transmit one or more messages to one or more neighbors on the network in order to find a certain target. This iterative process stops when the desired information is found and sent to the source of the request.

As already mentioned, when each vertex has exhaustive information about the other vertices and the connectivity structure of the network, messages can be passed along the shortest path of length ℓ_{st} from the source s to the target t. The delivery time T_N and number of messages exchanged (i.e. the traffic involved) are also given by ℓ_{st} which, in most real-world complex networks, scales on average only as log N (or slower) with the size N of the network (Chapter 2). Such low times and traffic are, however, obtained at the expense of storing in each node a potentially enormous amount of information: for example, in the case of the WWW, each server should keep the address and content of all existing web pages, a clearly impossible configuration. Another concern regards nodes with large betweenness centrality, through which many shortest paths transit, and which will therefore receive high traffic if many different queries are sent simultaneously. We will see in Chapter 11 how congestion phenomena can then arise and how various routing strategies can alleviate such problems.

In the opposite case, when no information is available on the position of the requested information, nor on the network's structure, the simplest strategy is the so-called *broadcast search* (see Figure 8.4): the source node sends a message to *all* its neighbors. If none of them has the requested information (or file), they iterate this process by forwarding the message to all their own neighbors (except the source node). This process is repeated, flooding the network until the information is retrieved. In order to avoid excessive traffic and a possibly never-ending process, a corresponding time-to-live (TTL) is assigned from the start to the message. At each transmission the TTL is decreased by one until it reaches 0 and the message is no longer transmitted. When the target vertex is reached by the message, it sends back to the source the requested item. Since the other nodes cannot be easily informed of the success of the query, the broadcast process continues in other parts of the network. The broadcast algorithm, akin to the breadth-first algorithm commonly used to find shortest paths, proceeds in parallel and is clearly able to

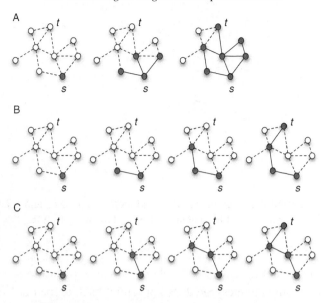

Fig. 8.4. Schematic comparison of various searching strategies to find the target vertex *t*, starting from the source *s*. A, Broadcast search; B, Random walk; C, Degree-biased strategy. The broadcast search finds the shortest path, at the expense of high traffic.

explore the entire network very rapidly. In particular, the desired target is reached after a number of steps equal to the shortest path from the source, and the delivery time is therefore equal to that obtained with full knowledge of the network. The obvious drawback consists in the large amount of traffic generated, since at each request all nodes within a shortest path distance ℓ of the source are visited, where ℓ is given by the TTL. The number of such nodes typically grows exponentially with ℓ and a large fraction of (or the whole) network receives a message at each search process. The traffic thus grows linearly with the size N of the network. Refined approaches have been put forward, in particular with the aim of obtaining efficient Peer-to-Peer search methods, such as the replication of information at various nodes, or iterative deepening, which consists of starting with a small TTL and increasing it by one unit at a time only if the search is not successful (see, for example, Yang and Garcia-Molina [2002]; Lv *et al.* [2002]). The amount of traffic can then be sensibly reduced but will still remain high.

Intermediate situations are generally found, in which each node possesses a limited knowledge about the network, such as the information stored in each of its neighbors. A straightforward and economical strategy is then given by the *random walk search*, illustrated in Figure 8.4. In this case, the source node starts by checking if any of its neighbors has the requested information. If not, it sends a

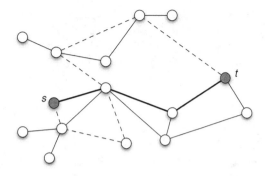

Fig. 8.5. Difference between the shortest path between nodes *s* and *t* (continuous thick line), of length 3, and an instance of a random walk (dashed line), here of length 6.

message to one randomly chosen neighbor,[2] which iterates the process (at each step, the node holding the message does not send it back to the node from which it was received), until a neighbor of the target is reached: the message is then passed directly to the target. Clearly, the delivery time T_N will be larger than for a broadcast strategy, since the random walk does not follow the shortest path from the source to the target (Figure 8.5). In fact, the random walk tends to visit the same nodes several times, and the probability of following exactly the shortest path is very low (Sneppen, Trusina and Rosvall, 2005; Rosvall, Minnhagen and Sneppen, 2005; Rosvall *et al.*, 2005). In generalized random networks with a scale-free degree distribution $P(k) \propto k^{-\gamma}$, with exponent $\gamma = 2.1$ corresponding to the value of Peer-to-Peer networks, Adamic *et al.* (2001) obtain for the random walk search $T_N \sim N^{0.79}$. This power-law behavior denotes a much worse behavior than the broadcast strategy in terms of delivery time. The traffic generated, however, is equal to T_N (since only one walker is generated at each request), and therefore remains smaller than the linear growth of the broadcast search.

We will see in the next sections how various strategies can be devised, depending on the network's structure and on the various levels of (local) knowledge available to the nodes.

8.3.2 Search in a small world

As illustrated by the Watts–Strogatz model (see Chapter 3), small-world networks are characterized by "shortcuts" which act as bridges linking together "far away" parts of the network and allow for a strong decrease of the network's diameter.

---

[2] The efficiency of the search process can be increased by the use of several random walkers in parallel, although this generates additional costs in terms of traffic (Lv *et al.*, 2002).

The concept of "far away" here refers to an underlying space (e.g. geographical) in which the nodes are located, as in Milgram's original experiment. In such cases, it is natural to try to use these shortcuts for the searching process. Such an approach is not as easy as it might seem: for example, in the Watts–Strogatz model, short-cuts are totally random and of arbitrary length, so that it is difficult to select the correct shortcut. Such bridges must therefore encode some information about the underlying structure in order to be used. Kleinberg (2000a) has formalized the conditions under which a Watts–Strogatz-like small-world network can be efficiently navigated (see also Kleinberg [2000b]). In his model, shortcuts are added to a D-dimensional hypercubic lattice in the following way: a link is added to node i, and the probability for this link to connect i to a vertex j at geographical distance r_{ij} is proportional to $r_{ij}^{-\alpha}$, where α is a parameter. Each node knows its own position and the geographical location of all its neighbors. The search process used is the simplest greedy one: a message has to be sent to a certain target node t whose geographical position is known. A node i receiving the message forwards it to the neighbor node j (either on the lattice or using the long-range link) that is geographically closest to the target, i.e., which minimizes r_{jt}. The idea is thus to send the message at each step closer and closer to the target. Kleinberg (2000a) shows a very interesting result: if $\alpha = D$, the delivery time scales as $\log^2(N)$ with the size N of the network. Various generalizations of this result can be found in the literature. For example, modified versions of the algorithm consider that nodes may consult some nearby nodes before choosing to which neighbor it forwards the message (see Lebhar and Schabanel [2004]; Manku, Naor and Wieder [2004] and Kleinberg [2006] for a review). On the contrary, as soon as the exponent α deviates from the space dimension D, the delivery time increases as a power of N, i.e., much faster (Figure 8.6).

The intuitive interpretation of this result is as follows: when α is too large, short-cuts connect nodes that are geographically not very distant, and therefore do not

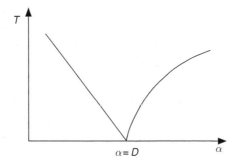

Fig. 8.6. Schematic evolution of the delivery time as a function of α in Kleinberg's model of a small-world network. Figure adapted from Kleinberg (2000a).

shorten efficiently the distances. When α is too small on the other hand, long-range shortcuts are available, but they are typically very long and are not useful to find a specific target. Only at $\alpha = D$ does one obtain shortcuts at all length scales so that the target can be found efficiently through the greedy algorithm.

The situation depicted by Kleinberg's result makes Milgram's experiment even more striking and intriguing. It means indeed that only very particular small-world networks can be searched in short times: the ones with the adequate distribution of shortcuts. A first objection that can come to mind is that we do not live on a hypercubic lattice and that the local population density is strongly heterogeneous. In order to take this aspect into account, Liben-Nowell *et al.* (2005) propose a rank-based friendship linking model, in which, for each node (individual) i, the geographical neighbors are ranked according to their distance to i, and a long-range link is established between i and j with probability inversely proportional to j's rank. The real length of such links will therefore depend on the local density, being shorter in more densely populated areas. Such a generalization of Kleinberg's model, which is in fact realized in the online community called Live-Journal (Liben-Nowell *et al.*, 2005), turns out to support the greedy search algorithm efficiently. A particular linking rule (using the inverse of the rank) is needed, however, just as a particular value of the exponent α is crucial in Kleinberg's result.

A very important point to understand how searching effectively works in social networks is that several distances may coexist, such as geographical but also based on social criteria or affinity: the creation of a link between two individuals may occur not only as a function of geographical distance (e.g. because two persons live in the same neighborhood), but also because they are members of the same group, because they share the same interests or hobby, or because they have the same profession.[3] Interestingly, Killworth and Bernard (1978) have shown that, in Milgram's experiment, the choice of the neighbor to whom the message was forwarded was made mostly on the basis of two criteria: geography and occupation. The participants were therefore taking advantage of the potential proximity of their contacts to the target in both geographical and social space. While geography can be represented by an embedding in a D-dimensional space ($D = 2$), a natural approximation of the space of occupations is a hierarchical structure. Following this idea, Watts, Dodds and Newman (2002) put forward a model in which each node exists at the same time in different hierarchies, each hierarchy corresponding to some dimension of social space. In each hierarchy, a network is constructed, favoring links between nearby nodes, and the union of these networks forms the global social network. The effective distance between nodes is then the

---

[3] Menczer (2002) also generalizes the idea of an underlying distance to the WWW, by using a semantic or lexical distance between web pages.

minimum possible distance over all the hierarchies. Numerical simulations reveal that the greedy algorithm is efficient if there exists a small number of different ways to measure proximity, and if the network's creation favors homophily, i.e. links exist in each social dimension preferentially among close nodes (see also Kleinberg [2001] for a rigorous approach on similar concepts of group membership). The more recent study of Adamic and Adar (2005) on a real social network (obtained through the email network of a firm) has also revealed that using the hierarchical structure increases the search efficiency (see also Şimşek and Jensen [2005]). Finally, Boguñá, Krioukov and claffy (2007) consider networks in which nodes reside in a hidden metric space, and study the greedy routing in this hidden space. They show that scale-free networks with small exponents for the degree distribution and large clustering are then highly navigable. Such approaches allow us to rationalize the apparent contrast between the success of Milgram's experiment and the restrictive result of Kleinberg.

8.3.3 Taking advantage of complexity

In Milgram's experiment, the identity and location of the target node were known. Such information is not always available, for example in networks without geographical or hierarchical structure. Moreover, when searching for a precise item, one may not even know the identity of the target node. In P2P applications for instance, requests for a specific file are sent without knowing which peers may hold it. In such cases, the greedy algorithm previously described cannot be applied, but the idea of using the shortcuts can still be exploited. In particular, the small-world character of many real-world networks is due to the presence of hubs that connect together many different parts of the network. While the probability of reaching a given target by a random walk is reduced when going through a hub, because of the large choice for the direction of the next step (Rosvall *et al.*, 2005), the fact that typically many shortest paths go through hubs makes them easily accessible (Dall'Asta *et al.*, 2005; Sneppen *et al.*, 2005). Search processes can thus take advantage of this property with a strategy that biases the routing of messages towards the nodes with large degree. This *degree-biased* searching approach (see Figure 8.4C) has been studied by Adamic *et al.* (2001) and Kim *et al.* (2002b), assuming that each node has information about the identity and contents of its neighbors, and on their degree. Using this purely local information, a node holding the message (search request) forwards it either to the target node if it is found among its neighbors, or to its most connected neighbor (see also Adamic, Lukose and Huberman [2003]). Numerical simulations allow comparing the efficiency of this method with the random-walk search. For example, on generalized random networks with a scale-free degree distribution $P(k) \sim k^{-\gamma}$ with $\gamma = 2.1$, Adamic *et al.* (2001) obtain a delivery time $T_N \sim N^{0.7}$ smaller than for the random walk. Such increased performance is confirmed by analytical calculations and can be understood by the following intuitive argument: if each node has a knowledge of the information stored in all its neighbors, hubs will naturally have more knowledge than low degree nodes, and it seems therefore a sensible choice to send them the request. Starting from a random node on the network, a message which follows the degree-biased strategy will clearly reach the most-connected vertex after a few steps. From this position, it will start in fact to explore the network by visiting the nodes in decreasing order of their degree. Interestingly, the deterministic character of the rule turns out to be crucial, and any amount of randomness (such as sending the message to a neighbor chosen with a probability proportional to its degree) yields much larger delivery times (Kim *et al.*, 2002b). Moreover, the strategy turns out to be quite inefficient in homogeneous networks such as Erdős–Rényi or Watts–Strogatz models where no hubs are present (Adamic *et al.*, 2001; Kim *et al.*, 2002b; Adamic *et al.*, 2003; Adamic and Adar, 2005). The clear drawback of the degree-biased strategy lies in the potentially high traffic imposed on the hubs (see Chapter 11 for a detailed discussion on traffic issues).

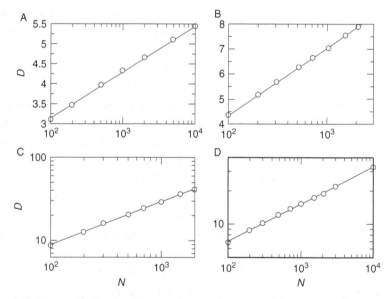

Fig. 8.7. Network effective diameter D as a function of size in a Barabási–Albert network for (A) shortest paths, (B) deterministic degree-biased strategy, (C) random walk strategy and (D) preferential degree-biased strategy (neighbor chosen with probability proportional to its degree). D scales logarithmically for A and B, and as a power-law for C and D. Data from Kim *et al.* (2002b).

Search strategies may also be used to find (short) paths between nodes, that can be subsequently re-used in future communications (Kim *et al.*, 2002b; Adamic *et al.*, 2003). With respect to the outcome of the search strategy, loops and possible backward steps are removed, yielding a path between the source and the target which is shorter than the real path followed by the message. The average path length obtained by selecting random source and target nodes defines an effective diameter of the network which depends on the search strategy. Numerical simulations on Barabási–Albert networks show that this effective diameter scales in very different ways with the network size N for the various strategies, from logarithmically for the degree-biased case to power laws for random walks (see Figure 8.7). Interestingly, the degree-biased strategy, even if it yields delivery times scaling as a power of the size, allows very short paths to be found once the loops have been taken out.

Various studies have recently aimed at improving the simple degree-biased search strategy. Thadakamalla, Albert and Kumara (2005) in particular consider the case of weighted networks in which a cost may be associated to the transmission of a message through an edge (see also Jeong and Berman [2007]). A *local betweenness centrality* can then be defined as the number of shortest (weighted) paths in the subnetwork formed by the neighbors and the neighbors' neighbors of a node (see also Trusina, Rosvall and Sneppen [2005]). Sending the message to the neighbor with largest local betweenness centrality then proves to be quite efficient, since it connects in an effective way information on the degree and weights. In general, complex networks are searchable thanks to their degree heterogeneity, and the combination of information on the degrees of neighboring node with knowledge of other attributes such as geography (Thadakamalla, Albert and Kumara, 2007), weights (Thadakamalla *et al.*, 2005; Jeong and Berman, 2007), social attributes (Şimşek and Jensen, 2005), information on the shortest path to the nearest hub (Carmi, Cohen and Dolev, 2006), or even the use of adaptative schemes for P2P applications (Yang and Garcia-Molina, 2002), allow the building of search algorithms that perform rather efficiently.

9

Epidemic spreading in population networks

The mathematical modeling of epidemics is a very active field of research which crosses different disciplines. Epidemiologists, computer scientists, and social scientists share a common interest in studying spreading phenomena and rely on very similar models for the description of the diffusion of viruses, knowledge, and innovation. Epidemic modeling has developed an impressive array of methods and approaches aimed at describing various spreading phenomena, as well as incorporating many details affecting the spreading of real pathogens. In particular, understanding and predicting an epidemic outbreak requires a detailed knowledge of the contact networks defining the interactions between individuals. The theoretical framework for epidemic spreading has thus to be widened with opportune models and methods dealing with the intrinsic system complexity encountered in many real situations.

In this chapter, we introduce the general framework of epidemic modeling in complex networks, showing how the introduction of strong degree fluctuations leads to unusual results concerning the basic properties of disease spreading processes. Using some specific examples we show how plugging in complex networks in epidemic modeling enables one to obtain new interpretative frameworks for the spread of diseases and to provide a quantitative rationalization of general features observed in epidemic spreading. We end the chapter by discussing general issues about modeling the spread of diseases in complex environments, and we describe metapopulation models which are at the basis of modern computational epidemiology.

9.1 Epidemic models

The prediction of disease evolution and social contagion processes, can be conceptualized with a variety of mathematical models of spreading and diffusion processes. These models evolved from simple compartmental approaches into structured frameworks in which the hierarchies and heterogeneities present at the

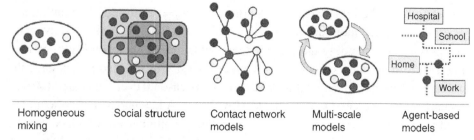

| Homogeneous mixing | Social structure | Contact network models | Multi-scale models | Agent-based models |

Fig. 9.1. Structures at different scales used in epidemic modeling. Circles represent individuals and each shade corresponds to a specific stage of the disease. From left to right: homogeneous mixing, in which individuals are assumed to interact homogeneously with each other at random; social structure, where people are classified according to demographic information (age, gender, etc.); contact network models, in which the detailed network of social interactions between individuals provide the possible virus propagation paths; multiscale models which consider subpopulations coupled by movements of individuals, while homogeneous mixing is assumed on the lower scale; agent-based models which recreate the movements and interactions of any single individual on a very detailed scale (a schematic representation of a part of a city is shown).

community and population levels are becoming increasingly important features (Anderson and May, 1992) (see Figure 9.1). As in many other fields, an interplay exists between the simplicity of the model and the accuracy of its predictions. An important question is how to optimize the predictive power of a model, which usually means reducing the number of tunable parameters. The challenging tasks of understanding and modeling the spread of diseases in countries, cities, etc., make this debate very real. At the geopolitical level, modeling approaches have evolved to explicitly include spatial structures and consist of multiple subpopulations coupled by movements among them, while the epidemics within each subpopulation are described according to approximations depending on the specific case studied (Hethcote, 1978; Anderson and May, 1984; May and Anderson, 1984; Bolker and Grenfell, 1993; Bolker and Grenfell, 1995; Lloyd and May, 1996; Grenfell and Bolker, 1998; Keeling and Rohani, 2002; Ferguson *et al.*, 2003). The modern versions of this patch or meta-population modeling scheme are multiscale frameworks in which different granularities of the system (country, inter-city, intra-city) are considered through different approximations and are coupled through interaction networks describing the flows of people and/or animals (Rvachev and Longini, 1985; Keeling *et al.*, 2001; Ferguson *et al.*, 2003; Hufnagel, Brockmann and Geisel, 2004; Longini *et al.*, 2005; Ferguson *et al.*, 2005; Colizza *et al.*, 2006a). Computers also serve as in-silico laboratories where the introduction of agent-based models (ABM) extends the modeling perspective, simulating propagation and contagion processes at the individual level (Chowell *et al.*, 2003; Eubank *et al.*, 2004).

Generally all these models are concerned with the evolution of the number and location of infected individuals in the population as a function of time. They therefore aim at understanding the properties of epidemics in the equilibrium or long time steady state, the existence of a non-zero density of infected individuals, the presence or absence of a global outbreak, the non-seasonal cycles that are observed in many infections, etc. (Anderson and May, 1992). A key parameter in the understanding of these properties is the basic reproductive number R_0, which counts the number of secondary infected cases generated by one primary infected individual. It is easy to understand that any epidemic will spread across a non-zero fraction of the population only for $R_0 > 1$. In this case the epidemics are able to generate a number of infected individuals larger than those who have recovered, leading to an increase of the total number of infected individuals $I(t)$ at time t. This simple consideration leads to the definition of a crucial epidemiological concept, namely the epidemic threshold. Indeed, if the spreading rate is not large enough to allow a reproductive number larger than 1, the epidemic outbreak will not affect a finite portion of the population and dies out in a finite time. The epidemic threshold is a central concept in epidemiology since it provides a reference frame to understand the evolution of contagion processes and to focus on policies aimed at an effective reduction of the reproductive number in order to stop epidemic outbreaks.

9.1.1 Compartmental models and the homogeneous assumption

The simplest class of epidemic models assumes that the population can be divided into different classes or compartments depending on the stage of the disease (Anderson and May, 1992; Bailey, 1975; Daley and Gani, 2000; Diekmann and Heesterbeek, 2000), such as susceptibles (denoted by S, those who can contract the infection), infectious (I, those who have contracted the infection and are contagious), and recovered (R, those who have recovered from the disease). Additional compartments could be considered in order to model, for example, people immune to the disease or people exposed to the infection but not yet infectious (latent). According to this minimal framework, within each compartment individuals are assumed to be identical and homogeneously mixed, and the larger the number of sick and infectious individuals among one individual's contacts, the higher the probability of transmission of the infection. If we consider the spread of a disease in a population of N individuals, it is possible to obtain a general formulation of epidemic models which governs the time evolution of the number of individuals $X^{[m]}(t)$ in the class $[m]$ at time t (with $N = \sum_m X^{[m]}(t)$). The dynamics of the individuals between the different compartments depends on the specific disease considered. In general, the transition from a compartment to the other is specified by a reaction rate that depends on the disease etiology, such as the infection transmission rate or the recovery or cure rate. In compartmental models there are two possible types of elementary processes ruling the disease dynamics. The first class of processes refers to the spontaneous transition of one individual from one compartment $[m]$ to another compartment $[h]$

$$X^{[m]} \to X^{[m]} - 1 \tag{9.1}$$

$$X^{[h]} \to X^{[h]} + 1. \tag{9.2}$$

Processes of this kind are the spontaneous recovery of infected individuals ($I \to R$) or the passage from a latent condition to an infectious one ($L \to I$) after the incubation period. In this case the variation in the number of individuals $X^{[m]}$ is simply given by $\sum_h v_h^m a_h X^{[h]}$, where a_h is the rate of transition from the class $[h]$ and $v_h^m = 1, 0$ or -1 is the change in the number of $X^{[m]}$ due to the spontaneous process from or to the compartment $[h]$.

The second class of processes refers to binary interactions among individuals such as the contagion of a susceptible individual in interaction with an infectious one

$$S + I \to 2I. \tag{9.3}$$

In this case, the variation of $X^{[m]}$ is given by $\sum_{h,g} v_{h,g}^m a_{h,g} N^{-1} X^{[h]} X^{[g]}$, where $a_{h,g}$ is the transition rate of the process and $v_{h,g}^m = 1, 0$ or -1 the change in the number of $X^{[m]}$ due to the interaction. The factor N^{-1}, where N is the number of individuals, stems from the fact that the above expression considers an homogeneous approximation in which the probability for each individual of class $[h]$ to interact with an individual of class $[g]$ is simply proportional to the density $X^{[g]}/N$ of such individuals. The homogeneous approximation is therefore equivalent to the mean-field one used for physical models and considers an effective interaction, a mass-action law, determining the force of infection in the same way for all individuals in the system. By using the above expressions it is possible to write the general deterministic reaction rate equations for the average number of individuals in the compartment $[m]$ as

$$\partial_t X^{[m]} = \sum_{h,g} v_{h,g}^m a_{h,g} N^{-1} X^{[h]} X^{[g]} + \sum_h v_h^m a_h X^{[h]}, \tag{9.4}$$

where now the quantities $X^{[m]}$ are continuous variables representing the average number of individuals in each class $[m]$. If the total number of individuals is constant, these equations must satisfy the conservation equation $\sum_m \partial_t X^{[m]} = 0$. It is also worth stressing that the deterministic continuous approximation neglects stochastic fluctuations that may be relevant in some cases, as we will see in the next sections.

This general framework easily allows us to derive the dynamical equations of the three basic models which are commonly used to illustrate the general properties of epidemic spreading processes. The simplest epidemiological model one can consider is the susceptible–infected (SI) model in which individuals can only exist in two discrete states, namely, susceptible and infected. The probability that a susceptible vertex acquires the infection from any given neighbor in an infinitesimal time interval dt is $\beta\, dt$, where β defines the pathogen *spreading rate*. Individuals that enter the infected class remain permanently infectious. The epidemics can only grow as the number of infectious individuals $I(t)$ is constantly increasing and the seed of infectious individuals placed at time $t = 0$ will therefore ultimately infect the rest of the population. The evolution of the SI model is therefore completely defined by the number of infected individuals $I(t)$ or equivalently the corresponding density $i(t) = I(t)/N$.

In the homogeneous assumption, the force of the infection (the per capita rate of acquisition of the disease for the susceptible individuals) is proportional to the average number of contacts with infected individuals, which for a total number of contacts k is approximated as ki. In a more microscopic perspective – which is at the basis of many numerical simulations of these processes – this can be understood by the following argument (see also Chapter 4). Since each infected individual attempts to infect a connected susceptible vertex with probability $\beta\, dt$, a susceptible vertex with n infected neighbors will have a total probability of getting infected during the time interval dt given by $1 - (1 - \beta\, dt)^n$. Neglecting fluctuations, each susceptible vertex with k connections will have on average $n = ki$ infected neighbors, yielding at the leading order in $\beta\, dt \ll 1$ an infection acquisition probability $1 - (1 - \beta\, dt)^{ki} \simeq \beta ki\, dt$ and the per capita acquisition rate βki.[1] This approach makes explicit the dependence of the spreading rate with the number of contacts k of each individual, which will be very useful later in extending the calculation in the case of heterogeneous systems. As a first approximation, let us consider that each individual/vertex has the same number of contacts/edges, $k \simeq \langle k \rangle$. The continuous and deterministic reaction rate equation describing the evolution of the SI model then reads as

$$\frac{di(t)}{dt} = \beta \langle k \rangle i(t) \left[1 - i(t) \right].   \tag{9.5}$$

The above equation states that the growth rate of infected individuals is proportional to the spreading rate $\beta \langle k \rangle$, the density of susceptible vertices that may become infected, $s(t) = 1 - i(t)$, where $s(t) = S(t)/N$, and the number of infected individuals in contact with any susceptible individual.

---

[1] Sometimes alternative definitions of the infection mechanism refer to $\beta\, dt$ as the probability of acquiring the infection if one or more neighbors are infected. In this case the total acquisition probability is given by $\beta\, dt[1 - (1 - i)^k]$, i.e. the spreading probability times the probability that at least one neighbor is infected. Also in this case, for $\beta\, dt \ll 1$ and $i \ll 1$ an acquisition rate βki is recovered at the leading order.

The susceptible–infected–susceptible (SIS) model is mainly used as a paradig-matic model for the study of infectious diseases leading to an endemic state with a stationary and constant value for the prevalence of infected individuals, i.e. the degree to which the infection is widespread in the population as measured by the density of those infected. In the SIS model, individuals exist in the suscepti-ble and infected classes only. The disease transmission is described as in the SI model, but infected individuals may recover and become susceptible again with probability μdt, where μ is the recovery rate. Individuals thus run stochastically through the cycle susceptible $\rightarrow$ infected $\rightarrow$ susceptible, hence the name of the model. The equation describing the evolution of the SIS model therefore contains a spontaneous transition term and reads as

$$\frac{di(t)}{dt} = -\mu i(t) + \beta \langle k \rangle i(t) \left[1 - i(t) \right]. \tag{9.6}$$

The usual normalization condition $s(t) = 1 - i(t)$ has to be valid at all times.

The SIS model does not take into account the possibility of an individ-ual's removal through death or acquired immunization, which would lead to the so-called susceptible–infected–removed (SIR) model (Anderson and May, 1992; Murray, 2005). The SIR model, in fact, assumes that infected individuals disap-pear permanently from the network with rate μ and enter a new compartment R of removed individuals, whose density in the population is $r(t) = R(t)/N$. The intro-duction of a new compartment yields the following system of equations describing the dynamics:

$$\frac{ds(t)}{dt} = -\beta \langle k \rangle i(t) \left[1 - r(t) - i(t) \right]$$

$$\frac{di(t)}{dt} = -\mu i(t) + \beta \langle k \rangle i(t) \left[1 - r(t) - i(t) \right] \tag{9.7}$$

$$\frac{dr(t)}{dt} = \mu i(t). \tag{9.8}$$

Through these dynamics, all infected individuals will sooner or later enter the recovered compartment, so that it is clear that in the infinite time limit the epi-demics must fade away. It is interesting to note that both the SIS and SIR models introduce a time scale $1/\mu$ governing the self-recovery of individuals. We can think of two extreme cases. If $1/\mu$ is smaller than the spreading time scale $1/\beta$, then the process is dominated by the natural recovery of infected to susceptible or removed individuals. This situation is less interesting since it corresponds to a dynamical process governed by the decay into a healthy state and the interaction with neigh-bors plays a minor role. The other extreme case is in the regime $1/\mu \gg 1/\beta$, i.e. a spreading time scale much smaller than the recovery time scale. In this case, as a first approximation, we can neglect the individual recovery that will

occur at a much later stage and focus on the early dynamics of the epidemic outbreak. This case corresponds to the simplified susceptible–infected (SI) model, for
which infected nodes remain always infective and spread the infection to susceptible neighbors with rate β. In this limit the infection is bound to affect the whole
population. The two limits therefore define two different regions in the parameters
space and a transition from one regime to the other must occur at a particular value
of the parameters, as we will see in the following.

9.1.2 The linear approximation and the epidemic threshold

All the basic models defined in the previous section can be easily solved at the
early stage of the epidemics when we can assume that the number of infected
individuals is a very small fraction of the total population; i.e. $i(t) \ll 1$. In this
regime we can use a linear approximation for the equation governing the dynamics
of infected individuals and neglect all i^2 terms. In the case of the SI model the
resulting equation for the evolution of the density of infected individuals reads as

$$\frac{di(t)}{dt} = \beta \langle k \rangle i(t), \tag{9.9}$$

yielding the solution

$$i(t) \simeq i_0 e^{\beta \langle k \rangle t}, \tag{9.10}$$

where i_0 is the initial density of infected individuals. This solution simply states
that the time scale $\tau = (\beta \langle k \rangle)^{-1}$ of the disease prevalence is inversely proportional
to the spreading rate β; the larger the spreading rate, the faster the outbreak will
be. In the SI model the epidemic always propagates in the population until all
individuals are infected, but the linear approximation breaks down when the density
of infected individuals becomes appreciable and the behavior of $i(t)$ no longer
follows a simple exponential form. For the SI model, it is in fact possible to derive
the full solution of Equation (9.5), to obtain

$$i(t) = \frac{i_0 \exp(t/\tau)}{1 + i_0(\exp(t/\tau) - 1)}. \tag{9.11}$$

As expected, this expression recovers (9.10) for $t \ll \tau$, and shows saturation
towards $i \to 1$ for $t \gg \tau$.

Along the same lines, it is possible to determine the early stage epidemic behavior of the SIS and SIR models. By neglecting the i^2 terms we obtain for both models
the same linearized equation

$$\frac{di(t)}{dt} = -\mu i(t) + \beta \langle k \rangle i(t), \tag{9.12}$$

where in the case of the SIR model the term $r(t)$ can be considered of the same order as $i(t)$. The solution of the above equation is straightforward, yielding

$$i(t) \simeq i_0 e^{t/\tau}, \tag{9.13}$$

with i_0 the initial density of infected individuals and where τ is the typical outbreak time (Anderson and May, 1992)

$$\tau^{-1} = \beta \langle k \rangle - \mu. \tag{9.14}$$

The above relation contains a striking difference from the SI equation. The exponential characteristic time is the combination of two terms and can assume negative values if the recovery rate is large enough. In this case the epidemic will not spread across the population but will instead fade away on the time scale $|\tau|$. The previous consideration leads to the definition of a crucial epidemiological concept, namely the *epidemic threshold*. Indeed, if the spreading rate is not large enough (i.e., if $\beta < \mu/\langle k \rangle$), the epidemic outbreak will not affect a finite portion of the population and will die out in a finite time. The epidemic threshold condition can be readily written in the form

$$\tau^{-1} = \mu(R_0 - 1) > 0, \tag{9.15}$$

where $R_0 = \beta \langle k \rangle / \mu$ identifies the basic reproductive rate in the SIS and SIR models, which has to be larger than 1 for the spreading to occur.

The simple linear analysis allows the drawing of a general picture for epidemic evolution consisting of three basic stages (see Figure 9.2). Initially, when a few infected individuals are introduced into the population we define a pre-outbreak stage in which the evolution is noisy and dominated by the stochastic effects that are extremely relevant in the presence of a few contagious events. This stage is not described by the deterministic continuous equations derived in the homogeneous approximation and would require a full stochastic analysis. This is a stage in which epidemics may or may not disappear from the population just because of stochastic effects. When the infected individuals are enough to make stochastic effects negligible, but still very few if compared with the whole population, we enter the exponential spreading phase described by the linearized equations. In this case, the epidemic will pervade the system according to the exponential growth (9.13) whose time scale depends on the basic reproductive rate. Below the epidemic threshold the epidemic will just disappear in a finite time.

The final stage of the epidemic is model-dependent. The decrease of susceptible individuals reduces the force of infection of each infected individual (the i^2 term in Equation (9.5) is negative and therefore slows down the dynamics) and the exponential growth cannot be sustained any longer in the population. The SI model will continue its spread at a lower pace until the total population is in the infected state.

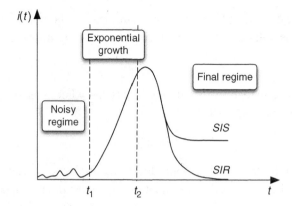

Fig. 9.2. Typical profile of the density of infected individuals $i(t)$ versus time on a given realization of the network. In the first regime $t < t_1$, the outbreak is subject to strong statistical fluctuations. In the second regime, $t_1 < t < t_2$ there is an exponential growth characterized by the reproductive number R_0. In the final regime ($t > t_2$), the density of infected either converges to a constant for the SIS model or to zero for the SIR model. In individuals the SI model the epidemics will eventually pervade the whole population.

This cannot happen in the SIR and SIS model where individuals recover from the disease. The SIR model will inevitably enter a clean-up stage, since the susceptible compartment becomes depleted of individuals that flow into the removed compartment after the infectious period, and the epidemics will ultimately disappear. The SIS model will enter a stationary state in which the infectious individuals density is fixed by the balance of the spreading and recovery rate. However, it is worth stressing that while the outbreak will occur with finite probability if the parameters poise the system above the epidemic threshold, this probability is not equal to 1. Actually the stochastic fluctuations may lead to the extinction of the epidemics even well above the epidemic threshold. In general it is possible to estimate that the extinction probability of an epidemic starting with n infected individuals is equal to R_0^{-n} (Bailey, 1975). For instance, in the case of a single infected individual, even for values of R_0 as high as 2 the outbreak probability is just 50%.

The concept of epidemic threshold is very general and a key property of epidemic models. For instance, the addition of extra compartments such as latent or asymptomatic individuals defines models whose epidemic thresholds can still be calculated from the basic transition rates among compartments. Also, clustered but homogeneous connectivity patterns between individuals, such as regular lattices, meshes and even the Watts–Strogatz network, do not alter this scenario and just provide a different scaling behavior of the prevalence close to the threshold (Anderson and May, 1992; Marro and Dickman, 1999; Moore and Newman, 2000; Kuperman and Abramson, 2001). The relevance of the epidemic

threshold is also related to the protection of populations by means of immunization programs. These correspond to vaccination policies aimed at the eradication
of the epidemics. The simplest immunization procedure one can consider consists
of the random introduction of immune individuals in the population (Anderson and
May, 1992), in order to get a *uniform* immunization density. In this case, for a fixed
spreading rate β, the relevant control parameter is the density of immune vertices
present in the network, the *immunity g*. At the mean-field level, the presence of a
uniform immunity will have the effect of reducing the spreading rate β by a factor
$1 - g$; indeed, the probability of infecting a susceptible and non-immune vertex will
be $\beta(1 - g)[1 - i(t)]$. For homogeneous networks we can easily see that, for a constant β, we can define a critical immunization value g_c above which the effective
system with spreading rate $\beta(1 - g)$ is pushed below the epidemic threshold:

$$g_c = 1 - \frac{\mu}{\beta \langle k \rangle}. \tag{9.16}$$

Thus, for a uniform immunization level larger than g_c, homogeneous networks
are completely protected and no large epidemic outbreaks or endemic states are
possible. The immunization threshold is very important in the prevention of epidemic outbreaks and in the clean-up stage. In practice, the aim of vaccination
program deployment is to achieve a density of immunized individuals that pushes
the population into the healthy region of the phase diagram.

9.2 Epidemics in heterogeneous networks

The general picture presented in the previous section is obtained in the framework
of the homogeneous mixing hypothesis and by assuming that the network which
describes the connectivity pattern among individuals is homogeneous; each individual in the system has, to a first approximation, the same number of connections
$k \simeq \langle k \rangle$. However, many networks describing systems of epidemiological relevance exhibit a very heterogeneous topology. Indeed, social heterogeneity and the
existence of "super-spreaders" have been known for a long time in the epidemics
literature (Hethcote and Yorke, 1984). Similarly, much attention has been devoted
to heterogeneous transmission rates and heterogeneous connectivity patterns. Generally, it is possible to show that the reproductive rate R_0 gets renormalized by
fluctuations in the transmissibility or contact pattern as $R_0 \rightarrow R_0(1 + f(\sigma))$ where
$f(\sigma)$ is a positive and increasing function of the standard deviation of the individual transmissibility or connectivity pattern (Anderson and May, 1992), pointing
out that the spreading of epidemics may be favored in heterogeneous populations.
Recently, empirical evidence has emphasized the role of heterogeneity by showing that many epidemiological networks are heavy-tailed and therefore the average

degree $\langle k \rangle$ is no longer the relevant variable. One then expects the fluctuations to play the main role in determining the epidemic properties. Examples of such networks relevant to epidemics studies include several mobility networks and the web of sexual contacts. Furthermore, computer virus spreading can be described in the same framework as biological epidemics (Kephart, White and Chess, 1993; Kephart *et al.*, 1997; Pastor-Satorras and Vespignani, 2001b; Aron *et al.*, 2001; Pastor-Satorras and Vespignani, 2004).

The presence of topological fluctuations virtually acting at all scales calls for a mathematical analysis where the degree variables explicitly enter the description of the system. This can be done by considering a degree block approximation that assumes that all nodes with the same degree are statistically equivalent. This assumption allows the grouping of nodes in the same degree class k, yielding the convenient representation of the system by quantities such as the density of infected nodes and susceptible nodes in the degree class k

$$i_k = \frac{I_k}{N_k}; \qquad s_k = \frac{S_k}{N_k}, \qquad (9.17)$$

where N_k is the number of nodes with degree k and I_k and S_k are the number of infected and susceptible nodes in that class, respectively. Clearly, the global averages are then given by the expressions

$$i = \sum_k P(k)i_k; \qquad s = \sum_k P(k)s_k. \qquad (9.18)$$

This formalism is extremely convenient in networks where the connectivity pattern dominates the system's behavior. When other attributes such as time or space become relevant, they must be added to the system's description: for instance, the time dependence is simply introduced through the time-dependent quantities $i_k(t)$ and $s_k(t)$.

9.2.1 The SI model

As a first assessment of the effect of the network heterogeneities, let us consider the simple SI model. In this case we know that the system is eventually totally infected whatever the spreading rate of the infection, but it is interesting to see the effect of topological fluctuations on the spreading velocity. In the case of the SI model the evolution equations read

$$\frac{di_k(t)}{dt} = \beta [1 - i_k(t)] k \Theta_k(t), \qquad (9.19)$$

where the creation term is proportional to the spreading rate β, the degree k, the probability $1 - i_k$ that a vertex with degree k is not infected, and the density Θ_k

of infected neighbors of vertices of degree k. The latter term is thus the average probability that any given neighbor of a vertex of degree k is infected. This is a new and unknown quantity that in the homogeneous assumption is equal to the density of infected nodes. In a heterogeneous network, however, this is a complicated expression that must take into account the different degree classes and their connections.

The simplest situation we can face corresponds to a complete lack of degree correlations. As already discussed in the previous chapters, a network is said to have no degree correlations when the probability that an edge departing from a vertex of degree k arrives at a vertex of degree k' is independent from the degree of the initial vertex k. In this case, the conditional probability does not depend on the originating node and it is possible to show that $P(k'|k) = k'P(k')/\langle k \rangle$ (see Chapter 1 and Appendix 1). This relation simply states that any edge has a probability of pointing to a node with degree k' which is proportional to k'. By considering that at least one of the edges of each infected vertex points to another infected vertex from which the infection has been transmitted, one obtains

$$\Theta_k(t) = \Theta(t) = \frac{\sum_{k'}(k'-1)P(k')i_{k'}(t)}{\langle k \rangle}, \tag{9.20}$$

where $\langle k \rangle = \sum_{k'} k'P(k')$ is the proper normalization factor dictated by the total number of edges. Thanks to the absence of correlations between the degrees of neighboring vertices, $\Theta_k(t)$ is then independent of k.

Combining Equations (9.19) and (9.20) one obtains the evolution equation for $\Theta(t)$. In the initial epidemic stages, neglecting terms of order $\mathcal{O}(i^2)$, the equations read

$$\frac{di_k(t)}{dt} = \beta k\Theta(t), \tag{9.21}$$

$$\frac{d\Theta(t)}{dt} = \beta\left(\frac{\langle k^2 \rangle}{\langle k \rangle} - 1\right)\Theta(t). \tag{9.22}$$

These equations can be solved and in the case of a uniform initial condition $i_k(t = 0) = i_0$, the prevalence of nodes of degree k reads as

$$i_k(t) = i_0\left[1 + \frac{k(\langle k \rangle - 1)}{\langle k^2 \rangle - \langle k \rangle}(e^{t/\tau} - 1)\right], \tag{9.23}$$

with

$$\tau = \frac{\langle k \rangle}{\beta(\langle k^2 \rangle - \langle k \rangle)}. \tag{9.24}$$

The prevalence therefore increases exponentially fast, and larger degree nodes display larger prevalence levels. The total average prevalence is also obtained as $i(t) = \sum_k P(k)i_k(t)$,

$$i(t) = i_0 \left[1 + \frac{\langle k \rangle^2 - \langle k \rangle}{\langle k^2 \rangle - \langle k \rangle} (e^{t/\tau} - 1) \right]. \tag{9.25}$$

The result (9.24) for uncorrelated networks implies that the growth time scale of an epidemic outbreak is related to the graph heterogeneity as measured by the heterogeneity ratio $\kappa = \langle k^2 \rangle / \langle k \rangle$ (see Chapter 2). In homogeneous networks with a Poisson degree distribution, in which $\kappa = \langle k \rangle + 1$, we recover the result $\tau = (\beta \langle k \rangle)^{-1}$, corresponding to the homogeneous mixing hypothesis. In networks with very heterogeneous connectivity patterns, on the other hand, κ is very large and the outbreak time scale τ is very small, signaling a very fast diffusion of the infection. In particular, in scale-free networks characterized by a degree exponent $2 < \gamma \leq 3$ we have that $\kappa \sim \langle k^2 \rangle \to \infty$ for networks of size $N \to \infty$. Therefore in uncorrelated scale-free networks we face a virtually instantaneous rise of the epidemic incidence. The physical reason is that once the disease has reached the hubs, it can spread very rapidly among the network following a "cascade" of decreasing degree classes (Barthélemy *et al.*, 2004; 2005).

9.2.2 The SIR and SIS models

The above results can be easily extended to the SIS and the SIR models. In the case of uncorrelated networks, Equation (9.19) contains, for both the SIS and the SIR models, an extra term $-\mu i_k(t)$ defining the rate at which infected individuals of degree k recover and become again susceptible or permanently immune (and thus removed from the population), respectively:

$$\frac{di_k(t)}{dt} = \beta k s_k(t) \Theta_k(t) - \mu i_k(t). \tag{9.26}$$

In the SIS model we have, as usual, $s_k(t) = 1 - i_k(t)$. In the SIR model, on the other hand, the normalization imposes that $s_k(t) = 1 - i_k(t) - r_k(t)$, where $r_k(t)$ is the density of removed individuals of degree k. The inclusion of the decaying term $-\mu i_k$, however, does not change the picture obtained in the SI model. By using the same approximations, the time scale for the SIR is found to behave as

$$\tau = \frac{\langle k \rangle}{\beta \langle k^2 \rangle - (\mu + \beta) \langle k \rangle}. \tag{9.27}$$

In the case of diverging fluctuations the time scale behavior is therefore still dominated by $\langle k^2 \rangle$ and the spreading is faster for higher network heterogeneity. This leads to a striking result concerning the epidemic threshold. In order to ensure an epidemic outbreak the basic condition $\tau > 0$ reads as[2]

$$\frac{\beta}{\mu} \geq \frac{\langle k \rangle}{\langle k^2 \rangle - \langle k \rangle}. \tag{9.28}$$

This implies that in heavy-tailed networks such that $\langle k^2 \rangle \rightarrow \infty$ in the limit of a network of infinite size we have *a null epidemic threshold*. While this is not the case in any finite size real-world network, larger heterogeneity levels lead to smaller epidemic thresholds. Also in this case, the parameter $\kappa = \langle k^2 \rangle / \langle k \rangle$ that defines the level of heterogeneity in the connectivity pattern is determining the properties of the physical processes occurring on the network. This is a very relevant result that, analogously to those concerning the resilience to damage (see Chapter 6), indicates that heterogeneous networks behave very differently from homogeneous networks with respect to physical and dynamical processes. In particular, the absence of any epidemic threshold makes scale-free networks a sort of ideal environment for the spreading of viruses, which even in the case of very weak spreading capabilities are able to pervade the network. We will see, however, that the picture is not so gloomy, as while the epidemic threshold is extremely reduced, the prevalence corresponding to small spreading rates is very small. Furthermore, we will see that the heterogeneity can be turned to our advantage by defining opportune vaccination strategies of great effectiveness.

9.2.3 The effect of mixing patterns

While we have so far restricted our study to the case of uncorrelated networks, it is worth noting that many real networks do not fulfil this assumption (Dorogovtsev and Mendes, 2003; Pastor-Satorras and Vespignani, 2004). In order to consider the presence of non-trivial correlations we have to fully take into account the structure of the conditional correlation function $P(k'|k)$. For the sake of simplicity let us consider the SI model. In this case, the equations we have written for the evolution of i_k can be stated as (Boguñá *et al.*, 2003b)

$$\frac{di_k(t)}{dt} = \beta \left[1 - i_k(t) \right] k \Theta_k(t)$$

$$\Theta_k = \sum_{k'} i_{k'} \frac{k' - 1}{k'} P(k'|k). \tag{9.29}$$

---

[2] In the SIS model the equation is slightly different because in principle each infected vertex does not have to point to another infected vertex since the vertex from which it received the infection can spontaneously become susceptible again. This leads to $\Theta(t) = \left[\sum_{k'} k' P(k') i_{k'}(t) \right] / \langle k \rangle$, and consequently the epidemic threshold is given by $\beta/\mu \geq \langle k \rangle / \langle k^2 \rangle$.

The function Θ_k takes into account explicitly the structure of the conditional probability that an infected vertex with degree k' points to a vertex of degree k, with any of the $k'-1$ free edges it has (not pointing to the original source of its infection). In the absence of correlations, $P(k'|k) = k'P(k')/\langle k \rangle$, recovering the results of Section 9.2.1. If the network presents correlations, measured by $P(k'|k)$, the situation is slightly more complex.

The evolution equation for $i_k(t)$ can be written at short times, neglecting terms of order $\mathcal{O}(i^2)$, as

$$\frac{di_k(t)}{dt} = \sum_{k'} \beta k \frac{k'-1}{k'} P(k'|k) i_{k'}(t)$$

$$\equiv \sum_{k'} C_{k,k'} i_{k'}(t), \qquad (9.30)$$

which is a linear system of differential equations given by the matrix $\mathbf{C} = \{C_{k,k'}\}$ of elements

$$C_{k,k'} = \beta k \frac{k'-1}{k'} P(k'|k). \qquad (9.31)$$

Elementary considerations from mathematical analysis tell us that the behavior of $i_k(t)$ will be given by a linear combination of exponential functions of the form $\exp(\Lambda_i t)$, where Λ_i are the eigenvalues of the matrix $\mathbf{C}$. Therefore, the dominant behavior of the average prevalence will be

$$i(t) \sim e^{\Lambda_m t}, \qquad (9.32)$$

where Λ_m is the largest eigenvalue of the matrix $\mathbf{C}$. The time scale governing the increase of the prevalence is thus given by $\tau \sim 1/\Lambda_m$. In the case of an uncorrelated network, the matrix $\mathbf{C}$, whose elements are $C_{k,k'}^{nc} = \beta k(k'-1)P(k')/\langle k \rangle$, has a unique eigenvalue satisfying

$$\sum_{k'} C_{k,k'} \Psi_{k'} = \Lambda_m^{nc} \Psi_k, \qquad (9.33)$$

where $\Lambda_m^{nc} = \beta(\langle k^2 \rangle/\langle k \rangle - 1)$, and where the corresponding eigenvector is $\Psi_k = k$, thus recovering the previous result Equation (9.24) of Section 9.2.1.

In the case of correlated networks, it has been shown using the Frobenius theorem (Gantmacher, 1974) that the largest eigenvalue of $\mathbf{C}$ is bounded from below (Boguñá, Pastor-Satorras, 2003)

$$\Lambda_m^2 \geq \min_k \sum_{k'} \sum_l (k'-1)(l-1)P(l|k)P(k'|l). \qquad (9.34)$$

This equation is very interesting since it can be rewritten as

$$\Lambda_m^2 \geq \min_k \sum_l (l - 1) P(l|k)(k_{nn}(l) - 1), \tag{9.35}$$

and it turns out (Boguñá *et al.*, 2003) that for scale-free networks with degree distribution $P(k) \sim k^{-\gamma}$ with $2 \leq \gamma \leq 3$, the average nearest neighbors degree $k_{nn}(l)$ diverges for infinite size systems ($N \to \infty$), which implies that Λ_m also diverges. Two particular cases have to be treated separately: it may happen that, for some k_0, $P(l|k_0) = 0$; then the previous limit for Λ_m^2 gives no information but it is possible to show with slightly more involved calculations that Λ_m still diverges for $N \to \infty$ (Boguñá *et al.*, 2003b). Another problem arises if $k_{nn}(l)$ diverges only for $l = 1$; this only happens in particular networks where the singularity is accumulated in a pathological way onto vertices with a single edge. Explicit examples of this situation are provided by Moreno and Vázquez (2003).

The previous result, which has been also checked numerically (Moreno, Gómez and Pacheco, 2003a), has the important consequence that, even in the presence of correlations, the time scale $\tau \sim 1/\Lambda_m$ tends to zero in the thermodynamic limit for any scale-free network with $2 < \gamma \leq 3$. It also underlines the relevance of the quantity k_{nn}, which gives a lower bound for Λ_m in finite networks. This result can be generalized to both the SIR and SIS model, recovering the conclusion of the previous section concerning the epidemic time scale and the epidemic threshold. In the next sections we will analyze numerically both the SI and SIS models in order to provide a full description of the dynamical properties that takes into account the network's complexity as well as finite size effects in the population.

As we stated in Chapter 1, the correlation function $P(k|k')$ provides only a first account of the network structure. Important structural properties might be encoded in higher order statistical correlations and there is no complete theory of the effect of these properties on epidemic behavior. Recent works have confirmed the general picture concerning the behavior of the epidemic threshold in heterogeneous networks also in the case of non-trivial clustering properties (Serrano and Boguñá, 2006), but these general results may be altered by specific constructions of the network that break the statistical equivalence of nodes with respect to their degree.

9.2.4 Numerical simulations

In order to illustrate the analytical predictions presented in the previous sections, we show numerical simulations obtained by using an agent-based modeling strategy for an SI model in which at each time step the stochastic disease dynamics is applied to each vertex by considering the actual state of the vertex and its neighbors

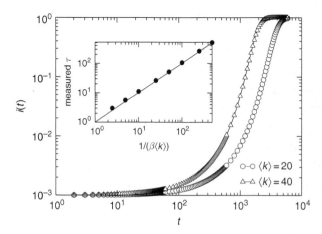

Fig. 9.3. Main frame: the symbols correspond to simulations of the SI model with $\beta = 10^{-4}$ on Erdős–Rényi networks with $N = 10^4$, $\langle k \rangle = 20, 40$; the lines are fits of the form of Equation (9.11). Inset: measured time scale τ, as obtained from fitting, versus the theoretical prediction for different values of $\langle k \rangle$ and β. From Barthélemy *et al.* (2005).

(see also Moreno, Gómez and Pacheco [2003a] for an alternative numerical method which directly solves the mean-field equations such as (9.29)). It is then possible to monitor the details of the spreading process, and to measure, for example, the evolution of the number of infected individuals. In addition, given the stochastic nature of the model, different initial conditions and network realizations can be used to obtain averaged quantities. We present simulations obtained for $N = 10^4$ and $\langle k \rangle$ ranging from 4 to 20. The results are typically averaged over a few hundred networks and for each network, over a few hundred different initial conditions. As an example of a homogeneous graph we consider the Erdős–Rényi network. In this case, Figure 9.3 shows the validity of Equation (9.11) and that the time scale of the exponential increase of the prevalence is given by $1/\beta\langle k \rangle$.

Needless to say, in the case of a homogeneous network, the hypothesis $k \simeq \langle k \rangle$ captures the correct dynamical behavior of the spreading. This is a standard result and we report the numerical simulations just as a reference for comparison with the following numerical experiments on heterogeneous networks. As a typical example of heterogeneous networks, the networks generated with the Barabási–Albert (BA) algorithm can be used (see Chapter 3). We report in Figure 9.4 the value of τ measured from the early time behavior of outbreaks in networks with different heterogeneity levels, as a function of $\langle k \rangle / \beta(\langle k^2 \rangle - \langle k \rangle)$. The numerical results recover the analytical prediction with great accuracy. Indeed, the BA network is a good example of uncorrelated heterogeneous networks in which the approximations used in the calculations are satisfied. In networks with correlations we expect

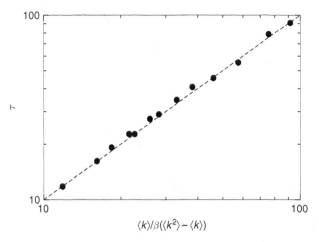

Fig. 9.4. Measured time scale τ in BA networks as obtained from exponential fitting to Equation (9.25) versus the theoretical prediction for different values of $\langle k \rangle$ and N corresponding to different levels of heterogeneity. From Barthélemy *et al.* (2005).

to find different quantitative results but qualitatively similar behavior, as happens in the case of the epidemic threshold evaluation (Boguñá *et al.*, 2003b).

While the present simulations are in very good agreement with the analytical results, the extent to which one expects such numerical simulations to recover exactly the analytical results is an open issue. The analytical results such as Equation (9.24) are obtained in the case of a linear expansion of the equations that is valid as long as $i(t) \ll 1$. On the other hand, the smaller the spreading time scale τ, the sooner the approximation breaks down. In very heterogeneous networks, the exponential regime may be extremely short or completely absent if the time scale of the epidemics is of the order of the unit time scale used in the simulations. In this case it is possible to show that the early time regime obeys a different behavior (Vázquez, 2006a; 2006c).

9.3 The large time limit of epidemic outbreaks

In the previous section we have limited ourselves to the analysis of the early time regime. It is, however, also interesting to study the opposite limit $t \to \infty$, in order to provide a more complete characterization of the epidemic outbreaks according to each model. The SI model is obviously uninteresting in the large time limit as it just asymptotically reaches the limit $i(t) \to 1$. The situation is very different if we consider the SIS and the SIR models. In these cases, the stationary prevalence or the total size of the epidemics depend upon the disease parameter of the model and the heterogeneity of the system.

9.3.1 The SIS model

The complete evolution equation for the SIS model on a network with arbitrary degree distribution can be written as

$$\frac{di_k(t)}{dt} = -\mu i_k(t) + \beta k [1 - i_k(t)] \Theta_k(t). \tag{9.36}$$

The creation term considers the density $1 - i_k(t)$ of susceptible vertices with k edges that might get the infection via a neighboring vertex. Let us first consider for the sake of simplicity the case in which the underlying network is a generalized random graph with no degree correlations. As already described in Sections 9.2.1 and 9.2.2, the calculation of Θ_k is then straightforward, as the average density of infected vertices pointed by any given edge that reads as

$$\Theta_k = \frac{1}{\langle k \rangle} \sum_{k'} k' P(k') i_{k'}, \tag{9.37}$$

which does not depend on k: $\Theta_k = \Theta$. Information on the $t \to \infty$ limit can be easily obtained by imposing the stationarity condition $di_k(t)/dt = 0$

$$i_k = \frac{k \beta \Theta}{\mu + k \beta \Theta}. \tag{9.38}$$

This set of equations shows that the higher a vertex degree, the higher its probability to be in an infected state. Injecting Equation (9.38) into (9.37), it is possible to obtain the self-consistent equation

$$\Theta = \frac{1}{\langle k \rangle} \sum_k k P(k) \frac{\beta k \Theta}{\mu + \beta k \Theta}, \tag{9.39}$$

whose solution allows the calculation of Θ as a function of the disease parameters β and μ (Pastor-Satorras and Vespignani, 2001a; 2001b).

The epidemic threshold can be explicitly calculated from Equation (9.39) by just noting that the condition is given by the value of β and μ for which it is possible to obtain a non-zero solution Θ^*. Using a geometrical argument, as for the analysis of percolation theory in random graphs in Chapter 6, the solution of Equation (9.39) follows from the intersection of the curves $y_1(\Theta) = \Theta$ and $y_2(\Theta) = (1/\langle k \rangle) \sum_k k P(k) \beta k \Theta/(\mu + \beta k \Theta)$. The latter is a monotonously increasing function of Θ between the limits $y_2(0) = 0$ and $y_2(1) = (1/\langle k \rangle) \sum_k k P(k) \beta k/(\mu + \beta k) < 1$. In order to have a solution $\Theta^* \neq 0$, the slope of $y_2(\Theta)$ at the point $\Theta = 0$ must be larger than or equal to 1 (see Figure 9.5). This condition can be written as

$$\frac{d}{d\Theta} \left(\frac{1}{\langle k \rangle} \sum_k k P(k) \frac{\beta k \Theta}{\mu + \beta k \Theta} \right) \Bigg|_{\Theta=0} = \frac{\beta}{\mu} \frac{\langle k^2 \rangle}{\langle k \rangle} \geq 1. \tag{9.40}$$

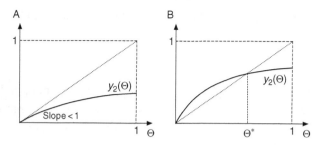

Fig. 9.5. Graphical solution of Equation (9.39). A, If the slope of the function $y_2(\Theta)$ at $\Theta = 0$ is smaller than 1, the only solution of the equation is $\Theta = 0$. B, When the slope is larger than 1, a non-trivial solution $\Theta^* \neq 0$ can be found.

The value of the disease parameters yielding the equality in Equation (9.40) defines the epidemic threshold condition that reads as

$$\frac{\beta}{\mu} = \frac{\langle k \rangle}{\langle k^2 \rangle}. \tag{9.41}$$

This condition recovers the results obtained from the linear approximation at short times and confirms that topological fluctuations lower the epidemic threshold.

It is moreover possible to compute explicitly the behavior of the stationary density of infected individuals $i_\infty = \lim_{t \to \infty} i(t)$ as a function of the disease parameters in random uncorrelated scale-free networks where the heavy-tailed character is modeled by a power-law degree distribution with arbitrary exponent γ (Pastor-Satorras and Vespignani, 2001b). Consider a network which, in the continuous k approximation, has a normalized degree distribution $P(k) = (\gamma - 1)m^{\gamma-1}k^{-\gamma}$ and average degree $\langle k \rangle = (\gamma - 1)m/(\gamma - 2)$, where m is the minimum degree of any vertex. According to the general result (9.41), the epidemic threshold for infinite networks depends on the second moment of the degree distribution and is given, as a function of γ, by

$$\frac{\beta}{\mu} = \begin{cases} \dfrac{\gamma - 3}{m(\gamma - 2)} & \text{if } \gamma > 3 \\ 0 & \text{if } \gamma \leq 3 \end{cases}. \tag{9.42}$$

The behavior of the density of infected individuals in the stationary state may be found by solving explicitly the self-consistent equation for Θ in the limit of β/μ approaching the epidemic threshold. The full calculation yields different cases as a function of the exponent γ (Pastor-Satorras and Vespignani, 2001a):

(a) $2 < \gamma < 3$

In this case the leading order terms in β/μ of the solution yields

$$i_\infty \sim \left(\frac{\beta}{\mu}\right)^{1/(3-\gamma)}. \tag{9.43}$$

As expected, this relation does not show any epidemic threshold and gives a non-zero prevalence for all values of β/μ. It is important to note that the exponent governing the behavior of the prevalence, $1/(3-\gamma)$, is larger than 1. This implies that for small β/μ the prevalence is growing very slowly, i.e. there exists a wide region of spreading rates in which $i_\infty \ll 1$.

(b) $\gamma = 3$

For this value of the degree exponent, logarithmic corrections dominate the scaling of the solution, yielding

$$i_\infty \sim e^{-\mu/m\beta}. \tag{9.44}$$

In this case, too, the absence of any epidemic threshold is recovered, and the prevalence approaches zero in a continuous way, exhibiting an exponentially small value for a wide range of spreading rates ($i_\infty \ll 1$).

(c) $3 < \gamma < 4$

The non-zero solution for Θ yields:

$$i_\infty \sim \left(\frac{\beta}{\mu} - \frac{\gamma-3}{m(\gamma-2)}\right)^{1/(\gamma-3)}. \tag{9.45}$$

That is, a power-law persistence behavior is observed. It is associated, however, to the presence of a non-zero threshold as given by Equation (9.42). Since $1/(\gamma-3) > 1$, the epidemic threshold is approached smoothly without any sign of the singular behavior associated to a critical point.

(d) $\gamma > 4$

The most relevant terms in the expansion of Θ now yield the behavior

$$i_\infty \sim \frac{\beta}{\mu} - \frac{\gamma-3}{m(\gamma-2)}. \tag{9.46}$$

That is, we recover the usual epidemic framework obtained for homogeneous networks.

In summary, the outcome of the analysis presented here is that the SIS model in scale-free uncorrelated random networks with degree exponent $\gamma \leq 3$ exhibits the absence of an epidemic threshold or critical point. Only for $\gamma > 4$ do epidemics on scale-free networks have the same properties as on homogeneous networks.

9.3.2 The SIR model

In the case of the SIR model, the number of infected individuals is ultimately zero and the epidemics die because of the depletion of the susceptible individuals that after the infection move into the removed compartment. One of the main pieces of information on the course of the epidemics is therefore provided by the total number of individuals affected by the infection which is equal to the number of recovered individuals if the starting population was composed only of susceptible individuals. Taking into account the degree heterogeneity, this number is expressed as $r_\infty = \lim_{t\to\infty} r(t)$, where $r(t) = \sum_k P(k) r_k(t)$. This quantity may be explicitly calculated using the SIR equation for the degree classes (May and Lloyd, 2001; Moreno, Pastor-Satorras and Vespignani, 2002b; Newman, 2002b; Boguñá et al., 2003b) which reads as

$$\frac{di_k(t)}{dt} = -\mu i_k(t) + \beta k s_k(t)\Theta_k(t), \tag{9.47}$$

$$\frac{ds_k(t)}{dt} = -\beta k s_k(t)\Theta_k(t), \tag{9.48}$$

$$\frac{dr_k(t)}{dt} = \mu i_k(t). \tag{9.49}$$

As in the previous sections, $\Theta_k(t)$ represents the average density of infected individuals at vertices pointed by any given edge and is given for an uncorrelated network by Equation (9.20). Equations (9.47), (9.48), (9.49), and (9.20), combined with the initial conditions $r_k(0) = 0$, $i_k(0)$, and $s_k(0) = 1 - i_k(0)$, completely define the SIR model on any random uncorrelated network with degree distribution $P(k)$. Let us consider the case of a homogeneous initial distribution of infected individuals, $i_k(0) = i_0$. In this case, in the limit of a very small number of initial infected individuals $i_0 \to 0$ and $s_k(0) \simeq 1$, Equations (9.48) and (9.49) can be directly integrated, yielding

$$s_k(t) = e^{-\beta k \phi(t)}, \qquad r_k(t) = \mu \int_0^t i_k(\tau)\, d\tau, \tag{9.50}$$

where we have defined the auxiliary function

$$\phi(t) = \int_0^t \Theta(\tau)\, d\tau = \frac{1}{\langle k \rangle}\mu^{-1}\sum_k (k-1)P(k)r_k(t). \tag{9.51}$$

In order to obtain a closed relation for the total density of infected individuals, it is more convenient to focus on the time evolution of $\phi(t)$. To this purpose, let us compute its time derivative

$$
\begin{aligned}
\frac{d\phi(t)}{dt} &= \frac{1}{\langle k \rangle} \sum_k (k-1) P(k) i_k(t) \\
&= \frac{1}{\langle k \rangle} \sum_k (k-1) P(k)[1 - r_k(t) - s_k(t)] \\
&= 1 - \frac{1}{\langle k \rangle} - \mu\phi(t) - \frac{1}{\langle k \rangle} \sum_k (k-1) P(k) e^{-\beta k \phi(t)},
\end{aligned} \tag{9.52}
$$

where we have used the time dependence of $s_k(t)$ obtained in Equation (9.50). For a general distribution $P(k)$, Equation (9.52) cannot be solved in a closed form, but it is however possible to get useful information on the infinite time limit; i.e. at the end of the epidemics. In particular, the total epidemic prevalence $r_\infty = \sum_k P(k) r_k(\infty)$ is obtained as a function of $\phi_\infty = \lim_{t \to \infty} \phi(t)$

$$
r_\infty = \sum_k P(k) \left(1 - e^{-\beta k \phi_\infty}\right), \tag{9.53}
$$

where we have used $r_k(\infty) = 1 - s_k(\infty)$ and Equation (9.50). Since $i_k(\infty) = 0$, and consequently $\lim_{t \to \infty} d\phi(t)/dt = 0$, we obtain from Equation (9.52) the following self-consistent equation for ϕ_∞:

$$
\mu\phi_\infty = 1 - \frac{1}{\langle k \rangle} - \frac{1}{\langle k \rangle} \sum_k (k-1) P(k) e^{-\beta k \phi_\infty}. \tag{9.54}
$$

The value $\phi_\infty = 0$ is always a solution. The non-zero ϕ_∞ solution, corresponding to finite prevalence $r_\infty > 0$, exists only if

$$
\frac{d}{d\phi_\infty} \left(1 - \frac{1}{\langle k \rangle} - \frac{1}{\langle k \rangle} \sum_k (k-1) P(k) e^{-\beta k \phi_\infty} \right) \Bigg|_{\phi_\infty = 0} \geq \mu. \tag{9.55}
$$

This condition, which can be obtained with a geometrical argument analogous to that used to obtain Equation (9.40) for the SIS model, is equivalent to

$$
\frac{\beta}{\langle k \rangle} \sum_k k(k-1) P(k) \geq \mu. \tag{9.56}
$$

This defines the epidemic threshold condition

$$
\frac{\beta}{\mu} = \frac{\langle k \rangle}{\langle k^2 \rangle - \langle k \rangle}, \tag{9.57}
$$

below which the epidemic prevalence is $r_\infty = 0$, and above which it reaches a finite value $r_\infty > 0$. This recovers the result (9.28) obtained by the linear approximation in the early stage of the dynamics and shows that the effects of topological fluctuations are consistently obtained in both limiting solutions.

The above equations for r_∞ cannot be solved explicitly for any network. As for the SIS model in the previous section, in the case of generalized random scale-free networks with degree distribution $P(k) = (\gamma - 1)m^{\gamma-1}k^{-\gamma}$ it is however possible to find an explicit solution that is an example of what happens in heavy-tailed networks (Moreno, Pastor-Satorras and Vespignani, 2002b; May and Lloyd, 2001; Newman, 2002b). Not surprisingly it is possible to show that r_∞ displays the same behavior as a function of β/μ as obtained for the stationary infected density i_∞ in the SIS model, and the various cases for the different ranges of γ values are recovered. In particular, for heavy-tailed networks of infinite size, and small values of β/μ, one obtains $r_\infty \sim (\beta/\mu)^{1/(3-\gamma)}$ and $r_\infty \sim e^{-\mu/m\beta}$ for $2 < \gamma < 3$ and $\gamma = 3$, respectively.

9.3.3 Epidemic models and phase transitions

The previous analysis allows us to draw an analogy between epidemic models and non-equilibrium continuous phase transitions. The SIS and SIR models are characterized by a threshold defining a transition between two very different regimes. These regimes are determined by the values of the disease parameters, and characterized by the global parameters i_∞ and r_∞, which are zero below the threshold and assume a finite value above the threshold. From this perspective we can consider the epidemic threshold as the critical point of the system, and i_∞ and r_∞ represent the order parameter characterizing the transition in the SIS and SIR, respectively. Below the critical point the system relaxes in a frozen state with null dynamics, the healthy phase. Above this point, a dynamical state characterized by a macroscopic number of infected individuals sets in, defining an infected phase. Finally we have explained how, in the case of strong heterogeneity, the epidemic threshold is reduced and eventually suppressed by topological fluctuations in the case of infinite size networks.

It is therefore possible to draw a qualitative picture of the phase diagram of the system as depicted in Figure 9.6. The figure also shows the difference between homogeneous and heterogeneous networks where the epidemic threshold is shifted to very small values. On the other hand, for scale-free networks with degree distribution exponent $\gamma \leq 3$ the associated prevalence i_∞ is extremely small in a large region of values of β/μ. In other words the bad news of the suppression (or very small value) of the epidemic threshold is balanced by the very low prevalence attained by the epidemics.

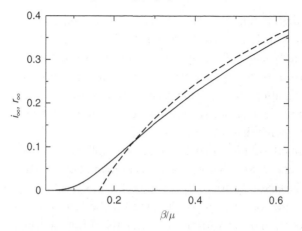

Fig. 9.6. Total prevalence (i_∞ for the SIS and r_∞ for the SIR model) in a heterogeneous network (full line) as a function of the spreading rate β/μ, compared with the theoretical prediction for a homogeneous network (dashed line). From Pastor-Satorras and Vespignani (2001a).

The mapping between epidemic models and non-equilibrium phase transitions has been pointed out in various contexts for a long time. As stressed by Grassberger (1983), the SIR model static properties can be mapped to an edge percolation process. Indeed, the epidemic threshold and the behavior close to threshold of the SIR model have the same form found for the percolation threshold in generalized networks (see Chapter 6). Analogously, it is possible to recognize that the SIS model is a generalization of the contact process model (Harris, 1974), widely studied as the paradigmatic example of an absorbing-state phase transition with a unique absorbing state (Marro and Dickman, 1999).

9.3.4 Finite size and correlations

As we discussed in the introductory chapters, real-world networks are composed of a number of elements that is generally far from the thermodynamic limit. This finite population introduces a maximum degree k_c, depending on the system size N or a finite connectivity capacity, which has the effect of restoring a bound to the degree fluctuations. The presence of the cut-off translates, through the general expression Equation (9.41), into an effective non-zero epidemic threshold due to *finite size effects*, as usually observed in non-equilibrium phase transitions (Pastor-Satorras and Vespignani, 2002a; May and Lloyd, 2001; Marro and Dickman, 1999). This positive epidemic threshold, however, is not an *intrinsic* property as in homogeneous systems, but an artifact of the limited system size that vanishes when increasing the network size or the degree cut-off.

To illustrate this point, let us focus on the SIS model in uncorrelated random networks with a scale-free degree distribution of the form $P(k) \simeq k^{-\gamma} \exp(-k/k_c)$. We define the epidemic threshold condition as

$$\frac{\beta}{\mu} = \rho_c, \tag{9.58}$$

and compute the effective nonzero epidemic threshold within the continuous k approximation (Pastor-Satorras and Vespignani, 2002a) as

$$\rho_c(k_c) = \frac{\langle k \rangle_{k_c}}{\langle k^2 \rangle_{k_c}} = \frac{\int_m^\infty k^{-\gamma+1} \exp(-k/k_c)}{\int_m^\infty k^{-\gamma+2} \exp(-k/k_c)} \equiv \frac{\Gamma(2-\gamma, m/k_c)}{\Gamma(3-\gamma, m/k_c)}. \tag{9.59}$$

Here m is the minimum degree of the network, and $\Gamma(x, y)$ is the incomplete Gamma function (Abramowitz and Stegun, 1972). For large k_c we can perform a Taylor expansion and retain only the leading term, obtaining for any $2 < \gamma < 3$ a threshold condition

$$\rho_c(k_c) \simeq \left(\frac{k_c}{m}\right)^{\gamma-3}. \tag{9.60}$$

The limit $\gamma \to 3$ corresponds instead to a logarithmic divergence, yielding at leading order

$$\rho_c(k_c) \simeq \frac{1}{m \ln(k_c/m)}. \tag{9.61}$$

It is interesting to compare the intrinsic epidemic threshold obtained in homogeneous networks with negligible degree fluctuations, $\rho_c^H = \langle k \rangle^{-1}$, with the non-zero effective threshold of bounded scale-free distributions. Figure 9.7 represents the ratio $\rho_c(k_c)/\rho_c^H$ as a function of k_c/m, for different values of the degree exponent γ. We can observe that, even for relatively small cut-offs ($k_c/m \sim 10^2 - 10^3$), for a reasonable value $\gamma \approx 2.5$ the effective epidemic threshold of finite scale-free networks is smaller by nearly an order of magnitude than the intrinsic threshold corresponding to a homogeneous network with the same average degree. This fact implies that the use of the homogeneity assumption would lead in scale-free networks to a serious over-estimation of the epidemic threshold, even for relatively small networks.

Another assumption that we have used in the calculation of the $t \to \infty$ properties of epidemic models is that the structure of the network is just characterized by its degree distribution. This is far from reality in most networks where structural properties and correlations are present. A full solution of the dynamical equations taking into account the general structural properties of the network is not possible, and we are left with approaches that progressively include higher order correlations. As we have seen for the early stage dynamics, a first step is

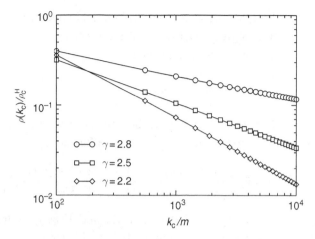

Fig. 9.7. Ratio between the effective epidemic threshold $\rho_c(k_c)$ in bounded scale-free networks with a soft exponential cut-off, and the intrinsic epidemic threshold ρ_c^H for homogeneous networks with the same average degree, for different values of γ. From Pastor-Satorras and Vespignani (2002a).

to include the correlations defined by the conditional probability $P(k'|k)$ that a vertex of degree k is connected to one of degree k'. Also, for the infinite time properties it is possible to show that the presence or lack of an epidemic threshold is directly related to the largest eigenvalue Λ_m of the connectivity matrix. This eigenvalue has been shown to diverge for all scale-free *unstructured networks* with infinite fluctuations,[3] for any kind of two-point correlations (Boguñá and Pastor-Satorras, 2002; Boguñá *et al.*, 2003). Recent results (Serrano and Boguñá, 2006) have shown that more complicate structural properties, such as three vertices correlations and high clustering coefficient, do not alter the general framework concerning the epidemic threshold existence. On the other hand it is always possible to imagine very specific epidemic behaviors dictated by a peculiar construction or particular engineering of the network not completely captured by the statistical properties of the degree distribution and degree correlations.

Let us finally insist on the fact that the results presented in this chapter have been obtained within mean-field approximations. It is worth a word of warning that the domain of validity of these approximations is not a priori known. The numerical simulations have shown that this approach allows one to grasp and understand how the behavior of epidemic models is altered and strongly influenced by heterogeneous connectivity patterns. There is no guarantee that similar approximations will

---

[3] The present result is only valid for networks with no internal structure, in which all the vertices with the same degree are statistically equivalent. It does not apply for regular lattices or structured networks (Klemm and Eguíluz, 2002b; Moreno and Vázquez, 2003), in which a spatial or class ordering constrains the connections among vertices.

be valid in the analytical study of any process taking place on networks. In particular, recent works have shown that mean-field predictions are contradicted by extensive numerical simulations in two particular models, namely the dynamical evolution of Ising spins located on the nodes of a network and interacting at zero temperature, and the contact process, in which particles located on the nodes of a network disappear or produce offspring according to particular rules (Castellano and Pastor-Satorras, 2006a; 2006b). It is therefore important to bear in mind that the mean-field approach, although usually very convenient and powerful, has to be systematically complemented by numerical investigations. The understanding of the limits of validity of the mean-field approximations represents, moreover, an interesting open problem.

9.4 Immunization of heterogeneous networks

The weakness of scale-free networks to epidemic attacks presents an extremely worrying scenario. The conceptual understanding of the mechanisms and causes for this weakness, however, allows us to develop new defensive strategies that take advantage of the heterogeneous topology. Thus, while random immunization strategies are utterly inefficient, it is yet possible to devise targeted immunization schemes which are extremely effective.

9.4.1 Uniform immunization

In heavy-tailed networks the introduction of a random immunization is able to depress the infection's prevalence locally, but it does so too slowly, being unable to find any critical fraction of immunized individuals that ensures the eradication of the infection. An intuitive argument showing the inadequacy of random immunization strategies is that they give the same importance to very connected vertices (with the largest infection potential) and to vertices with a very small degree. Because of the large fluctuations in the degree, heavily connected vertices, which are statistically very significant, can overcome the effect of the uniform immunization and maintain the endemic state.

In more mathematical terms, and as already stated in Section 9.1.2, the introduction of a density g of immune individuals chosen at random is equivalent to a simple rescaling of the effective spreading rate as $\beta \rightarrow \beta(1 - g)$, i.e. the rate at which new infected individuals appear is depressed by a factor proportional to the probability that they are not immunized. On the other hand, the absence of an epidemic threshold in the thermodynamic limit implies that any rescaling of the spreading rate does not bring the epidemic into the healthy region except in the case $g = 1$. Indeed, the immunization threshold g_c is obtained when the rescaled spreading rate is set equal to the epidemic threshold. For instance, for uncorrelated networks we obtain

$$\frac{\beta}{\mu}(1 - g_c) = \frac{\langle k \rangle}{\langle k^2 \rangle}. \tag{9.62}$$

In heavy-tailed networks with $\langle k^2 \rangle \to \infty$ only a complete immunization of the network ensures an infection-free stationary state in the thermodynamic limit (i.e. $g_c = 1$). The fact that uniform immunization strategies are less effective has been noted in the biological context in several cases of spatial heterogeneity (Anderson and May, 1992). In heavy-tailed networks, however, we face a limiting case due to the extremely high (virtually infinite) heterogeneity in the connectivity properties.

9.4.2 Targeted immunization

Although heavy-tailed networks hinder the efficiency of naive uniform immunization strategies, we can take advantage of their heterogeneity by devising immunization procedures that take into account the inherent hierarchy in the degree distribution. In fact, we know that heavy-tailed networks possess a noticeable resilience to *random* connection failures (Chapter 6), which implies that the network can resist a high level of accidental damage without losing its global connectivity properties, i.e. the possibility of finding a connected path between almost any two vertices in the system. At the same time, scale-free networks are strongly affected by *targeted* damage; if a few of the most connected vertices are removed, the network suffers a dramatic reduction of its ability to carry information. Applying this argument to the case of epidemic spreading, we can devise a *targeted* immunization scheme in which we progressively make immune the most highly connected vertices, which are the ones more likely to spread the disease. While this strategy is the simplest solution to the optimal immunization problem in heterogeneous populations (Anderson and May, 1992), its efficiency is comparable to the uniform strategies in homogeneous networks with finite degree variance. In heavy-tailed networks, it produces a striking increase of the network tolerance to infections at the price of a tiny fraction of immune individuals.

An approximate calculation of the immunization threshold in the case of a random scale-free network (Pastor-Satorras and Vespignani, 2002b) can be pursued along the lines of the analysis of the intentional attack of complex networks (see Section 6.5). Let us consider the situation in which a fraction g of the individuals with the highest degree have been successfully immunized. This corresponds, in the limit of a large network, to the introduction of an upper cut-off $k_c(g)$ – which is obviously a function of the immunization g – such that all vertices with degree $k > k_c(g)$ are immune. At the same time, the infective agent cannot prop-agate along all the edges emanating from immune vertices, which translates into a probability $r(g)$ of deleting any individual contacts in the network. The elimi-nation of edges and vertices for the spreading purposes yields a new connectivity pattern whose degree distribution and relative moments $\langle k \rangle_g$ and $\langle k^2 \rangle_g$ can be com-puted as a function of the density of immunized individuals (see the analogous calculation for the targeted removal of vertices in Section 6.5). The protection of the network will be achieved when the effective network on which the epidemic spreads satisfies the inequality $\langle k \rangle_g / \langle k^2 \rangle_g \geq \beta / \mu$, yielding the implicit equation for the immunization threshold

$$\frac{\langle k \rangle_{g_c}}{\langle k^2 \rangle_{g_c}} = \frac{\beta}{\mu}. \tag{9.63}$$

The immunization threshold is therefore an implicit function $g_c(\beta/\mu)$ and its analytic form will depend on the original degree distribution of the network.

In order to assess the efficiency of the targeted immunization scheme it is pos-sible to perform the explicit calculation for an uncorrelated network with degree exponent $\gamma = 3$ (Pastor-Satorras and Vespignani, 2002b). In this case the leading order solution for the immunization threshold in the case of targeted immunization reads as

$$g_c \sim \exp(-2\mu/m\beta), \tag{9.64}$$

where m is the minimum degree of the network. This clearly indicates that the targeted immunization program is extremely convenient, with a critical immu-nization threshold that is exponentially small in a wide range of spreading rates. This theoretical prediction can be tested by performing direct numerical simula-tions of the SIS model on Barabási–Albert networks in the presence of targeted immunization. In Figure 9.8 the results of the targeted immunization are com-pared with simulations made with a uniform immunization (Pastor-Satorras and Vespignani, 2002b). The plot shows the reduced prevalence i_g/i_0, where i_g is the stationary prevalence in the network with immunization density g and i_0 is the stationary prevalence in the non-immunized network, at a fixed spreading rate $\beta/\mu = 0.25$. This plot indicates that, for uniform immunization, the prevalence decays very slowly when increasing g, and will be effectively null only for $g \to 1$, as predicted by Equation (9.62).[4] On the other hand, for the targeted immunization scheme, the prevalence shows a very sharp drop and exhibits the onset of a sharp immunization threshold above which the system is infection-free. A linear regres-sion from the largest values of g yields an approximate immunization threshold

---

[4] The threshold is not exactly 1 because of the usual finite size effects present even in the simulations which are performed on networks of size $N = 10^7$.

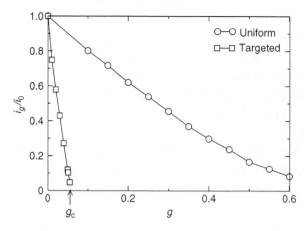

Fig. 9.8. Reduced prevalence i_g/i_0 from numerical simulations of the SIS model in the Barabási–Albert network (with $m = 2$) with uniform and targeted immunization, at a fixed spreading rate $\beta/\mu = 0.25$. A linear extrapolation from the largest values of g yields an estimate of the threshold $g_c \simeq 0.06$ for targeted immunization. From Pastor-Satorras and Vespignani (2002b).

$g_c \simeq 0.06$, that definitely proves that scale-free networks are very sensitive to the targeted immunization of a very small fraction of the most connected vertices.

Let us finally mention that, in a similar spirit, Dezső and Barabási (2002) propose a level of safety and protection policy, which is proportional to the importance of the vertex measured as a function of its local degree. This implies that high degree vertices are cured with a rate proportional to their degree, or more generally to k^α. At the theoretical level it is possible to show that any $\alpha > 0$ reintroduces a finite epidemic threshold.

9.4.3 *Immunization without global knowledge*

While the targeted strategy is very effective, it suffers from a practical drawback in its real-world application. Its implementation requires a *complete* knowledge of the network structure in order to identify and immunize the most connected vertices. For this reason, several strategies to overcome this problem have been proposed, mainly relying just on a local, rather than a global, knowledge of the network. In particular, an ingenious immunization strategy was put forward by Cohen, Havlin and ben Avraham (2003), levering on a local exploration mechanism (see also Madar *et al.* [2004]). In this scheme, a fraction g of vertices are selected at random and each one is asked to point to one of its neighbors. The neighbors, rather than the selected vertices, are chosen for immunization. Since by following edges at random it is more probable to point to high degree vertices which by definition have many links pointing to them, this strategy allows effective immunization of hubs without having any precise knowledge of the network connectivity. This strategy therefore manages to take advantage of the very same property that renders the network prone to infections. Variations on this idea have been subsequently proposed. A possibility consists in immunizing the vertex with highest degree found within shortest-path distance ℓ of randomly selected nodes (Gómez-Gardeñes, Echenique and Moreno, 2006). As ℓ increases, the necessary knowledge about the network's properties goes from local ($\ell = 1$) to global (when ℓ reaches the network's diameter). Holme (2004) also investigates *chained* versions of the immunization strategies, in which the nodes to be vaccinated are chosen successively as neighbors of previously vaccinated vertices. Such procedures turn out to be even more efficient, except in the case of very assortative or clustered networks. In the same spirit, it is possible to use the propagation properties of the heterogeneous networks to diffuse a vaccine on the network, starting from a random node, and using local heuristic propagation rules in which the probability that a node i sends the vaccine to a neighbor j increases with j's degree and decreases with i's, in order to preferentially immunize nodes with large degree (Stauffer and Barbosa, 2006).

In the particular case of the Internet, the propagation of viruses or worms is extremely fast. The development of vaccines occurs on a similar time scale, and dynamical and possibly collaborative response strategies have then to be developed in order to detect the virus and propagate the vaccine quickly.[5] When viruses and antiviruses are competing on the same network, viruses have the inherent advantage of an earlier diffusion starting time. A way to mitigate this advantage consists of disseminating a random set of "honey-pots" in the network (the name honey-pot originates from their function as traps), and an extra set of special edges that can be traversed only by the immunizing agents and connect all honey-pots in a complete graph topology (Goldenberg *et al.*, 2005). Any virus spreading through the network rapidly reaches one of the honey-pots, triggering the defense mechanism: a vaccine is developed and spread immediately to all the other honey-pots, each of which acts as a seed for the propagation of the vaccine. This multiple propagation from the "superhub" formed by the clique of honey-pots transforms the vulnerability of scale-free networks into an advantage, since it uses both the possibility of a fast detection and the fast subsequent propagation. It could moreover be further improved by choosing particular nodes for their location, such as nodes obtained through the process of acquaintance immunization described above (Goldenberg *et al.*, 2005).

---

[5] A detailed analysis of the computer science literature on this topic goes beyond the scope of the present book. (See Costa *et al.* [2005] for a recent work on such approaches.)

9.5 Complex networks and epidemic forecast

In recent years an impressive amount of study has focused on the effect of topological fluctuations on epidemic spreading. After the analysis of the SIS and SIR models, a long list of variations of the basic compartmental structure and the disease parameters have been proposed and studied in both the physics and mathematical epidemiology literature. While a detailed review of these papers goes beyond the scope of this book, a non-exhaustive list of recent works includes Sander *et al.* (2002); Joo and Lebowitz (2004); Dodds and Watts (2004); Olinky and Stone (2004); Petermann and De Los Rios (2004b); Dodds and Watts (2005); Crépey, Alvarez and Barthélemy (2006); Ahn *et al.* (2006); Karsai, Juhász and Iglói (2006); Vázquez (2006b); Zhou *et al.* (2006).

Most studies have mainly focused on systems in which each node of the network corresponds to a single individual; i.e. the network represents a single structured population. Only in recent years has the effect of heterogeneous connectivity patterns been studied for the case in which each node of the system may be occupied by any number of particles and the edges allow for the displacement of particles from one node to the other. Examples in which such a framework turns out to be relevant can be found in reaction–diffusion systems used to model a wide range of phenomena in chemistry and physics and mechanistic metapopulation epidemic models where particles represent people moving between different locations, such as cities or urban areas and the reaction processes between individuals simultaneously present at the same location are governed by the infection dynamics (Anderson and May, 1984; May and Anderson, 1984; Bolker and Grenfell, 1993; 1995; Lloyd and May, 1996; Grenfell and Bolker, 1998; Keeling and Rohani, 2002; Ferguson *et al.*, 2003; Watts *et al.*, 2005).

The development of such approaches and models, especially at the mechanistic level, is based on the detailed knowledge of the spatial structure of the environment, and of transportation infrastructures, movement patterns, and traffic networks. Most of these networks exhibit very heterogeneous topologies as in the case of the airport network among cities, the commuting patterns in inter- and intra-urban areas, and several info-structures (see Chapter 2). This clearly calls for a deeper understanding of the effect of heterogeneous connectivity patterns on these processes.

In more detail, metapopulation models describe spatially structured interacting subpopulations, such as cities, urban areas, or defined geographical regions (Hanski and Gaggiotti, 2004; Grenfell and Harwood, 1997). Individuals within each subpopulation are divided into the usual classes denoting their state with respect to the modeled disease (Anderson and May, 1992) and the compartment dynamics accounts for the possibility that individuals in the same location may get in contact

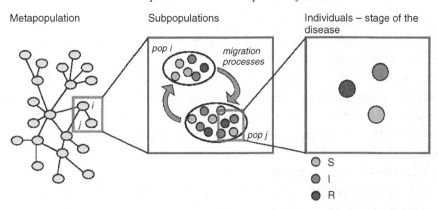

Fig. 9.9. Schematic representation of a metapopulation model. The system is composed of a heterogeneous network of subpopulations or patches, connected by migration processes. Each patch contains a population of individuals who are characterized with respect to their stage of the disease (e.g. susceptible, infected, removed), and identified with a different shade in the picture. Individuals can move from a subpopulation to another on the network of connections among subpopulations.

and change their state according to the infection dynamics. The interaction among subpopulations is the result of the movement of individuals from one subpopulation to the other (see Figure 9.9). A key issue in such a modeling approach lies in the accuracy with which we can describe the commuting patterns or traveling of people. In many instances even complicated mechanistic patterns can be accounted for by effective couplings expressed as a force of infection generated by the infectious individuals in subpopulation j on the individuals in subpopulation i (Bolker and Grenfell, 1995; Lloyd and May, 1996; Earn, Rohani and Grenfell, 1998; Rohani, Earn and Grenfell, 1999; Keeling, 2000; Park, Gubbins and Gilligan, 2002). An explicit mechanistic approach considers a detailed rate of traveling/commuting that defines the mixing subpopulation N_{ij} denoting the number of individuals of the subpopulation i present in the subpopulation j (Keeling and Rohani, 2002; Sattenspiel and Dietz, 1995). To these subpopulations are associated specific traveling and return rates that define different mobility patterns and stationary states, as in the work by Sattenspiel and Dietz (1995).

On the other hand, a simplified mechanistic approach uses a Markovian assumption in which at each time step the movement of individuals is given according to a matrix $p_{j\ell}$ that expresses the probability that an individual in the subpopulation j is traveling to the subpopulation ℓ. The Markovian character corresponds to the fact that we do not label individuals according to their original subpopulation (e.g. *home* in a commuting pattern framework) and at each time step the same traveling probability applies to all individuals in the subpopulation without having memory of their origin. This approach is extensively used for very large populations when the traffic $w_{j\ell}$ among subpopulations is known, by stating that $p_{j\ell} \sim w_{j\ell}/N_j$ where N_j is the number of individuals in node j. Several modeling approaches to the large-scale spreading of infectious diseases use this mobility process based on transportation networks for which it is now possible to obtain detailed data (Baroyan *et al.*, 1969; Rvachev and Longini, 1985; Longini, 1988; Flahault and Valleron, 1991; Grais, Ellis and Glass, 2003; Grais *et al.*, 2004; Hufnagel *et al.*, 2004; Colizza *et al.*, 2007a; Colizza, Pastor-Satorras and Vespignani, 2007b).

In the case of a simple SIR model for the evolution of the disease, the metapopulation approach amounts to writing, for each subpopulation, equations such as (here for the variation of infected individuals)

$$\Delta I_j(t) = K(I_j, S_j, R_j) + \Omega_j(I), \tag{9.65}$$

where the first term of the r.h.s. of the equation represents the variation of infected individuals due to the infection dynamics within the subpopulation j and the second term corresponds to the net balance of infectious individuals traveling in and out of city j. This last term, the transport operator Ω_j, depends on the probability $p_{j\ell}$ that an infected individual will go from city j to city ℓ, and can be simply written as

$$\Omega_j(I) = \sum_{\ell} \left(p_{\ell j} I_\ell - p_{j\ell} I_j \right), \tag{9.66}$$

representing the total sum of infectious individuals arriving in the subpopulation j from all connected subpopulations ℓ, minus the amount of individuals traveling in the opposite directions. Similar equations can be written for all the compartments included in the disease model, finally leading to a set of equations where the transport operator acts as a coupling term among the evolution of the epidemics in the various subpopulations. It is clear that the previous equations represent a solid challenge to an analytical understanding. One way to tackle the issue is to look into simplified versions in which the coupling among the subpopulations is defined by a homogeneous diffusion rate p among all subpopulations linked in the network. The connectivity pattern of the metapopulation network is therefore simply described as a random graph characterized by a degree distribution $P(k)$ with given first and second moments $\langle k \rangle$ and $\langle k^2 \rangle$, respectively. By using a mechanistic approach it is possible to show that, along with the usual threshold condition $R_0 > 1$ on the local reproductive number within each subpopulation, the system exhibits a global invasion threshold that provides a critical value for the diffusion rate p below which the epidemics cannot propagate globally to a relevant fraction of subpopulations. Furthermore the global invasion threshold is affected by the topological fluctuations of the metapopulation network. In particular, the larger the network heterogeneity, the smaller the value of the diffusion rate above which the epidemic may globally invade the metapopulation system. The relevance of these results stems from the fact that invasion thresholds typical of metapopulation models are also affected by heterogeneity and fluctuations in the underlying network, opening the path to a general discussion of threshold effects due to mobility fluxes in real-world networks.

Beyond the analysis of simplified metapopulation models, several approaches are focusing on data-driven implementation, aiming to explore the possibility of developing a computational infrastructure for reliable epidemic forecast and spreading scenario analysis. Taking into account the complexity of real systems in epidemic modeling has proved to be unavoidable, and the corresponding approaches have already produced a wealth of interesting results. But it is clear that many basic theoretical questions are still open. How does the complex nature of the real world affect our predictive capabilities in the realm of computational epidemiology? What are the fundamental limits on epidemic evolution predictability with computational modeling? How do they depend on the accuracy of our description and knowledge of the state of the system? Tackling such questions requires several techniques and approaches. Complex systems and networks analysis, mathematical biology, statistics, non-equilibrium statistical physics and computer science are all playing an important role in the development of a modern approach to computational epidemiology.

10

Social networks and collective behavior

The study of collective behavior in social systems has recently witnessed an increasing number of works relying on computational and agent-based models. These models use very simplistic schemes for the micro-processes of social influence and are more interested in the emerging macro-level social behavior. Agent-based models for social phenomena are very similar in spirit to the statistical physics approach. The agents update their internal state through an interaction with their neighbors and the emergent macroscopic behavior of the system is the result of a large number of these interactions.

The behavior of all of these models has been extensively studied for agents located on the nodes of regular lattices or possessing the ability to interact homogeneously with each other. But as described in Chapter 2, interactions between individuals and the structure of social systems can be generally represented by complex networks whose topologies exhibit many non-trivial properties such as small-world, high clustering, and strong heterogeneity of the connectivity pattern. Attention has therefore recently shifted to the study of the effect of more realistic network structures on the dynamical evolution and emergence of social phenomena and organization. In this chapter, we review the results obtained in four prototypical models for social interactions and show the effect of the network topology on the emergence of collective behavior. Finally, the last section of the chapter is devoted to recent research avenues taking into account that network topology and social interactions may be mutually related in their evolution.

10.1 Social influence

Social influence is at the core of social psychology and deals with the effect of other people on individuals' thoughts and behaviors. It underpins innovation adoption, decision-making, rumor spreading, and group problem solving which all unfold at a macro-level and affect disciplines as diverse as economics, political science, and anthropology. The overarching question in these phenomena is how the micro-processes between individuals are related to the macro-level behavior of groups or whole societies.

In particular, an important issue is the understanding of diversity or uniformity of attitudes and beliefs in a large number of interacting agents. If agents interact via linear assimilative influence, meaning that on average the recipient of influence moves some percentage toward the influencer's position, we immediately face the inevitable outcome of a complete uniformity of attitude in the system. This is not what we observe in reality, as minority opinions persist and we often see polarization of various kinds in politics and culture. The uniformity collapse, however, may be avoided by considering several other features of real-world social systems. First of all, social influence is not always a linear mechanism. An attitude that we could imagine as a continuous variable is not the same as a behavior, which is often a discrete variable. This is exemplified by the continuum of political attitudes with respect to the electoral behavior that often results, in the end, in a binary choice. Environmental influences can also be extremely relevant. Sometimes, social influence can generate a contrast that opposes assimilation. A typical example is provided by social identity and the motive to seek distinctiveness of subgroups. Finally, the patterns of connectivity among individuals may be very complex and foster or hinder the emergence of collective behavior and uniformity.

Several pioneering works use the agent-based models to explore how macro-level collective behavior emerges as a function of the micro-level processes of social influence acting among the agents of the system (Granovetter, 1978; Nowak, Szamrej and Latané, 1990; Axelrod, 1997b). These papers adopt an approach that is akin to the statistical physics approach, and the incursions by statistical physicists into the area of social sciences have become frequent. Nowadays, a vast array of agent-based models aimed at the study of social influence have been defined (see the recent review by Castellano, Fortunato and Loreto [2007]). A first class of models is represented by behavioral models where the attributes of agents are binary variables similar to Ising spins (see Chapter 5) as in the case of the Voter model (Krapivsky, 1992), the majority rule model (Galam, 2002; Krapivsky and Redner, 2003a), and the Sznajd model (Sznajd-Weron and Sznajd, 2000). In other instances additional realism has been introduced. Continuous opinion variables have been proposed by Deffuant *et al.* (2000) (see also Ben-Naim, Krapivsky and Redner [2003]) or by Hegselmann and Krause (2002). Along the path opened by Axelrod (1997b), models in which opinions or cultures are represented by vectors of cultural traits have introduced the notion of bounded confidence: an agent will not interact with any other agent independently of their opinions, but only if they

are close enough. Finally, a large class of models bears similarities with the propagation of epidemics described in Chapter 9. These models aim at understanding
the spread of rumors, information and the sharing of a common knowledge and can
also be used to describe data dissemination or marketing campaigns. In all these
models, the connectivity pattern among agents is extremely important in determining the macro-level behavior. A complete review of all the models is, however,
beyond the scope of this book and in the next sections we focus on some classic
models of social influence where the role of fluctuations and heterogeneities in the
connectivity patterns has been studied in detail.

10.2 Rumor and information spreading

Rumors and information spreading phenomena are the prototypical examples of
social contagion processes in which the infection mechanism can be considered
of psychological origin. Social contagion phenomena refer to different processes
that depend on the individual propensity to adopt and diffuse knowledge, ideas, or
simply a habit. The similarity between social contagion processes and epidemiological models such as those described in Chapter 9 was recognized quite a long
time ago (Rapoport, 1953; Goffman and Newill, 1964; Goffman, 1966). We also
refer the reader interested in further details to the reviews by Dietz (1967) and
Tabah (1999).[1] A simple translation from epidemiological to social vocabulary is
in order here. A "susceptible" individual is an agent who has not yet learned the
new information, and is therefore called "ignorant". An "infected" individual in
epidemiology is a "spreader" in the social context, who can propagate rumors,
habits, or knowledge. Finally "recovered" or "immunized" individuals correspond
to "stiflers" who are aware (adopters) of the rumor (knowledge) but who no longer
spread it. As in the case of epidemic modeling, it is possible to include other
compartments at will such as latents or skeptics, and, when data is available, to
compare the models with real propagation of ideas (Bettencourt *et al.*, 2006). It
is also worth stressing the relevance of these modeling approaches in technological and commercial applications where rapid and efficient spread of information is
often desired. To this end, epidemic-based protocols for information spreading may
be used for data dissemination and resource discovery on the Internet (Vogels, van
Renesse and Birman, 2003; Kermarrec, Massoulie and Ganesh, 2003) or in marketing campaigns using the so-called viral marketing techniques (Moreno, Nekovee
and Pacheco, 2004a; Leskovec, Adamic and Huberman, 2006).

---

[1] An interesting modeling approach which instead considers the propagation of an idea as a cascade phenomenon
similar to those of Chapter 11 can be found in Watts (2002).

Social and physiological contagion processes differ, however, in some important features. Some straightforward dissimilarities can be understood at a qualitative level: first of all, the spread of information is an intentional act, unlike a pathogen contamination. Moreover, it is usually advantageous to access a new idea or information and being infected is no longer just a passive process. Finally, acquiring a new idea or being convinced that a new rumor or information is grounded may need time and exposure to more than one source of information, which leads to the definition of models in which memory has an important role (Dodds and Watts, 2004; 2005). Such differences, which are important at the level of interpretation, do not necessarily change the evolution equations of the spreading model. On the other hand, a non-trivial modification to epidemiological models was proposed by Daley and Kendall (1964) in order to construct a stochastic model for the spread of rumors. This important modification takes into account the fact that the transition from a spreader state to a stifler state is not usually spontaneous: an individual will stop spreading the rumor if he encounters other individuals who are already informed. This implies that the process, equivalent to the recovery process in infectious diseases, is no longer a spontaneous state transition, but rather an interaction process among agents.

Following the parallel between disease and information spreading, the rumors model of Daley and Kendall (1964) (see also Daley and Kendall [1965]; Maki and Thompson [1973]; Daley and Gani [2000]) considers that individuals are compartmentalized in three different categories, ignorant, spreaders, and stiflers, described by their respective densities $i(t) = I(t)/N$, $s(t) = S(t)/N$, and $r(t) = R(t)/N$, where N is the number of individuals in the system and $i(t) + s(t) + r(t) = 1$. As in epidemic spreading, a transition from the ignorant (susceptible) to the spreader (infected) compartment is obtained at a rate λ when an ignorant is in contact with a spreader. On the other hand, and in contrast with basic epidemic models, the transition from spreader to stifler is not spontaneous. On the contrary, the recovery occurs only because of the contact of a spreader with either other spreaders or stiflers according to a transition rate α, introducing a key difference with respect to standard epidemic models.[2] The above spreading process can be summarized by the following set of pairwise interactions

$$
\begin{cases}
I + S \xrightarrow{\lambda} 2S \\
S + R \xrightarrow{\alpha} 2R \\
S + S \xrightarrow{\alpha} R + S.
\end{cases}
\tag{10.1}
$$

---

[2] The inverse of the rate α can be also seen as a measure of the number of communication attempts with other individuals already knowing the rumor before the spreader turns into a stifler.

Note that these reaction processes correspond to the version of Maki and Thompson (1973) in which the interaction of two spreaders results (at rate α) in the conversion of the first spreader into a stifler; in the version of Daley and Kendall (1964), both are turned into stiflers (the last line is then $S + S \overset{\alpha}{\to} 2R$).

In order to provide some insight on the model's behavior, we first consider the homogeneous hypothesis in which all individuals are interacting at each time step with a fixed number of individuals $\langle k \rangle$ randomly selected in the population. The network of contacts is therefore assumed to have no degree fluctuations and the evolution equations of the densities of ignorants, spreaders and stiflers can be written as

$$\frac{di}{dt} = -\lambda \langle k \rangle i(t) s(t) \tag{10.2}$$

$$\frac{ds}{dt} = +\lambda \langle k \rangle i(t) s(t) - \alpha \langle k \rangle s(t) [s(t) + r(t)] \tag{10.3}$$

$$\frac{dr}{dt} = \alpha \langle k \rangle s(t) [s(t) + r(t)]. \tag{10.4}$$

These equations are derived by using the usual homogenous assumption as shown in Chapters 4 and 9. The terms $i(t)s(t)$ and $s(t)[s(t) + r(t)]$ are simply the mass action laws expressing the force of infection and recovery in the population. As anticipated in the model's description, the difference from a Susceptible–Infected–Removed model (see Chapter 9) lies in the non-linear decay rate $s(t)[s(t)+r(t)]$ of Equation (10.3), or equivalently in the right-hand side of Equation (10.4). Despite this non-linearity, the infinite time limit of these equations can be obtained. In the stationary regime, we can set the time derivatives equal to 0 which implies that $\lim_{t \to \infty} s(t) = 0$: individuals either have remained ignorants or are aware of the rumor but have stopped spreading it. The value of $r_\infty \equiv \lim_{t \to \infty} r(t)$ allows us to understand whether the information propagation process has reached a finite fraction of the population ($r_\infty > 0$) or not ($r_\infty = 0$), starting from e.g. one single initial spreader. This quantity defines the *reliability* r_∞ of the process. Given the initial conditions $s(0) = 1/N$, $i(0) = 1 - s(0)$ and $r(0) = 0$, the equation for the density of ignorants can be formally integrated, yielding

$$i(t) = i(0) \exp \left[-\lambda \langle k \rangle \int_0^t d\tau s(\tau) \right]. \tag{10.5}$$

The insertion of $s(t) + r(t) = 1 - i(t)$ and of Equation (10.2) into Equation (10.4) yield the following relation valid for any time t

$$\int_0^t dt \frac{dr}{dt} = \alpha \langle k \rangle \int_0^t d\tau s(\tau) + \frac{\alpha}{\lambda} \int_0^t dt \frac{di}{dt}, \tag{10.6}$$

or equivalently

$$\alpha \langle k \rangle \int_0^t \mathrm{d}\tau s(\tau) = r(t) - r(0) - \frac{\alpha}{\lambda}[i(t) - i(0)].$$ (10.7)

Equation (10.5) leads then straightforwardly to

$$i_\infty \equiv \lim_{t \to \infty} i(t) = \exp(-\beta r_\infty),$$ (10.8)

with $\beta = 1 + \lambda/\alpha$ and where we have used $r(0) = 0$, $i(0) \approx 1$ and $i_\infty + r_\infty = 1$. The transcendental equation for the density of stiflers at the end of the spreading process finally reads (Sudbury, 1985)

$$r_\infty = 1 - e^{-\beta r_\infty}.$$ (10.9)

The solutions of this equation can be obtained similarly to those of Equation (6.10) for the percolation phenomenon. It can indeed be written as $r_\infty = F(r_\infty)$, where $F(x) = 1 - \exp(-\beta x)$ is a monotonously increasing continuous function with $F(0) = 0$ and $F(1) < 1$. A non-zero solution can therefore exist if and only if the first derivative $F'(0) > 1$, leading to the inequality

$$\left. \frac{\mathrm{d}}{\mathrm{d}r_\infty}(1 - e^{-\beta r_\infty}) \right|_{r_\infty=0} > 1,$$ (10.10)

which yields the condition on the spreading rate for the obtention of a finite density of stiflers at large times

$$\frac{\lambda}{\alpha} > 0.$$ (10.11)

Interestingly, and in strong contrast with the case of epidemic spreading, this condition is always fulfilled for any arbitrary positive spreading rate. The introduction of the non-linear term by Daley and Kendall (1964) in the description of the propagation has therefore the far-reaching consequence of removing the epidemic threshold: the rumor has always a non-zero probability of pervading a macroscopic fraction of the system whatever the values of the rates α and λ.

Information, like diseases, propagates along the contacts between individuals in the population. The transmission of information or rumors uses various types of social or technological networks (collaboration or friendship networks, email networks, telephone networks, WWW, the Internet...), which typically are far from being homogeneous random networks. Therefore the definition and structure of the contact network will be of primary importance in determining the properties of the spreading process. The first studies of rumors models in complex networks have focused on the effect of clustering and small-world properties. Zanette (2001) (see also Zanette [2002]) has investigated the spreading phenomenon (with $\lambda = \alpha = 1$) in Watts–Strogatz (WS) small-world networks through numerical simulations. In particular, the final fraction of stiflers, r_∞, has been measured as a function of the disorder in the network structure, quantified by the rewiring probability p of the WS network. As p increases, more and more shortcuts exist between distant regions of the initial one-dimensional chain, and one can expect that the information will propagate better. In fact, Zanette (2001) provided evidence of a striking transition phenomenon between a regime, for $p < p_c$, in which the rumor does not spread out of the neighborhood of its origin, and a regime where it is finally known to a finite fraction of the population, as shown in Figure 10.1. The value of p_c, which is close to 0.2 for a network with $\langle k \rangle = 4$, decreases when $\langle k \rangle$ increases. The rationale behind the appearance of this non-equilibrium phase transition is the following: at small p, the network is highly locally clustered (just as the one-dimensional chain at $p = 0$), so that many interactions take place on triangles. A newly informed spreader will therefore interact with large probability with other neighbors of the individual who has informed him, thereby easily becoming a stifler and stopping the propagation. The "redundant" links forming triangles therefore act as an inhibiting structure and keep the information localized. When enough links are instead devoted to long-range random connections, the information can spread out of a given neighborhood before the spreaders become stiflers. It is particularly interesting to notice that this dynamic transition occurs at a finite value of the disorder p, in contrast with the case of equilibrium dynamics described in Chapter 5, for which any strictly positive p leads to a transition. The dynamics at $p > p_c$ presents an additional feature which is also present in epidemics: for a given network and a given initial condition, different stochastic realizations of the spreading can lead either to a rapid extinction ($r_\infty = 0$), or to a propagation affecting a macroscopic

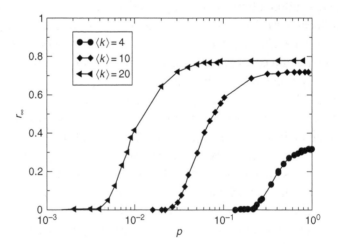

Fig. 10.1. Reliability of the rumor spreading process as a function of p, for Watts–Strogatz networks of size $N = 10^5$ and various values of the average degree $\langle k \rangle$. Data from Zanette (2001).

fraction of the population. The average over all realizations of r_∞ is therefore positive but in fact hides a bimodal distribution where the fraction of realizations with finite r_∞ naturally grows as p increases.

In order to consider how the model of Maki and Thompson (1973) is affected by the possible degree heterogeneity of the network on which the spreading takes place, the densities of ignorants, spreaders, and stiflers in each class of degree k have to be introduced, namely i_k, s_k and r_k. The evolution equations for these densities can be written as

$$\frac{di_k}{dt} = -\lambda k i_k(t) \sum_{k'} s_{k'}(t) \frac{k'-1}{k'} P(k'|k)$$

$$\frac{ds_k}{dt} = +\lambda k i_k(t) \sum_{k'} s_{k'}(t) \frac{k'-1}{k'} P(k'|k)$$

$$\qquad\qquad - \alpha k s_k(t) \sum_{k'} [s_{k'}(t) + r_{k'}(t)] P(k'|k)$$

$$\frac{dr_k}{dt} = \alpha k s_k(t) \sum_{k'} [s_{k'}(t) + r_{k'}(t)] P(k'|k),$$

(10.12)

where $P(k'|k)$ is the conditional probability that an edge departing from a vertex of degree k arrives at a vertex of degree k' and the terms $(k'-1)/k'$ take into account that each spreader must have one link connected to another spreader, from which it received the information. Even in the case of uncorrelated networks with $P(k'|k) = k' P(k')/\langle k \rangle$, the non-linear term in the evolution of the stifler density makes these equations much more involved than their counterparts for the SIR model and an analytical solution has not yet been derived.[3] Numerical simulations (Moreno, Nekovee and Vespignani, 2004b; Moreno, Nekovee and Pacheco, 2004a) nonetheless allow the spread of information to be characterized in this model. The final density of individuals who are aware of the rumor, i.e. the global reliability r_∞ of the information/rumor diffusion process, increases as expected if α decreases. More interestingly, homogeneous networks allow for larger levels of reliability than heterogeneous ones, for the same value of the average degree and of the parameters λ and α (Liu, Lai and Ye, 2003; Moreno, Nekovee and Vespignani, 2004b). This result may seem surprising as epidemic propagation is usually enhanced by the presence of hubs. For information spreading, however, hubs produce conflicting effects: they potentially allow a large number of nodes to be reached, but lead as well to many spreader–spreader and spreader–stifler interactions, thus turning themselves into stiflers before being able to inform all their neighbors. Owing to

---

[3] An additional linear term corresponding to the fact that spreaders could spontaneously become stiflers can be added (Nekovee *et al.*, 2007), leading to the appearance of an epidemic threshold in homogeneous networks, but this modification does not allow for an analytical solution of Equations (10.12).

the percolation properties of heterogeneous networks, the inhibition of propagation for a small fraction of the hubs will fragment the network in non-communicating components, isolating many nodes that will therefore never be informed. Numerical simulations also allow the final density of ignorants in each degree class to be measured, showing that it decays exponentially with the degree. This result implies that hubs acquire the rumor or information very efficiently: from the point of view of the hubs, high reliability is obtained even at large α.

While the reliability of the information spreading process is clearly important in technological applications, its scalability represents a crucial issue as well (Vogels *et al.*, 2003). Information systems and infrastructures have become very large networks and, while the number of messages may not be a concern in direct social contacts, the load L (defined as the average number of messages per node sent during the spreading process) imposed on a technological network when transmitting information should be kept at a minimum. The trade-off between a large reliability r_∞ and a small load L is measured through the *efficiency* $E = r_\infty/L$. From this perspective, scale-free networks lead to slightly more efficient processes than homogeneous random networks (see Figure 10.2). Additionally, epidemic-like spreading process achieves a better efficiency, for a broad range of α-values, than

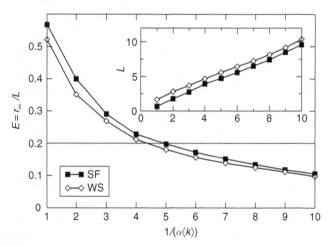

Fig. 10.2. Efficiency of the rumor spreading process as a function of $1/(\alpha\langle k\rangle)$, for networks of size $N = 10^4$ and average degree $\langle k\rangle = 6$. The homogeneous Watts–Strogatz network with rewiring probability $p = 1$ (diamonds) yields a lower efficiency than the heterogeneous one (SF, black squares). Larger α corresponds to smaller load and smaller reliability but larger efficiency. The straight horizontal line corresponds to the efficiency of a broadcast algorithm, for which each node passes the message deterministically to all its neighbors (except the one from which it received the information): in this case $r_\infty = 1$ and $L = \langle k\rangle - 1$. The inset shows the generated load as a function of $1/(\alpha\langle k\rangle)$. Data from Moreno, Nekovee and Vespignani (2004b).

the simplest broadcast algorithm in which each node transmits the information deterministically to all its neighbors. Finally, the hubs also play a role if they are taken as seeds, i.e. as the initial source of information. The degree of the initially informed node does not affect the final reliability of the process, but a larger degree yields a faster growth of $r(t)$. For intermediate stages of the process, larger densities of stiflers are therefore obtained. These results highlight the importance of well-connected hubs in the efficiency of rumor/information spreading processes: if the process aims at a given level of reliability, starting from well-connected nodes allows this level to be reached quickly, i.e. at smaller costs or load.

10.3 Opinion formation and the Voter model

As mentioned earlier in this chapter, numerous models have been devised to describe the evolution of opinions and cultural traits in a population of interacting agents. A classic example of collective behavior is given by the emergence of consensus in a population of individuals who can a priori have contradictory opinions and interact pairwise. While several models have been put forward to study this phenomenon, the Voter model represents the simplest modeling framework for the study of the onset of consensus due to opinion spreading.

The Voter model is defined on a population of size N in which each individual i has an opinion characterized by a binary variable $s_i = \pm 1$: only two opposite opinions are here allowed (for example a political choice between two parties). The dynamical evolution, starting from a random configuration of opinions, is the following: at each elementary step, an agent i is randomly selected, chooses one of his neighbors j at random and adopts his opinion, i.e. s_i is set equal to s_j (one time step corresponds to N such updates). This process therefore mimics the homogenization of opinions but, since interactions are binary and random, does not guarantee the convergence to a uniform state. When the connectivity pattern of individuals can be modeled as a regular D-dimensional lattice, the update rules lead to a slow coarsening process, i.e. to the growth of spatially ordered regions formed by individuals sharing the same opinion: large regions tend to expand while small ones tend to be "convinced" by the larger neighboring groups of homogeneous agents (see Figure 10.3). The dynamics is therefore defined by the evolution of the frontiers or "interfaces" between these ordered regions. For $D < 2$, the density of interfaces decays with time t (measured as the number of interactions per individual) as $t^{(D-2)/2}$ (Krapivsky, 1992; Frachebourg and Krapivsky, 1996). A logarithmic decay is observed for $D = 2$ (Dornic *et al.*, 2001), while a stationary active state with coexisting domains is reached for $D > 2$. In this last case, consensus (i.e. a global homogeneous state) is, however, obtained for finite systems; in a population of size N, fluctuations lead to a consensus after a typical time $\tau \propto N$.

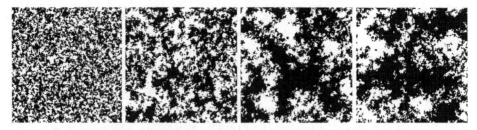

Fig. 10.3. Evolution of the Voter model for agents located on the sites of a two-dimensional square lattice of linear size $L = 200$. Black dots represent opinions $+1$, while empty spaces are left where opinions take the value -1. From left to right the system, starting from a disordered configuration, is represented at times $t = 5, 50, 500, 1000$. Adapted from Dornic *et al.* (2001).

As most interactions in the social context do not take place on regular lattices, recent studies have investigated the dynamics of the Voter model for agents interacting according to various complex connectivity patterns (Castellano *et al.*, 2003; Vilone and Castellano, 2004; Suchecki, Eguíluz and San Miguel, 2005a; Wu and Huberman, 2004; Sood and Redner, 2005; Castellano *et al.*, 2005; Suchecki, Eguíluz and San Miguel, 2005b; Castellano, 2005; Antal, Redner and Sood, 2006). Because of the lack of a Euclidean distance defining spatial domains and interfaces, coarsening phenomena are not well defined on general networks, in contrast with the case of regular lattices. The evolution of the system can, however, be measured by the fraction $n_A(t)$ of active bonds at time t, i.e. of bonds connecting sites with opposite values of the variable s_i, and by the survival probability $P_S(t)$ of a process, which is the probability that the fully ordered state has not been reached up to time t. The evolution of the time to reach a completely ordered state (time to consensus) with the size of the population is also clearly of great interest.

In this context, Castellano *et al.* (2003) have investigated the Voter model for individuals interacting along a Watts–Strogatz small-world network. After an initial transient, $n_A(t)$ exhibits a plateau whose length increases with the system size N, and which is followed for any finite N by an exponential decrease to 0. In the end the net result is an ordering time scale smaller than in the one-dimensional lattice, for which $n_A(t)$ decreases steadily as a power-law. Moreover, the height of the plateau increases as the parameter p of the Watts–Strogatz model increases, i.e. as the randomness of the network is increased. This dynamical process can be understood along the lines of Section 5.3.1. In the one-dimensional lattice, ordering is obtained through random diffusion of active bonds, which annihilate when they meet, resulting in $n_A(t) \sim 1/\sqrt{t}$. At finite p, the shortcuts inhibit this random diffusion through the influence of distant nodes with different opinions. This creates a "pinning" phenomenon, where the shortcuts act like obstacles to the homogenization process. The crossover is reached when the typical size of an ordered domain, namely $1/n_A(t)$, reaches the typical distance between shortcuts $\mathcal{O}(1/p)$ (see Chapter 3). The crossover time is thus given by $t^* \sim p^{-2}$, with $n_A(t^*) \sim p$. In addition, the various curves giving $n_A(t, p)$ can be rescaled by

$$n_A(t, p) = p\mathcal{G}(tp^2), \tag{10.13}$$

where $\mathcal{G}$ is a scaling function behaving as $\mathcal{G}(x) \sim \sqrt{x}$ for $x \ll 1$ and $\mathcal{G}(x) = \text{const}$ for large x (see Figure 10.4).

Further insight is given by the study of the survival probability $P_S(t)$, which decreases as $\exp[-t/\tau(N)]$, with $\tau(N) \propto N$ (Castellano *et al.*, 2003), as shown also analytically in the annealed version of the small-world networks (Vilone and Castellano, 2004).[4] On the other hand, the fraction of active bonds *averaged only over surviving runs*, $n_A^S(t)$, reaches a finite value at large times. In fact, the long-time decay of $n_A(t) = P_S(t)n_A^S(t)$ is solely due to the exponential decay of $P_S(t)$, showing that the system never orders in the thermodynamic limit, but retains a finite density of active links. The picture is therefore very different from the usual coarsening occurring in finite dimensions for which $n_A^S(t)$ steadily decays as a power-law of time, showing that the small-world effect created by shortcuts strongly affects the behavior of the Voter model.

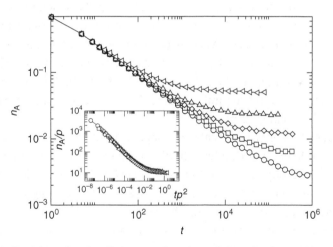

Fig. 10.4. Density of active links as a function of time for Watts–Strogatz networks of size $N = 10^5$ and various rewiring probabilities p: $p = 0.0002$ (circles), $p = 0.0005$ (squares), $p = 0.001$ (diamonds), $p = 0.002$ (triangles up) and $p = 0.005$ (triangles left). The inset displays $n_A(t, p)/p$ vs tp^2, showing the validity of Equation (10.13). Data from Castellano *et al.* (2003).

---

[4] In the annealed small-world graphs, the long-range links are rewired randomly at each time step.

Various studies, following these first investigations, have focused on the effect of the network heterogeneity and on the behavior of the time needed to reach consensus as a function of the network size N. As noted by Suchecki *et al.* (2005a) and Castellano (2005), the definition of the Voter model has to be made more precise if the individuals are situated on the nodes of a heterogeneous network. A randomly chosen individual has a degree distribution given by the network's distribution $P(k)$, while the *neighbor* of a randomly chosen node has degree distribution $kP(k)/\langle k \rangle$, and hence a higher probability of being a hub. Since, in the update rule, the two interacting nodes do not play symmetric roles, such a difference can be relevant, and leads to the following possible rules: (i) in the original definition, a randomly selected node picks at random one of its neighbors and adopts its state or opinion; (ii) in the link update rule, a link is chosen at random and the update occurs in a random direction along this link; (iii) in the *reverse* Voter model, known as Moran model, a randomly chosen *neighbor j* of a randomly chosen node i adopts the opinion of i. These three definitions are equivalent on a lattice, but may induce relevant changes in heterogeneous networks since the probability for a hub to update its state will vary strongly from one rule to the other (Suchecki *et al.*, 2005a; Castellano, 2005).

As shown in Figure 10.5, the essential features of the dynamical process are similar on heterogeneous and homogeneous networks: $P_S(t)$ decays exponentially with a characteristic time scale $\tau(N)$ depending on the system size, while $n_A^S(t)$ reaches a plateau: the system shows incomplete ordering, in contrast with the case of finite-dimensional lattices.[5] Various behaviors for $\tau(N)$ are, however, obtained depending on the updating rules. For agents interacting on Barabási–Albert networks, numerical simulations yield $\tau(N) \sim N^\nu$ with $\nu \approx 0.88$ for the original Voter model, and $\tau(N) \sim N$ for the link-update rule (Suchecki *et al.*, 2005a; Castellano *et al.*, 2005; Castellano, 2005).

For arbitrary (uncorrelated) scale-free networks with degree distribution $P(k) \sim k^{-\gamma}$, it is possible to obtain analytically the behavior of $\tau(N)$ through a mean-field approach which groups the nodes in degree classes (Sood and Redner, 2005), in the same way as seen in previous chapters. We start by defining ρ_k as the fraction of nodes having opinion $+1$ among the nodes of degree k. This density evolves because of the probabilities $P(k; - \rightarrow +)$ and $P(k; + \rightarrow -)$ that a node of degree k changes state, respectively from -1 to $+1$ or the opposite. In the original Voter model, the change $- \rightarrow +$ occurs in an update if the first randomly selected node has degree k and opinion -1 (which has probability $P(k)(1 - \rho_k)$) and if its randomly chosen neighbor has opinion 1. In an uncorrelated network, this

---

[5] Suchecki *et al.* (2005b) have also observed that the Voter model can even display a coarsening dynamics on a particular scale-free structured network, which however is known to have an effective dimension equal to 1 (Klemm and Eguíluz, 2002b).

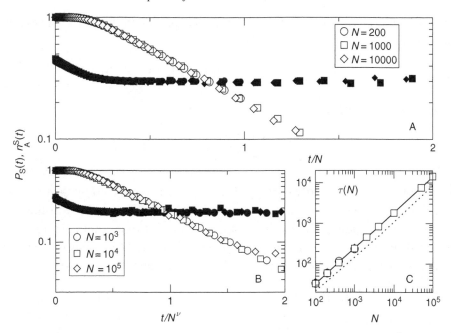

Fig. 10.5. Survival probability $P_S(t)$ (empty symbols) and fraction $n_A^S(t)$ of active bonds in surviving runs (filled symbols) for the voter model on (A) a random homogeneous graph with $\langle k \rangle = 10$ and (B) a Barabási–Albert (BA) network with $\langle k \rangle = 6$, for various sizes. Time is rescaled (A) by N and (B) by N^ν with $\nu \approx 0.88$. The plot (C) shows $\tau(N)$ vs N for the BA network; the continuous line corresponds to N^ν, the dotted one to $N/\ln(N)$: both scalings are compatible with the data over three orders of magnitude. Data from Castellano *et al.* (2005).

probability reads

$$\sum_{k'} \frac{k' P(k')}{\langle k \rangle} \rho_{k'}, \tag{10.14}$$

since the probability for the neighbor to have degree k' is $k' P(k')/\langle k \rangle$, and we sum over all possible degrees k'. We thus obtain

$$P(k; - \to +) = P(k)(1 - \rho_k) \sum_{k'} \frac{k' P(k')}{\langle k \rangle} \rho_{k'}, \tag{10.15}$$

and similarly

$$P(k; + \to -) = P(k)\rho_k \sum_{k'} \frac{k' P(k')}{\langle k \rangle} (1 - \rho_{k'}). \tag{10.16}$$

The number N_k^+ of nodes of degree k and state $+1$ thus evolves according to $dN_k^+/dt = N(P(k; - \to +) - P(k; + \to -))$. Since $\rho_k = N_k^+/N_k$, we obtain the evolution equation for ρ_k in the mean-field approximation and for uncorrelated networks

$$\frac{d\rho_k}{dt} = \frac{N}{N_k}[P(k; - \to +) - P(k; + \to -)], \qquad (10.17)$$

which can be rewritten as

$$\frac{d\rho_k}{dt} = \sum_{k'} \frac{k'P(k')}{\langle k \rangle}(\rho_{k'} - \rho_k). \qquad (10.18)$$

This shows that the quantity $\omega = \sum_k k\rho_k P(k)/\langle k \rangle$ is conserved by the dynamics, and that a stationary state with $\rho_k = \omega$ is obtained. In contrast with the study of epidemic spreading detailed in Chapter 9, the long time limit of ρ cannot be obtained self-consistently, but a recursion equation for the consensus time $\tau(N)$ can be written as a function of the densities ρ_k (Sood and Redner, 2005):

$$\tau(\{\rho_k\}) = \sum_k P(k; - \to +)[\tau(\rho_k + \delta_k) + \delta t]$$

$$+ \sum_k P(k; + \to -)[\tau(\rho_k - \delta_k) + \delta t]$$

$$+ Q(\{\rho_k\})[\tau(\{\rho_k\}) + \delta t], \qquad (10.19)$$

where $Q(\{\rho_k\}) = 1 - \sum_k P(k; - \to +) - \sum_k P(k; + \to -)$ is the probability that no update takes place during δt, and $\delta_k = 1/(NP(k))$ is the change in ρ_k when a node of degree k changes opinion. Using the expressions of $P(k; - \to +)$, $P(k; + \to -)$ and expanding to second order in δ_k yields

$$\sum_k (\rho_k - \omega)\partial_k \tau - \frac{1}{2N} \sum_k \frac{1}{P(k)}(\rho_k + \omega - 2\omega\rho_k)\partial_k^2 \tau = 1, \qquad (10.20)$$

where we have $\delta t = 1/N$ as the elementary time-step, and $\partial_k \equiv \partial/\partial\rho_k$. The first term vanishes at long times, thanks to the convergence of ρ_k to ω. Using $\partial_k = kP(k)/\langle k \rangle \partial_\omega$, Equation (10.20) becomes

$$\frac{\langle k^2 \rangle}{\langle k \rangle^2}\omega(1 - \omega)\partial_\omega^2 \tau = -N, \qquad (10.21)$$

which can be integrated with the conditions that τ vanishes for both $\omega = 0$ and $\omega = 1$ (these values of ω indeed correspond to the two possible consensus states, from which the system cannot evolve). We finally obtain

$$\tau(N) = -N\frac{\langle k \rangle^2}{\langle k^2 \rangle}[\omega \ln \omega + (1 - \omega) \ln(1 - \omega)]. \qquad (10.22)$$

Table 10.1 Scaling of the time to consensus $\tau(N)$ for uncorrelated scale-free networks with distribution $P(k) \sim k^{-\gamma}$, as a function of the exponent γ.

γ	$\gamma > 3$	$\gamma = 3$	$3 > \gamma > 2$	$\gamma = 2$	$\gamma < 2$
$\tau(N)$	N	$N/\ln N$	$N^{(2\gamma-4)/(\gamma-1)}$	$(\ln N)^2$	$\mathcal{O}(1)$

For an initial random uncorrelated configuration $\rho_k(0) = \rho(0)$, $\omega = \rho(0)$ and

$$\tau(N) = -N \frac{\langle k \rangle^2}{\langle k^2 \rangle} [\rho(0) \ln \rho(0) + (1 - \rho(0)) \ln(1 - \rho(0))]. \qquad (10.23)$$

The dependence of the convergence time on the system size N therefore depends on the fluctuations of the degree distribution and can be computed by using Equation (2.3) and the results of Appendix 1. In Table 10.1 we report the scaling of the convergence time as a function of the system size N in random scale-free networks with power-law exponent γ. The result obtained for $\gamma = 3$ indicates a logarithmic scaling $\tau(N) \sim N/\ln N$, apparently in contradiction to the power-law best fit $\tau(N) \sim N^{\nu}$ with $\nu \approx 0.88$ obtained by Suchecki et al. (2005a) and Castellano et al. (2005). On the other hand, the logarithmic fit is still compatible and the small logarithmic corrections are hard to validate on the three orders of magnitude accessible from simulations (see Figure 10.5).

A similar analysis can be carried out for the reverse Voter model, confirming the relevance of the updating rule in the case of heterogeneous networks (Castellano, 2005). The analytical mean-field approach leads indeed to

$$\tau(N) = -N \langle k \rangle \left\langle \frac{1}{k} \right\rangle [\rho(0) \ln \rho(0)(1 - \rho(0)) \ln(1 - \rho(0))], \qquad (10.24)$$

and thus to a completely different scaling with respect to the exponents of the network degree distribution: $\tau(N) \sim N$ for $\gamma > 2$, $\tau(N) \sim N \ln N$ for $\gamma = 2$, and $\tau(N) \sim N^{1/(\gamma-1)}$ for $1 < \gamma < 2$ (Castellano, 2005).

In summary, the dynamics of the Voter model is strongly different for agents interacting on the nodes of a network with respect to the case of regular lattices. Interestingly, the possible heterogeneous structure of the interaction network only marginally affects the dynamical process: the presence of hubs does not modify the absence of ordering in the thermodynamic limit but only the scaling of the ordering time for finite sizes. In this respect, the Voter model is very different from processes such as epidemics spreading for which the divergence of $\langle k^2 \rangle$ has strong consequences, as explained in Chapter 9.

Let us finally note that interesting extensions of the Voter model have recently been studied. Castelló, Eguíluz and San Miguel (2006) introduce the possibility of an intermediate state '±1', through which an individual must pass when changing opinion from +1 to −1 or the opposite.[6] This modification is in the spirit of the *Naming Game* model, in which an agent can have an arbitrary number of possible opinions at the same time, and which has been recently extensively studied for agents interacting on networks with various topologies (Steels, 1996; Lenaerts *et al.*, 2005; Baronchelli *et al.*, 2006; Dall'Asta and Baronchelli, 2006). It turns out that such a modification of the Voter model leads to a faster convergence to consensus for agents interacting on a small-world network, with $\tau \sim \ln N$, essentially by avoiding the "pinning" phenomenon due to shortcuts.

10.4 The Axelrod model

In the Voter model, each agent possesses a unique opinion which can take two discrete values only. While an interesting extension consists of considering opinions as continuous variables (Deffuant *et al.*, 2000; Ben-Naim *et al.*, 2003), it is clear that social influence and interaction do not act on a single dimensional space. In general, social influence applies to a wide range of cultural attributes such as beliefs, attitudes and behavior, which cannot be considered in isolation. Social influence is more likely when more than one of these attributes are shared by the interacting individuals, and it may act on more than a single attribute at once. It is in this spirit that Axelrod (1997b) has proposed a simple but ambitious model of social influence that studies the convergence to global polarization in a multi-dimensional space for the individual's attributes.

In the Axelrod model each agent is endowed with a certain number F of cultural features defining the individual's attributes, each of those assuming any one of q possible values. The Voter model is thus a particular case of Axelrod's model, with $F = 1$ and $q = 2$. The model takes into account the fact that agents are likely to interact with others only if they already share cultural attributes, and that the interaction then tends to reinforce the similarity. The precise rules of the model are therefore the following: at each time step, two neighboring agents are selected. They interact with probability proportional to the number of features (or attributes) for which they share the same value (among the q possible ones).[7] In this case, one of the features for which they differ is chosen, and one of the agents selected at random adopts the value of the other. The F different attributes are therefore not

---

[6] See also Dall'Asta and Castellano (2007) for a slightly different mechanism which also takes into account a memory effect.

[7] In some versions of the model, the probability of interaction is simply 1 if at least one feature is shared, and 0 otherwise.

completely independent in that the evolution of each feature depends on the values of the others (and in particular the possible agreement with other agents).

Numerical simulations performed for agents connected as in a regular two-dimensional grid show the influence of the number of features F per agent and of the cultural variability q on the final state of the system. The local convergence rules can lead either to global polarization or to a culturally fragmented state with coexistence of different homogeneous regions. As F increases, the likelihood of convergence towards a globally homogeneous state is enhanced, while this likelihood decreases when q increases (Axelrod, 1997b). Indeed, when more features are present (at fixed q), there is a greater chance that two neighboring agents share at least one of them, and therefore the probability of interaction is enhanced. At larger q instead, it is less probable for each feature to have a common value in two agents, and they interact with smaller probability. Castellano, Marsili and Vespignani (2000) have studied in detail the phase diagram of the model, uncovering a non-equilibrium phase transition between the ordered (homogeneous opinions) and disordered (culturally fragmented) phases. The dynamics evolves through a competition between the initially disordered situation and the tendency of the interaction to homogenize the agents' attributes, leading to a coarsening phenomenon of regions in which the agents have uniform features. The system finally reaches an absorbing state in which all links are inactive: each link connects agents who either have all features equal or all different, so that no further evolution is possible. At small q, the absorbing state reached by the system is expected to be homogeneous, while at large q one expects a highly fragmented state. The order parameter which determines the transition is defined by the average size of the largest homogeneous region in the final state, $\langle S_{max} \rangle$. At fixed F, a clear transition is observed as q crosses a critical value q_c: for $q < q_c$, $\langle S_{max} \rangle$ increases with the system size L and tends to L^2 (for a two-dimensional lattice of linear size L, the number of sites or agents is equal to $N = L^2$), while $\langle S_{max} \rangle / L^2 \to 0$ for $q > q_c$. The transition is continuous for $F = 2$, discontinuous for $F > 2$, and can be analyzed through mean-field approaches (Castellano *et al.*, 2000). For agents interacting on a one-dimensional chain, a mapping to a reaction–diffusion process shows that the critical value q_c is equal to F if the initial distribution of values taken by the features is uniform, and to F^2 for an initial Poisson distribution (Vilone, Vespignani and Castellano, 2002).

Interestingly, cultural drift and global interactions can also be introduced into the model in a simple way. Random changes in the attributes' values at a certain rate r, or the addition of an interaction of the agents with a global field, account for these phenomena (Klemm *et al.*, 2003a; Klemm *et al.*, 2005; González-Avella *et al.*, 2006). An interesting result in this context is that even a very small rate of random changes drives the population of agents to an homogeneous state for any value of F and q by potentially allowing two neighboring agents who did not share any feature to interact again.

The possible effect of long-distance interactions perturbing the grid ordering of agents was also briefly discussed in the original work of Axelrod (1997b). The evolution of Axelrod's model for agents interacting on complex networks with various topological characteristics has subsequently been investigated in detail through numerical simulations by Klemm *et al.* (2003b). In the case of Watts–Strogatz (two-dimensional) small-world networks, the disorder is shown to favor the ordered (homogeneous) state: the critical value q_c grows as the disorder p is increased (see Figure 10.6). On random scale-free networks, the hubs enhance the spreading of cultural traits so much that the value of q_c diverges. At any finite size, an effective transition from the ordered to the fragmented state is observed at a pseudo-critical value $q_c(N)$, which diverges as N^β (with $\beta \sim 0.4$ for Barabási–Albert networks, see Figure 10.7) with the size N of the network (Klemm *et al.*, 2003b). In other words, the existence of a culturally fragmented phase is hindered by the presence of hubs and is no longer possible in the thermodynamic limit. Once again, this behavior is reminiscent of the results obtained for the Ising model for which the critical temperature diverges for scale-free networks (see Section 5.3.2), owing to the strong polarization role of the hubs. Finally, we refer to Centola *et al.* (2007) and to Section 10.6 for the interesting case in which the network of interaction itself may evolve on the same time scale as the agents' features.

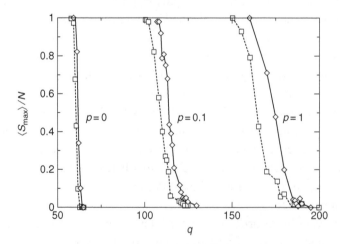

Fig. 10.6. Average order parameter $\langle S_{\max} \rangle / N$ as a function of q for three different values of the small-world parameter p. System sizes are $N = 500^2$ (squares) and $N = 1000^2$ (diamonds), number of features $F = 10$. Data from Klemm *et al.* (2003b).

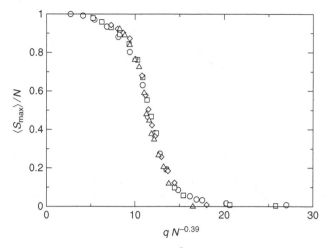

Fig. 10.7. Plot of $\langle S_{max}\rangle/N$ versus $qN^{-\beta}$ in random scale-free networks for $F = 10$, for different system sizes: 1000 (circles), 2000 (squares), 5000 (diamonds), and 10000 (triangles). Data from Klemm *et al.* (2003b).

10.5 Prisoner's dilemma

Another interesting and often puzzling social behavior that has attracted the attention of scientists across different research areas lies in the emergence of cooperation between a priori selfish interacting agents. One of the best-known paradigms of a system able to display both cooperative and competitive behaviors is given by the prisoner's dilemma game (Rapoport and Chammah, 1965; Axelrod and Hamilton, 1981). Indeed, the basic two-players iterated prisoners dilemma has been described as the *Escherichia coli* of the social sciences and it is one of the basic framework for the study of evolution and adaptation of social behavior (Axelrod, 1997a). In this model, agents (players) interact pairwise, and have at each interaction (or round of the game) the option either to cooperate (C strategy) or to defect (D strategy). They obtain various payoffs according to these choices. If both players cooperate, they earn a quantity R ("reward"), if they both defect they earn P ("punishment") while if they use different strategies, the defector earns T ("temptation" to defect) and the cooperator S ("sucker's payoff"), as sketched in Figure 10.8.

The language and payoff matrix have their inspiration in the following situation from which the name of the model derives. Two suspects A and B are arrested by the police. The police do not have sufficient evidence for a conviction and separate them in the hope that each one of them may testify for the prosecution of the other. If both A and B remain silent (they cooperate with each other) they both receive a sentence for minor charges. If one of the two betrays the other, the betrayer goes

Player A / Player B	Cooperation	Defection
Cooperation	R $\quad$ R	R $\quad$ T $\quad$ S
Defection	S $\quad$ T	S $\quad$ P $\quad$ P

Fig. 10.8. Sketch of the prisoner's dilemma rules for two players A and B. If the two players cooperate (C strategy), they both receive the reward R. If they both defect (D strategy), they are both punished and receive only P. If one betrays the other by defecting while the other cooperates, the traitor receives T (temptation to defect) and the cooperator gets S.

free while the other receives a heavy sentence. This is the case in which one of the two agents has defected the other while the second one is still cooperating. Finally both of them could betray and get a sentence intermediate between the two previous cases. The main question is therefore what would be the best strategy for the two suspects. If $T > R > P > S$, an interesting paradox arises. In any given round, it is in each agent's interest to defect, since the D strategy is better than the C one, regardless of the strategy chosen by the other player. On the other hand, on average, if both players systematically use the D strategy, they earn only P while they would receive R if they both chose to cooperate: in the long run, cooperation is therefore favored.

While the initial definition involves two players engaged in successive encounters, Nowak and May (1992) have considered the evolution of a population of players situated on the nodes of a two-dimensional square lattice, thus introducing local effects. For simplicity, the various payoffs are taken as $P = S = 0$, $R = 1$, and the only parameter is $T > 1$, which characterizes the advantage of defectors against cooperators. Moreover, at each round (time) t agents play with their neighbors, and each agent tries to maximize its payoff by adopting in the successive round $t + 1$ the strategy of the neighbor who scored best at t. Agents have otherwise no memory of past encounters. A large variety of patterns is generated by such rules (Nowak and May, 1992; 1993). In particular, for $T < 1.8$ a cluster of defectors in a background of cooperators shrinks, while it grows for $T > 1.8$. A cluster of cooperators instead grows for $T < 2$ and shrinks for $T > 2$. In the parameter region $1.8 < T < 2$, many chaotically evolving patterns of C and D regions are observed.

If agents interact on more realistic networks, a plethora of rich and interesting behaviors emerges. On Watts–Strogatz small-world networks, the fraction of defectors rises from 0 at small values of T to 1 at large T, but with intriguing features such as a non-monotonic evolution of this fraction (at fixed T) when the small-world parameter p changes (Abramson and Kuperman, 2001). The introduction of an "influential node" which has links to a large number of other agents, in an otherwise small-world topology, leads moreover to sudden drops (breakdowns) of the cooperation when this node becomes a defector, followed by gradual recovery of the cooperation (Kim *et al.*, 2002a).

Santos and Pacheco (2005) (see also Santos, Rodrigues and Pacheco [2005]; Santos, Pacheco and Lenaerts [2006a]; [2006b]; Santos and Pacheco [2006]) show on the other hand that cooperation is more easily achieved in heterogeneous topologies than in homogeneous networks (Figure 10.9). Gómez-Gardeñes *et al.* (2007a) analyze the transition, as T is increased, from cooperation to defection, in heterogeneous and homogeneous networks. Interestingly, the paths are quite different in the two topologies. In homogeneous networks, all individuals are cooperators for $T = 1$. As T increases, some agents remain "pure cooperators", i.e. always cooperate, while others fluctuate. Pure defectors arise as T is further increased, and finally take over at large T. In heterogeneous networks, the hubs remain pure cooperators until the cooperators' density vanishes; thanks to their percolation properties, the subgraph of all pure cooperators therefore remains connected, and the pure defectors start by forming various small clusters which finally merge and invade the whole network. For homogeneous networks, on the other hand, the cooperator core splits easily (as T grows) in several clusters surrounded by a sea of fluctuating agents. The comparison of these two behaviors has the following outcome: for T close to 1, a larger fraction of cooperators is observed in homogeneous topologies, but heterogeneous networks yield larger levels of cooperation as the temptation to defect, T, is increased to intermediate or large values.

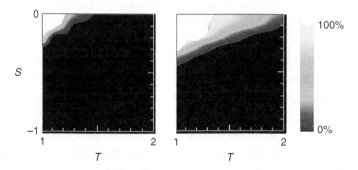

Fig. 10.9. Evolution of the density of cooperators in the prisoner's dilemma for individuals with different networks of contact, as a function of the parameters T and S. Left: homogeneous network; right: scale-free network. Adapted from Santos *et al.* (2006b).

Cooperation on *empirical* social networks has also been studied: the density of cooperators then exhibits a complex non-stationary behavior corresponding to transitions between various quasi-stable states (Holme *et al.*, 2003). Lozano, Arenas and Sanchez (2006) also investigate the role of networks' community structures by comparing the prisoner's dilemma model on empirical social networks with very different community properties: the possible presence of local hubs in communities has a strong stabilizing role and favors cooperation, while more homogeneous communities can more easily change strategy, with all agents becoming defectors. Strong community structures (with sparse connectivity between communities) also contribute to the stability of a community composed of cooperators.

The various static topological features of real social networks thus have a very strong impact on the emergence of cooperative phenomena between selfish and competing agents. Interestingly, taking into account the *plasticity* of the network, i.e. the possibility for (social) links to disappear, be created, or rewired, leads to further interesting effects, as described in the next section.

10.6 Coevolution of opinions and network

Most studies of the dynamical phenomena taking place on complex networks have focused on the influence of the network's topological features on the dynamics. Many networks, however, have a dynamical nature, and their evolution time scales may have an impact on the dynamical processes occurring between the nodes. Such considerations are particularly relevant for social networks which continuously evolve a priori on various time scales (both fast and slow).

Let us first consider the case in which the dynamical evolution of the nodes' (or agents') states is very fast with respect to the network topological evolution. This is the point of view adopted by Ehrhardt, Marsili and Vega-Redondo (2006a) (see also Ehrhardt, Marsili and Vega-Redondo [2006b]). In this approach, each agent carries an internal variable (which can describe an opinion, a level of knowledge, etc), which is updated through interaction with its neighbors, for example through diffusion processes (mimicking knowledge diffusion phenomena) or opinion exchanges. Each link between agents decays spontaneously at a certain rate λ, and new links are created at rate ξ *only between agents whose internal variables are close enough*. The topology thus has an impact on the evolution of the agents' states, which in its turn determines how the topology can be modified. When the knowledge or opinion update (rate ν) is fast with respect to the link's update process, the competition between link decay and creation rates leads to a phase transition between a very sparse phase in which the population is divided into many small clusters, and a denser globally connected network with much larger average degree. This transition, which can be studied through mean-field approaches in the limit $v \gg 1$, turns out to be sharp, and to display hysteresis phenomena (Ehrhardt *et al.*, 2006a; 2006b).

More generally, however, the time scale on which opinions and interactions evolve can be of the same order. Links can be broken more easily if the two interacting agents differ in their opinions but new contacts do generally appear owing to random events in the social life of the individuals. A model integrating the breaking of social relations is given by a population of agents interacting through the Voter model paradigm (Zanette and Gil, 2006). The interaction between two agents who do not share the same opinion can lead either to a local consensus (one agent adopts the opinion of the other), or to a breaking of the link if the agents fail to reach an agreement (this occurs with a certain probability p). Starting from a fully connected population, and depending on the model's parameters, the final state can be formed of one or more separated communities of agents sharing the same opinion. In order to mimic the introduction of new social relations, another natural hypothesis consists in considering that links do not "disappear" but are simply rewired by the agent who decides to change interaction partner. Holme and Newman (2006) uncover an interesting out-of-equilibrium phase transition in the coevolution of the network of contacts of agents interacting through a Voter-like model. In this model, starting from a random network of agents with randomly chosen opinions, an agent is selected at each time step; with probability p, he adopts the opinion of one randomly selected neighbor, and with probability $1 - p$ one of his links is rewired towards another agent who shares the same opinion. The total number of edges is thus conserved during the evolution. When p is smaller than a certain critical value, the system evolves towards a set of small communities of agents who share the same opinion. At large p on the other hand, opinions change faster than the topology and consensus is obtained in a giant connected cluster of agents.[8]

The coevolution of agents, opinions, or strategies and of their interaction network also has a strong impact on the Axelrod model. For example, it is possible to consider that the links between agents who do not share any attributes can be broken. If such links are then rewired at random, the critical value of q (number of possible values of each attribute) above which the system becomes disordered is largely increased (Centola *et al.*, 2007). In other words, the parameter range leading to global consensus is strongly enlarged. The structure of the network itself is also affected, breaking into small components of agents with homogeneous opinions at

---

[8] Strong effects of the network adaptive character are also observed for a model of agents with continuous opinions and bounded confidence who can break the contacts with other agents who have too diverse an opinion (Kozma and Barrat, 2008).

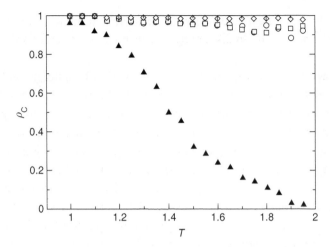

Fig. 10.10. Average fraction of cooperators ρ_C in the steady state as a function of the temptation to defect T, for various rewiring probabilities p: $p = 0$ (triangles), $p = 0.01$ (circles), $p = 0.1$ (squares) and $p = 1$ (diamonds). Data from Zimmermann *et al.* (2004).

large q. Various studies have also shown that cooperation is favored in the prisoner's dilemma case when agents can change their neighborhood. More precisely, links between cooperators are naturally maintained; in the case of a link between a cooperator and a defector, the cooperator may want to break the interaction, but not the defector, so that these reactions are balanced, and for simplicity only links between defectors are assumed to be rewired, with a certain probability p at each time step (Zimmermann, Eguíluz and San Miguel, 2001; 2004; Zimmermann and Eguíluz, 2005; Eguíluz *et al.*, 2005b). Defectors are thus effectively competitive, since they rewire links at random until they find a cooperative neighbor. A rich phenomenology follows from these dynamical rules, yielding a stationary state with a large number of cooperators as soon as $p > 0$ (see Figure 10.10), exploited by a smaller number of defectors, who have on average a larger payoff. Random perturbations to this stationary state can either increase the number of cooperators, or on the contrary trigger large-scale avalanches towards the absorbing state in which all agents are defectors (Zimmermann and Eguíluz, 2005). The rate of topological changes plays also an important role: faster responses of individuals to the nature of their ties yield easier cooperation (Santos *et al.*, 2006a).

While the interactions and feedback between dynamical processes *on* networks and *of* their topologies have been described here from the point of view of socially motivated models, continuous topological reshaping of networks by the dynamics which they support has a broad range of applications. The evolution of the World Wide Web, for example, is certainly strongly influenced by the performance and popularity of search engines. Infrastructure networks may also be affected by rapid evolutions during cascading failures (see Chapter 6), or may have to be reshaped because of the traffic they carry (Chapter 11). While the coevolution of networks and dynamical behavior according to the feedback between dynamical processes and topology is clearly a key issue in the understanding of many systems, its study in the context of complex networks is still at an early stage.

11

Traffic on complex networks

The main role of many networks is to provide the physical substrate to the flow of some physical quantity, data, or information. In particular, this is the case for the transportation and technological infrastructures upon which the everyday functioning of our society relies. In this context, congestion phenomena, failures, and breakdown avalanches can have dramatic effects, as witnessed in major blackouts and transportation breakdowns. Also, in technological networks, if the single element does not have the capacity to cope with the amount of data to be handled, the network as a whole might be unable to perform efficiently. The understanding of congestion and failure avalanches is therefore a crucial research question when developing strategies aimed at network optimization and protection.

Clearly the study of network traffic and the emergence of congestion and large scale failures cannot neglect the specific nature of the system. The huge literature addressing these questions is therefore domain oriented and goes beyond the scope of this book. On the other hand, large-scale congestion and failure avalanches can also be seen as the emergence of collective phenomena and, as such, studied through simple models with the aim of abstracting the general features due to the basic topology of the underlying network. This chapter is devoted to an overview of results concerning traffic, congestions, and avalanches and is more focused on the general and ubiquitous features of these phenomena than on system-specific properties. However, some of the general results highlight properties and phenomenologies that can be applied in a wide range of phenomena and therefore provide insight on real-world systems.

11.1 Traffic and congestion

It is impossible to provide a general and complete account of the study of traffic in networks, since in each domain we find a specific research area devoted to the subject, each worth a chapter of this book.

Traffic and congestion in transportation networks have a clear impact on the economy and the environment. Understanding individuals' movements is also crucial in epidemiology. This has led to a substantial body of work that concerns the problem of how to determine the traffic on every link of the network once the outgoing and incoming flows are known for all locations (the so-called origin–destination matrix). This problem, known as the "traffic assignment problem," was addressed a long time ago by Wardrop (1952) in the equilibrium perspective. It is usually assumed that every user will try to minimize her or his travel cost computed as a function of distance, transportation cost, travel time, etc. The main assumption formulated by Wardrop is that the equilibrium is reached when no user can improve travel time by changing routes unilaterally. This user equilibrium condition is akin to the Nash equilibrium in economics. The interaction between users occurs because of congestion effects which increase the travel time when the traffic is too large. Despite the large number of studies on this equilibrium principle, it is still unclear how the different heterogeneities observed in transportation networks will affect the traffic patterns.

Indeed, several studies have reported evidence in transportation networks of complex topological properties encoded in large-scale heterogeneities and heavy-tailed statistical distributions on different systems such as the airline system (Guimerà and Amaral, 2004; Barrat *et al.* 2004a), subways or railways (Latora and Marchiori, 2002; Sen *et al.*, 2003) and roads and city streets (Cardillo *et al.*, 2006; Buhl *et al.*, 2006; Crucitti *et al.*, 2006; Scellato *et al.*, 2006; Kalapala *et al.*, 2006). Although complex networks made their appearance in transportation research through the empirical measures, the impact of complex topologies on traffic is still to be understood. In addition, even if the separation between these different processes – the evolution of transportation networks and the dynamical traffic patterns – allows for a simple presentation of this field, they are ultimately coupled (see for example the paper by Yerra and Levinson [2005]). The growth of urban centers is facilitated by their accessibility, which in turn will create high demand in terms of transportation means. This simple fact means that modeling approaches should consider the coupling between various socio-economical factors, an effort which has already been undertaken by geographers, economists, and transportation researchers, but in which complex networks still play a minor role.

In the context of information technology, the behavior of internet traffic has been analyzed since the early times of computer networks.[1] An impressive biblio graphical guide to the literature in the early years can be found in Willinger, Taqqu and Erramilli (1996), but work in this field is still rapidly progressing (Crovella,

---

[1] This kind of analysis has been supported by the timely availability of packet traffic tools such as `tcpdump`, developed by V. Jacobson, C. Leres, and S. McCanne in 1989.

Lindemann and Reiser, 2000; Crovella and Krishnamurthy, 2006). The largest part
of these studies generally focuses on a *local* view of the network, by analyzing
a single link or a small collection of them, providing just the properties of the
local traffic. On the other hand, the vision obtained from a single link is often
the outcome of a global dynamic in which very distant regions of the network
also cooperate. Indeed, the characterization of traffic in technological networks
has contributed enormously to the development of the theory of self-similar processes that traces back to the early work of Mandelbrot (1969), and is obviously
related to fractal and scale-free phenomena (Mandelbrot, 1982). In general, traffic
time series on internet connections are scale-free, a property which has correspon-
dences in many different research fields, ranging from surface growth (Barabási
and Stanley, 1995) to economics (Mantegna and Stanley, 1999) or geophysics
(Rodriguez-Iturbe and Rinaldo, 1997). Moreover, the scale-free behavior is not
only observed in traffic series. The traffic self-similarity also affects network performance and queueing time (Park, Kim and Crovella, 1997), inter-arrival time,
and end-to-end delay (Huffaker *et al.*, 2000; Percacci and Vespignani, 2003) which
displays statistical distributions with heavy tails and power spectrum of $1/f$ type
(Csabai, 1994; Paxson and Floyd, 1995; Fowler, 1999; claffy, 1999). Finally, self-
similarity is a general property of the large majority of traffic generated on the
Internet, including WWW traffic from file requests on web servers (Crovella and
Bestavros, 1997).

A global characterization of internet traffic and performance adds a new dimension to the problem that requires a large amount of resources and the solutions
to several technical problems. Traffic and performance analysis on a global scale
implies the collection of traffic traces and performance estimators which, even on
a single link, represent a noticeable volume of data. It is easy to realize that gathering data on hundreds or thousands of links and routers poses great problems
in data handling and storage, as well as in the traffic generated by the measurement process itself. Additionaly, on the global level, these data must be correlated
with the detailed physical and geographical structure (bandwidth, link length, etc.)
of the Internet. Despite these technical difficulties, an increasing body of work
focuses on the Internet as a whole specially aimed to forecast future performance
trends. For instance, interdomain traffic can be studied at a global level by looking at data representing the whole traffic received by specific service providers
(Uhlig and Bonaventure, 2001). Modern measurement infrastructure also allows
the construction of traffic matrices representing the traffic flow between pairs of
service providers (claffy, 1999; Huffaker *et al.*, 2000). In this case, traffic flows
are aggregated on the basis of the sources and destination addresses. These traffic
matrices can also be correlated with the geographical location of traffic sources
and destinations in order to obtain information on regional or national policies

and connectivity. Finally, measurements also focus on internet performance at the global level (Huffaker *et al.*, 2000; Lee and Stepanek, 2001; Percacci and Vespignani, 2003) and on its resilience to failures (Paxson, 1997; Labovitz and Malan, 1998; Labovitz, Ahuja and Jahanian, 1999; Magnasco, 2000).

While these previous investigations constitute a large body of work, the investigation of network traffic with the aim of abstracting general properties independent of the detailed system properties is less frequent. This approach, which focuses on *universal* properties, naturally finds its roots in statistical physics and has been pursued in several cases by explicitly considering the complex topological properties of many networks. A series of works along these lines focused on the study in five different large-scale natural and technological networks of the relationship between average flux or traffic $\langle f_i \rangle$ on a node and the corresponding fluctuations σ_i (de Menezes and Barabási, 2004a; Barabási *et al.*, 2004). Data were obtained for the daily traffic on internet routers, for the activity of logic gates in a microprocessor, for the number of daily visits of websites, for the traffic on highways, and also for the stream flow of a river network. Strikingly, power-law relations $\sigma \sim \langle f \rangle^{\alpha}$ were observed, with $\alpha = 1/2$ for the Internet and the microchip, and $\alpha = 1$ for the other cases. While the value $\alpha = 1/2$ is obtained in the case of a very simple model of independent random walkers visiting the sites of the network, the value $\alpha = 1$ arises when *externally induced* fluctuations are large (i.e., in the case of the simple model of random walkers, if the number of these walkers fluctuates strongly). As a confirmation of this prediction, de Menezes and Barabási (2004b) and Barabási *et al.* (2004) propose a method of separating the internal and external contributions to the dynamics. Namely, if $f_i(t)$ is the signal measured on node i ($i = 1, \ldots, N$) at time t, during a time window $[1, T]$, one can introduce the ratio of the total traffic through i during the observation time and the total overall traffic in the network

$$A_i = \frac{\sum_{t'=1}^{T} f_i(t')}{\sum_{t'=1}^{T} \sum_{j=1}^{N} f_j(t')}, \tag{11.1}$$

so that the expected amount of traffic through i can be written as

$$f_i^{\text{ext}}(t) = A_i \sum_{j=1}^{N} f_j(t). \tag{11.2}$$

Indeed, $\sum_{j=1}^{N} f_j(t)$ is the total traffic on the network at time t, and is assumed to be dispatched onto the various nodes in proportion to the values of A_i.

This expected traffic f_i^{ext} is thus a quantity which fluctuates because of externally induced time changes in the global network traffic. On the other hand, $f_i^{\text{int}} = f_i - f_i^{\text{ext}}$ captures the deviations from the traffic average through node i.

The systematic measure of $(f_i^{\text{ext}}, f_i^{\text{int}})$ and of the corresponding variances $(\sigma_i^{\text{ext}}, \sigma_i^{\text{int}})$ shows that $\sigma_i^{\text{ext}} \ll \sigma_i^{\text{int}}$ in the cases $\alpha = 1/2$ while $\sigma_i^{\text{ext}} \sim \sigma_i^{\text{int}}$ when $\alpha = 1$.[2] The Internet and the microchip are thus networks with internally robust dynamics, while the activity of the other three studied networks (WWW, the highway network, and the river network) is driven by the external demand.

Notably, such studies allow us to gain some understanding of global features of the traffic without requiring precise knowledge of the network structure. On the other hand, a theoretical modeling of traffic in large-scale networks cannot neglect the overall topology of the system. In this area most of the modeling activity is concerned with the self-similarity of traffic and the emergence of congestion in technological networks such as the Internet. These approaches propose that the presence of phase transition and emergent phenomena in simple models mimicking the routing of information packets is what lies behind the appearance of congestions and traffic self-similarity (Ohira and Sawatari, 1998; Takayasu, Fukuda and Takayasu, 1999; Fukuda, Takayasu and Takayasu, 2000; Solé and Valverde, 2001; Valverde and Solé, 2002). In general, the phase transition mechanism appears as an elegant explanation that arises as a complex emergent phenomenon due to the global dynamics of the network. Many other mechanisms which depend, however, on the specific nature of the system can be put forward and validated by specific experimental work. In contrast, the very abstract nature of statistical mechanics models does not often find clear-cut support from experimental data.[3] In the following we focus on a series of works that consider the effect of a network's complexity on routing algorithms and congestion phenomena clearly inspired by problems in the information technology world. These models are naturally fitting in the present book, but we warn the reader about the existence of a large body of literature that approaches these problems in the orthogonal perspective of a very detailed account of the microscopic processes as dictated by the specific technology under study.

11.2 Traffic and congestion in distributed routing

Most studies about traffic and congestion phenomena on complex networks focus on technological networks, particularly on packet switched dynamics such as the one used in the Internet to route data packets. In these attempts, the Internet is modeled as a network whose vertices are assumed to be hosts that generate a certain number of packets (the traffic to be routed) or routers that forward packets.

---

[2] Duch and Arenas (2006) show, however, that for models in which packets perform random walks on the network, varying the number of steps each packet remains in the network and taking into account the finite capacity of the nodes can lead to continuously varying exponents α between $1/2$ and 1.

[3] An interesting discussion of the models' validation issues is provided by Willinger *et al.* (2002).

These packets travel to their destination following routing dynamics that, in general, involve a nearest neighbor passing of packets. Most studies consider for the sake of simplicity that all nodes can play both roles of hosts and routers, and only a few distinguish, in a more realistic way, between hosts which can generate and receive packets, and routers which can only forward them to another node (Ohira and Sawatari, 1998; Solé and Valverde, 2001). In this configuration the network's behavior strongly depends upon the injection rate of new packets. This is generally modeled by the creation of packets at the rate R on randomly chosen nodes of the network. This simply implies that an average of NR packets are created at each unitary time step. Each packet is also assigned a random destination at which it is annihilated upon arrival. The packets then travel from one node to another: at each time step a node can treat and forward a certain number of packets to its neighbors. Such a capacity is usually taken as uniform (Ohira and Sawatari, 1998; Arenas, Díaz-Guilera and Guimerà, 2001), but can also depend on the node's characteristics (Zhao *et al.*, 2005; Lin and Wu, 2006). Each vertex is also supposed to have a buffer (a queue) in which packets can be stored if they arrive in a quantity larger than the handling limit. Packets may accumulate indefinitely or can be discarded if they exceed the buffer limit. The forwarding rules, from a node to its neighbor, define the *routing policy*: they can either follow the shortest path on the network (Ohira and Sawatari, 1998; Solé and Valverde, 2001; Arenas *et al.*, 2001; Guimerà *et al.*, 2002a), or perform a random walk until they reach their destination or its neighborhood (Tadić, Thurner and Rodgers, 2004; Tadić and Thurner, 2004; Valverde and Solé, 2004), as shown schematically in Figure 11.1.

In the context of packet routing models, the total network load or traffic is given by the number of packets traveling simultaneously on the network. As shown in

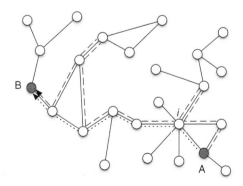

Fig. 11.1. Schematic comparison of shortest-path (dotted line) routing and random walk (dashed line) routing for a packet generated in A and sent to B. Node i has a large degree and betweenness centrality and will therefore typically receive high traffic.

Figure 11.2, this load fluctuates around a steady-state value if the creation rate is not too large. If, on the other hand, the routing strategy does not allow the delivering of packets quickly enough, they start accumulating in the network and the load steadily increases with time, defining a congested (sometimes called jammed) phase.

In order to distinguish between the free flow and congested phases, Arenas *et al.* (2001) define a specific global parameter (see also Guimerà *et al.* [2002a]) given by

$$\eta = \lim_{t \to \infty} \frac{n(t + \Delta t) - n(t)}{NR\Delta t}, \tag{11.3}$$

where $n(t)$ is the number of packets in the network of size N at time t and the $t \to \infty$ limit is taken in order to ensure that the transient initial state does not influence the result. The quantity $NR\Delta t$ is the number of packets injected into the system in the interval time Δt, and η is equal to or smaller than zero if the system is able to route to destination a number of packets larger than the generated traffic. Otherwise, if $\eta > 0$ the system enters a congested phase with packets accumulating in the network. Therefore, if the system is in a free flow phase, $\eta = 0$, while the congestion phase corresponds to $\eta > 0$. In practice, an average over a large enough Δt is performed, as fluctuations may create a negative η as well as transient positive values.

Analogous to phase transitions, the study of the congestion phenomena consists of identifying the critical R_c separating the free flow from the jammed phase, finding the mechanisms leading to the congestion, and understanding how the various ingredients of the model (network structure, routing policy) modify the congestion transition. Implicitly, the congestion transition also depends on the topology of the

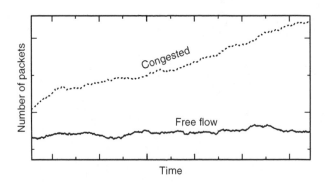

Fig. 11.2. Evolution in time of the number of data packets in the network: if the creation rate is small enough, this number fluctuates around an average value. If the creation rate is too large, the system becomes congested and the number of packets increases (roughly) linearly with time.

network as the routing dynamics is clearly affected by the underlying connectivity pattern. Indeed, most routing policies can be approximated by a shortest path routing from the origin to the destination node. This assumes that a routing table is associated to each node which indicates, according to all possible final destinations, to which neighbor the packet should be routed. Since each packet follows the shortest path to its destination, it is evident that the concept of betweenness centrality will play a central role in the study of traffic properties. We have seen in Chapter 1 that the betweenness of the vertex i is defined as

$$b_i = \sum_{h \neq j \neq i} \frac{\sigma_{hj}(i)}{\sigma_{hj}}, \tag{11.4}$$

where σ_{hj} is the total number of shortest paths from h to j and $\sigma_{hj}(i)$ is the number of these shortest paths that pass through the vertex i. Equation (11.4) is in fact very general, and can be used for any *static routing protocol* (SRP), i.e. assignment of routes between all $N(N-1)/2$ pairs of nodes.[4] In this case σ_{hj} and and $\sigma_{hj}(i)$ are the numbers of paths corresponding to the routing dynamics implemented in the network. In other words, the betweenness centrality as defined by the routing protocol is a first indicator of the relative traffic that a node may eventually handle.

Let us now consider a given SRP. Packets are generated at random, with random destinations (chosen among the $N-1$ other nodes) so that for each source h, the traffic generated toward each destination j is $R/(N-1)$. Therefore, the average number of packets handled at a node i at each time step is on average $Rb_i/(N-1)$, where b_i is the betweenness centrality due to the routing dynamics. Moreover, if each node i can handle c_i packets at each time step, the condition for the system not to be congested reads as

$$R \frac{b_i}{N-1} \leq c_i \ \forall i, \tag{11.5}$$

which, in the simple case of a uniform capacity $c_i = 1$, gives the congestion threshold (Arenas *et al.*, 2001)

$$R_c = \frac{N-1}{\max_i b_i} = \frac{N-1}{b^*}, \tag{11.6}$$

where b^* is the largest value of centrality in the network; i.e. the value belonging to the most central node in the network which is also the first one to become congested. The above equation clearly shows the role of network topology in the onset of congestions as the maximum betweenness is determined by both the routing

---

[4] Newman (2005a) also defines the random walk betweenness centrality as the number of times a random walk between h and j passes through i, averaged over all pairs (h,j).

dynamics and the network connectivity pattern. The larger the maximal between-ness of the system, the smaller the traffic injection rate R_c at which the network results in congestion. In addition, the evolution of the congestion threshold with the system size can also be evaluated in terms of the scaling of the maximal betweenness with the number of nodes N. The simple expression (11.6) explains why homogeneous networks have typically much larger congestion thresholds than heterogeneous graphs which display broad distributions of both degrees and betweennesses. In most heterogeneous networks, the betweenness typically grows with the degree, so that the hubs become easily congested, leading to a small R_c even if most nodes are far from being overloaded (Echenique, Gómez-Gardeñes and Moreno, 2005). On the other hand, at small load, the presence of hubs can lead to particularly short paths, and therefore small transit times from one point of the network to another. The combination of these observations shows that the optimal structure of a traffic carrying network depends on the imposed global load (Guimerà *et al.*, 2002b). A star-like network is very efficient (with paths of length at most 2) if the number of packets is small. In the case of large traffic, on the other hand, the minimization of b^* corresponds to the most homogeneous possible network, in which the load is balanced between all nodes.

Once we understand the effect of the network topology on traffic, we have to cope with the fact that, in most cases, the structure of networks on which traf-fic takes place is given and cannot be changed at will,[5] and obtaining optimized topologies in a technologically efficient manner turns out to be quite challeng-ing (Krause *et al.*, 2006). Given a fixed network topology, we will see in the following different routes to improve the traffic properties therefore need to be followed: refined static routing protocols may be defined in order to modify the betweenness properties; alternatively, adaptive protocols have been proposed, which dynamically change routing such that the most loaded nodes are avoided.

11.2.1 Heterogeneity and routing policies

Most networks are characterized by large heterogeneities in the nature of their constituent elements. This implies different capacity c_i of the elements in traffic handling. In the case of heterogeneous capacities of the nodes, Equation (11.6) can be rewritten as

$$R_c = \frac{(N-1)C^*}{b^*},$$ (11.7)

---

[5] A noteworthy exception concerns the case of wireless ad hoc networks, whose topologies can be changed by modifying the transmission power of the nodes (Krause *et al.*, 2004; Glauche *et al.*, 2004; Krause, Scholz and Greiner, 2006).

where C^* is the capacity of the node with the largest betweenness. Since hubs are typically the first nodes to become overloaded, Zhao *et al.* (2005) propose to delay their congestion by considering a capacity which increases linearly either with the degree (see also Lin and Wu [2006]) or with the betweenness. In the second case in particular, the congestion threshold becomes independent of the topology: if $c_i = \beta b_i/N$ then $R_c = \beta$. The extension of the nodes' capacities may, however, not be easy, especially for large networks, since b^* may diverge with the network size as N^γ with $\gamma > 1$ (Danila *et al.*, 2006a,b; Sreenivasan *et al.*, 2007). Various attempts have thus been directed towards the optimization of the routing policy.

First, and not surprisingly, a comparison between random walks and deterministic shortest paths shows that the latter leads to better performances, although the former is of course easier to implement. In particular, Tadić and Thurner (2004) (see also Tadić *et al.* [2004]) consider that packets perform random walks on the network, but switch to a deterministic shortest path if they reach a node which is next-to-nearest neighbor of their destination. This small modification yields a large improvement, especially in terms of transit times. The crossover between random and deterministic paths has been further investigated by Valverde and Solé (2004) who consider that when a randomly diffusing packet reaches a node which is at distance m from its destination, it follows the shortest path to this destination. The parameter m determines the change from a fully random walk at $m = 0$ to a fully deterministic routing using only shortest paths at values of m larger than the network diameter. Although congestions are not considered by Valverde and Solé (2004) (the packets creation rate is tuned below the congestion threshold) they have shown that large values of m determine a much better performance with smaller load imposed on the network.

Since pure shortest path routing leads to rapid overloading of the hubs, while pure random walks are rather inefficient (and tend to overload hubs, see Chapter 8), routing strategies aimed at obtaining larger congestion thresholds naturally try to avoid paths through hubs, at the possible cost of increasing the path length for the transmitted packets. Such *hub avoidance strategies* can be defined in various ways. Yan *et al.* (2006), for instance, propose to use between any source h and destination j the path $\mathcal{P}_{hj}$ which minimizes $\sum_{i \in \mathcal{P}_{hj}} k_i^\beta$, where β is a parameter of the routing strategy. For $\beta = 0$ the usual shortest path routing is recovered. For $\beta > 0$ the paths, when possible, go around the hubs and R_c is noticeably increased. While computing such optimal paths implies the knowledge of the whole topology of the network and may be computationally very expensive and hardly feasible in real-world applications, local stochastic strategies can be devised. A node can forward a packet to a neighbor i with a probability $\propto k_i^\alpha$; i.e. depending on the degree of the neighbor itself. The value $\alpha = -1$ turns out to be optimal in increasing R_c, as

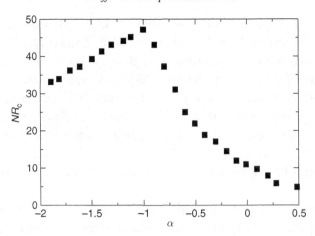

Fig. 11.3. Congestion threshold R_c as a function of the hub avoidance parameter α: each packet sitting at a node is forwarded to one of its neighbors i with probability $\propto k_i^{\alpha}$. An additional rule which forbids the packet to travel back and forth on the same link is used to obtain reasonable values of R_c. Here the network used is a Barabási–Albert model with $m = 5$, $N = 1000$. Each node can treat and forward 10 packets at each time step. Data from Wang *et al.* (2006a).

displayed in Figure 11.3 (Wang *et al.*, 2006a). Indeed, it can be shown (for uncorrelated networks) that the average number of packets located at a node i depends on its degree according to the form $n_i \propto k_i^{1+\alpha}$. The load is thus most evenly balanced between various degree classes when $n_i = $ const, yielding the condition $1 + \alpha = 0$.

The quest for increasing the congestion threshold R_c has led to the proposal of several other routing strategies that consider explicitly the network topology. Sreenivasan *et al.* (2007) put forward another hub avoidance strategy: (i) starting from a given network, a certain number of hubs are "removed", which typically will disconnect the network in various clusters (see Chapter 6); (ii) in each of these clusters, routing between each pair of nodes is assigned through shortest paths; some of these shortest paths may differ from (and be longer than) the shortest paths on the original network, which may pass through the removed hubs; (iii) the hubs are put back into the network and all pairs of nodes which do not yet have a routing path are joined through shortest paths. In this way, some of the paths assigned are not the shortest paths on the full network and avoid hubs. Danila *et al.* (2006b) alternatively, study a heuristic iterative algorithm that, given a network structure, minimizes the difference between the maximum and the average betweenness, i.e. obtains a distribution of betweenness as narrow as possible. Such a goal is obtained by iteratively assigning larger "costs" to the edges connected to the node with largest betweenness. The growth of b^* with N is substantially low-ered, but the algorithm computational time scales as $\mathcal{O}(N^4)$, which is too large for most real networks.

An extremely valuable result has been derived by Sreenivasan *et al.* (2007) who found that the network topology in fact induces a lower bound for b^*, whatever the routing protocol. Different static routing protocols can lead to different nodes having maximal betweenness, and different betweenness distributions, but b^* can-not be decreased indefinitely. For a scale-free network with degree distribution $P(k) \sim k^{-\gamma}$, this bound is estimated as $\mathcal{O}(N^{\gamma/(\gamma-1)})$: as $\gamma \to 2$, the network becomes increasingly star-like and the betweenness of the most central node scales as N^2 while, as $\gamma \to \infty$, one obtains a homogeneous network for which the best possible scaling of b^* with N is obtained. Consequently, the congestion threshold for deterministic routing algorithms has an upper bound due to the topology. Such reasoning does not, however, apply to adaptive strategies, considered in the next section, in which each node decides in real time the forwarding routes according to the local information about the most loaded nodes.

11.2.2 Adaptive (traffic-aware) routing policies

The fundamental feature of the traffic-aware routing policies is the possibility for each node to gather information about the load status of its neighbors. The decision of where to forward a packet can thus be taken in order to avoid overloading nodes with large queues. In the context of complex networks, such a scheme has been proposed by Echenique, Gómez-Gardeñes and Moreno (2004). In this work, the routing dynamics incorporates a tunable amount of information about traffic into the shortest-path-based routing. Let us consider that each node can treat and for-ward only one packet at each time step (see Chen and Wang [2006] for the case of a capacity depending on the degree) and that the packet held by a certain node l has destination j. For each neighbor i of l, the effective distance to j is then defined as

$$\delta_i = hd_i + (1 - h)q_i, \tag{11.8}$$

where d_i is the shortest-path distance between i and j, and q_i is the number of packets held by i (the length of its queue). The parameter h controls the interplay between traffic and shortest paths and, for $h < 1$, packets may be diverted to longer but less congested paths. The forwarding decision of node l can then either be probabilistic, each neighbor i having probability $\exp(-\beta\delta_i)$ of being chosen (Echenique *et al.*, 2004), or deterministic (Echenique *et al.*, 2005). According to the latter rule the packet is then sent to the neighbor i which minimizes the effective distance δ_i. For $h = 1$, the usual transition to congestion is observed as shown in Figure 11.4. The order parameter η increases continuously from 0 at $R < R_c$

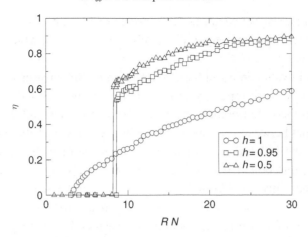

Fig. 11.4. Jamming transitions in the traffic-aware scheme proposed by Echenique *et al.* (2005), for various values of the parameter h (see Equation (11.8)). The order parameter η is given by Equation (11.3). The case $h = 1$ corresponds to the standard shortest path routing with no traffic awareness. Data from Echenique *et al.* (2005).

to positive values at $R > R_c$. On the other hand, as soon as some amount of traffic-awareness is included, $h < 1$, the appearance of congestion is retarded, i.e. R_c increases. It must be stressed that the parameter η presents a discontinuous jump at the transition and takes values larger than in the case $h = 1$. The study of congestion levels as a function of the node's characteristics and as a function of time, displayed in Figure 11.5, allows an understanding of how congestion arises and how the traffic-aware policy improves the situation. For $h = 1$ (usual shortest path routing), the nodes with large betweenness are congested while the nodes with small betweenness are not overloaded. In the case of $h < 1$, the nodes with large betweenness become congested first, but the traffic-awareness then leads the packets to follow paths which avoid these congested nodes, slowly spreading the congestion to the rest of the network. The interest of the traffic-aware scheme with respect to the shortest path routing thus actually depends on a trade-off between the value of R_c and the way in which congestion arises (continuously or abruptly).

Variations on the basic idea of dynamically avoiding the most loaded nodes have proposed routing mechanisms biased by the value of the queue q_i of each network element. In particular, each node l can decide stochastically to forward a packet, each neighbor i being chosen with probability $\propto (1 + q_i)^{-\beta}$ (Danila *et al.*, 2006a), or $\propto k_i (1 + q_i)^{-\beta}$ (Wang *et al.*, 2006b). It turns out that an optimal value for β is obtained (namely $\beta = 1$ for the first case, $\beta = 3$ for the second case), with a strong increase in the observed values of R_c. While $\beta > 0$ allows the most loaded nodes to be avoided, an excessively high β, i.e. too strict a load avoidance policy,

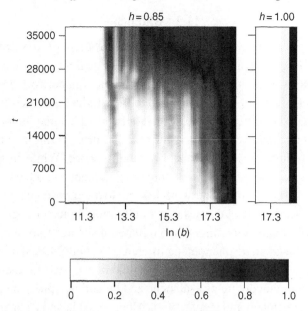

Fig. 11.5. Congestion levels in the congested phase as a function of time t and nodes' betweenness b. At each time step, the gray scale is normalized by the number of packets in the queue of the node with the largest congestion. Two radically distinct behaviors are obtained for the standard ($h = 1$, right panel) and the traffic-aware ($h = 0.85$, left panel) protocols. In both cases $R > R_c$. From Echenique *et al.* (2005).

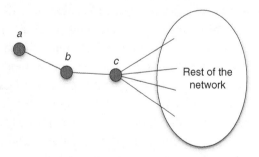

Fig. 11.6. "Traps" in the congestion aware scheme of Danila *et al.* (2006a). At large β, many packets get trapped between nodes b and a.

is counter-productive since it leads to the appearance of "traps" such as the one depicted in Figure 11.6 (Danila *et al.*, 2006a). The reason is that nodes a and b in Figure 11.6, having small betweenness, will typically receive few packets and have small queue lengths. Node c would then tend to send packets to b, and these packets would then get stuck for a long time traveling back and forth between a, b, and c. A value of β that is not too large allows the avoidance of such blocking phenomena.

11.3 Avalanches

In the previous section, we have considered the problem of how congestion arises (and may be delayed thanks to clever routing strategies) when the rate of packet creation increases in an otherwise intact and functioning network. Once congestion arises, the overloading of nodes may lead to their failure and their traffic will have to be redistributed to other elements of the network. This may induce the overload of other elements and eventually trigger other failures, giving rise to a cascade or avalanche of failures which propagates in the network. While in some cases the avalanche may just involve a limited number of elements, in other occasions it may affect a sizeable portion of the network leading to a macroscopic breakdown of the entire system. Avalanche dynamics is not only found in technological networks. A wide variety of phenomena ranging from material science to biological and social systems exhibit avalanche phenomena triggered by a progressive rearrangement of the system under the failure of some of its elements. Two ingredients appear as fundamental in the modeling of avalanche and breakdown phenomena in heterogeneous media (Herrmann and Roux, 1990; Alava, Nukala and Zapperi, 2006). First, the existence of local thresholds ensures that, above a certain level of stress, a local element will fail. The second important point lies in the existence of disorder (due to structural heterogeneities): a model in which all elements fail at the same time would be neither realistic nor very interesting. In the context of complex networks, the determination of the heterogeneity of the thresholds is a non-trivial point, since little is known about the real capacity of each element of real-world networks. In the absence of reliable information, one possibility is to distribute randomly the thresholds of the links, without any a priori correlations with the underlying topology. This corresponds to models akin to the fiber bundle model (FBM). Another choice corresponds instead to the hypothesis that the *actual* capacity of each link or node is strongly linked to its topological centrality as given by the betweenness centrality. The rationale behind this approach comes from the fact that the betweenness centrality gives the amount of all-to-all traffic (i.e. each node sends one data packet to each other node of the network) if shortest paths are used for transmitting the data, as seen in the previous section.

In all cases, the initial failure may be caused either by the progressive increase of an externally imposed force or load (such as the increasing number of data packets in a congested network), which will typically lead to the failure of the weakest or most overloaded part of the network, or by the sudden (accidental) failure of a given node or link; the cascading consequences of this link or node removal will then depend on its topological characteristics, just as in the case of random failures versus targeted attacks explored in Chapter 6.

11.3.1 Breakdown models

Classical models for avalanches and network breakdowns were first put forward by physicists in the context of disordered material science. While these models were formulated with physics problems in mind, they represent simple abstractions of the physical phenomena and as such can be easily generalized to other systems. These models share the presence of an external load or force which is acting globally on the system and can trigger the failure of network elements.

The random fuse network (RFN) is a simple model (de Arcangelis, Redner and Herrmann, 1985) which describes the breakdown of materials and electric grids. In this model, fuses form a regular lattice on which an electrical current flows. The fuses have a threshold which can fluctuate from one link to the other reflecting the heterogeneity in the system. When the local current reaches the fuse threshold, the corresponding fuse breaks and the current is then redistributed among the surviving links according to Kirchhoff's laws. This current redistribution may lead to an overload of some other fuses which fail in turn, triggering a failure cascade with the ability to break the system into disconnected parts. The RFN model thus demonstrates how the failure of a single element may trigger a cascade leading to a macroscopic breakdown. Among a number of general results, the existence of a macroscopic breakdown current I_c is of particular interest. When the applied external current I is increased, more and more fuses break, some causing small avalanches, until a global failure avalanche breaks the system for $I = I_c$. The value of this breakdown current I_c depends on different parameters such as the threshold distribution and the geometry of the lattice.

Some simple analytical arguments for the RFN on regular lattices might be helpful for the understanding of more involved cases. Let us consider a very large rectangular lattice of fuses with the same conductivity, and thresholds distributed according to $\phi(i_c)$ which we will assume to be uniform in the interval $[\langle i_c \rangle - w, \langle i_c \rangle + w]$. When the applied current is increased very slowly, the weakest fuse with threshold $\langle i_c \rangle - w$ fails first. The current is then redistributed on the lattice (Clerc *et al.*, 1990); for example, the current flowing through the neighbors of the broken link is multiplied by some factor $\alpha > 1$, equal to $\alpha = 4/\pi$ for an infinite rectangular lattice (the multiplication factor decreases with the distance of the links from the breaking point). If α is small or if w is large enough, these nearest neighbors will probably be able to cope with the current increase and will not break. On the other hand, a sufficient condition in order to trigger an avalanche leading to a global breakdown of the system is that the redistributed current is larger than the largest threshold in the system:

$$\alpha(\langle i_c \rangle - w) > \langle i_c \rangle + w. \tag{11.9}$$

This equation implies an important qualitative result. If the disorder is weak (small w) then the failure of the first fuses will very likely trigger a very large avalanche ("brittle" behavior). In contrast, if the threshold disorder is large, the increase of the applied current will break more and more fuses leading to microcracks which will grow and lead to global breakdown when they coalesce ("ductile" behavior).

In a similar spirit, the fiber bundle model represents a simplification of the RFN (Herrmann and Roux, 1990): instead of recalculating the currents in all the network when a link (or fiber) breaks, one assumes that the current is redistributed equally either on all the remaining links, or simply on the nearest neighbors of the failure. As in the RFN, each of the initially intact N fibers has a local threshold i_c distributed according to some distribution $\phi(i_c)$; if the applied current (or force) I is equally divided on all the fibers, a fiber breaks when $I/n(I)$ exceeds its threshold i_c, where $n(I)$ is the number of remaining fibers, which thus obeys

$$n(I) = N\text{Prob}(I/n(I) < i_c) = N \int di_c \phi(i_c)\theta\left(i_c - \frac{I}{n(I)}\right), \qquad (11.10)$$

where θ is the Heaviside function. Depending on the disorder distribution, the model exhibits different regimes of abrupt rupture or more progressive damage leading to global breakdown.

Although these models have been extensively studied on regular lattices, they can also describe avalanches in complex networks. In particular, Moreno, Gómez and Pacheco (2002a) consider a version of the fiber bundle model in which each node (rather than each link) of a Barabási–Albert network represents a fiber, with a random threshold distributed according to a given probability distribution. For definiteness, a Weibull distribution is taken, with a variance depending on a parameter ρ (the larger ρ, the smaller the variance and the narrower the range of threshold values). A global force or load F is applied to the network and initially equally divided among the nodes. When the load on a node is larger than its threshold, the node breaks and the load is equally redistributed among its neighbors, possibly causing an avalanche. As the externally applied force slowly increases, more and more nodes break, leading finally to a global destruction of the network at a critical load σ_c which depends on ρ: for larger ρ, i.e., for narrower distribution of thresholds, the rupture of the network becomes more abrupt. In this respect, the behavior of the scale-free network is no different from a regular lattice. A more detailed inspection shows that when the critical load is approached, the fraction of overloaded nodes is an increasing function of the degree; the macroscopic breakdown is reached when the hubs start to fail. When the last hubs are eventually overloaded, the network fails. This result is of course linked to the vulnerability of heterogeneous networks with respect to the removal of the most connected nodes (see Chapter 6), and appears here as a consequence of the local dynamics: the hubs

have more neighbors and thus have a higher probability of receiving additional load
from the breaking of a neighbor (see also Kim, Kim and Jeong [2005]).

11.3.2 Avalanches by local failures

As previously mentioned, the failures in a network can also occur because of the
random breakdown of one element, whose load has to be redistributed. Other elements may become overloaded and fail, leading to a new redistribution, and so
on. This cascade of events stops when all the elements of the network have a load
below their respective capacity. The size of the avalanche is given by the number of
nodes which have been broken in the process, and the damage is quantified by the
size of the largest connected component remaining in the network, in a way similar
to Chapter 6. In this case we are not driving the system with an external quantity
(force, generated traffic etc.), and the dynamics of the avalanche propagation and
the extent of the damage, depend on the specific system properties and dynamics
such as the capacities of the nodes or the redistribution rules.

In the absence of precise knowledge about the traffic flow on real networks, the
simplest hypothesis considers that in the initial state the load or traffic on each
link (i, j) is a random variable g_{ij} drawn from a probability distribution $U(g)$
(Moreno *et al.*, 2003b), and that the local threshold or capacity is uniform ($c = 1$
for each link). A failure can be simulated by selecting a link at random and overloading it, by raising its traffic to $g_{ij} = c$. At this point the traffic of the link is
redistributed among the (non-overloaded) links which depart from the end nodes
of the congested link. This model is once again thought for networks carrying
physical quantities or information packets such as power grids and the Internet.
Two types of redistribution can be considered. The first process attempts to deterministically redistribute the load equally among the links. In contrast, "random
redistribution" consists of randomly distributing the corresponding load across the
neighbors. When a failing link has no active neighbors, its load is equally shared
among the remaining functioning links ("conserved dynamics") or can be considered as lost ("dissipative" case). Moreno *et al.* (2003b) compute the phase diagram
of such a model when the initial system is a scale-free Barabási–Albert network:
the order parameter is given by the probability P_G of having a giant component
of connected nodes (i.e. with $g_{ij} < c$) of extensive size *after the avalanche*. If this
quantity is equal to one, the communication or transport capabilities of the network
still function with probability 1. In contrast, when P_G is zero, then no information
or quantity can propagate from one node of the network to another. The numerical
results distinguish three different regimes depending on the value of the average
initial load $\langle g \rangle$, as shown in Figure 11.7. For $\langle g \rangle < g_C^l$ where g_C^l depends on
the system details and on the redistribution process, one obtains $P_G = 1$. For

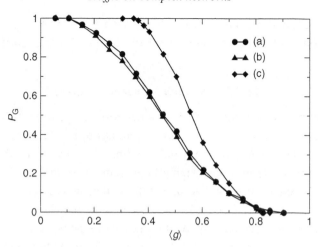

Fig. 11.7. Probability P_G of having a giant component of connected nodes of extensive size after the avalanche caused by the overloading of one link, as a function of the average initial load $\langle g \rangle$, in the model of Moreno *et al.* (2003b). Three redistribution rules are shown: (a) random and dissipative, (b) random and conserved, (c) deterministic and conserved. In all cases the critical values g_C^I and g_C^{II} are easily identified. The initial network has size $N = 10^4$. Data from Moreno *et al.* (2003b).

$g_C^I < \langle g \rangle < g_C^{II}$, P_G becomes smaller than 1 which signals that with a non-zero probability the network can develop a macroscopic breakdown. This region can be considered as fragile with respect to a local congestion. Finally, for $\langle g \rangle > g_C^{II}$, any small instability leads to a completely congested network i.e. $P_G = 0$. A small value is obtained for g_C^I, and quite a large one for g_C^{II}, defining a wide region of load values in which a small instability can propagate and lead to a global congestion with a finite probability. This is indeed what is observed in real congestion or failure phenomena in which apparently similar local disturbances lead either to a global failure or only to a very localized perturbation.

11.3.3 Avalanche and routing dynamics

In Section 11.2, the congestion phenomena have been studied using various routing protocols that are often based on shortest paths. A natural hypothesis considers therefore that the betweenness centrality represents a first order approximation of the real traffic in a technological network. If the capacities of the various elements are uniform, the large betweennesses of the hubs lead to a small congestion threshold. Moreover, the mere fact that a network evolves and grows leads to the creation of more and more shortest paths, i.e. to a traffic increase, which tends to accumulate on certain nodes, leading to a high fragility, as shown by Holme and Kim (2002b) in the case of a growing Barabási–Albert network. In this case, the nodes with large degree (and therefore large betweenness centrality) easily become overloaded and their removal yields both an important damage and an important redistribution phenomenon, leading eventually to a fragmented network. Another possibility is to assign, together with an initial traffic given by the betweenness centrality, a maximal capacity c_i for each node i *proportional* to its betweenness centrality. The parameter of the model is given by the ratio between capacity and initial traffic g_i: $c_i = (1 + \alpha)g_i$, where the constant $\alpha \geq 0$ is called *tolerance parameter* of the network (Motter and Lai, 2002; Crucitti, Latora and Marchiori, 2004; Kinney et al., 2005). If a node fails, the betweenness centralities of all other nodes change, since shortest paths have to be redirected. The load of some nodes may then be increased above their capacity, triggering an avalanching phenomenon. The extent of damage depends on the tolerance parameter: for large α the cascade remains local with few nodes being overloaded, while as α goes to zero the removal of a single node may lead to the entire breakdown of the system. Moreover, the topological characteristics of the initially broken node determine the extent of damage (Motter and Lai, 2002). As can be intuitively understood from Chapter 6, the removal of a node with large centrality is much more likely to produce significant damage than the removal of a low-centrality node. In the framework considered here, this is because the amount of load to redistribute is then large. In fact, the study of various types of networks shows that global cascades are triggered if the centrality is broadly distributed and if nodes with large centrality fail (Motter and Lai, 2002). In the case of networks with highly correlated degrees and centralities, the failure of a highly connected node easily leads to large damage. For homogeneous networks with narrow betweenness distributions, on the other hand, all nodes are essentially equivalent and large damages are obtained only if the tolerance α is very small.

11.3.4 Partial failures and recovery

The reaction of a network's properties to congestion or to a breakdown of one element may be less drastic than the simple failure of nodes. For instance, Crucitti et al. (2004) consider that each link (i, j) has a certain efficiency e_{ij}, which is simply decreased if the link's extremities are congested. The capacity of each node is supposed to be proportional to the initial betweenness centrality ($c_i = (1+\alpha)b_i$) and the links' efficiencies are taken initially uniform ($e_{ij}(0) - 1$ for all links). The efficiency of a *path* between any two nodes of the network is then defined as the inverse of the weighted distance along this path, where the weight of each link is the inverse of the efficiency: the more efficient a link (i, j), the smaller the "distance" between i and j. If ϵ_{lm} denotes the efficiency of the most efficient path between nodes l and m, then

$$\epsilon_{lm} = \max_{\mathcal{P}_{lm}} \left(\sum_{(i,j)\in\mathcal{P}_{lm}} \frac{1}{e_{ij}} \right)^{-1}, \qquad (11.11)$$

where $\mathcal{P}_{lm}$ are paths between l and m, and the global efficiency of the network is simply

$$E = \frac{1}{N(N-1)} \sum_{l \neq m} \epsilon_{lm}. \qquad (11.12)$$

The load of node i at any time is assumed to be equal to the number of the most efficient paths going through i. The failure of one node leads to a redistribution of the most efficient paths and possibly to the overloading of some nodes. In contrast with the cascades considered in the previous subsection, overloaded nodes are not removed from the network but become less efficient: if a node i has at time t a load $g_i(t)$ larger than its capacity c_i, the efficiency of all links between i and its neighbors j is reduced according to $e_{ij}(t) = e_{ij}(0)c_i/g_i(t)$. The global efficiency of the network drops in a way depending both on the tolerance parameter α and on the initial node removed. Interestingly, the system can either reach a steady value of efficiency or display oscillations due to the fact that overloaded nodes can become efficient again at later times. As α is decreased, the efficiency of the network after the cascading event undergoes a sharp decrease at a certain value α_c which depends on the type of removed nodes (see Figure 11.8): α_c is smaller for random removals than in the case where the node with the largest initial load is removed. The two

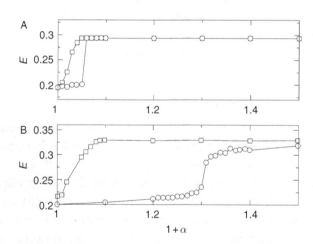

Fig. 11.8. Effect of a cascading failure in (A) an homogeneous ER graph and (B) a BA network, triggered by the removal of either a randomly chosen node (squares) or the node with the largest load (circles). The figure shows the final efficiency of the network vs. the tolerance parameter α. Both networks have size $N = 2000$. Data from Crucitti *et al.* (2004).

critical values are moreover very close for homogeneous graphs, but quite different for BA networks: a range of tolerance parameters in which the damage on the BA network is small for random removal but strong for targeted removal is obtained. Similar results are obtained for internet maps or the US power grid, and in fact for any network displaying a broad distribution of betweennesses. A similar but more detailed study in the case of the US power grid shows the existence of a large number of nodes whose removal causes almost no damage to the network, while there exists a small number of nodes whose removal has a very strong impact on the efficiency of the network (Kinney *et al.*, 2005).

11.3.5 Reinforcement mechanisms

The study of congestion phenomena and possible subsequent avalanches helps to rationalize the behavior of infrastructure networks: such networks typically can daily support a large number of random failures without any global damage, while dramatic consequences such as electrical blackouts may be triggered by apparently small initial events. In heterogeneous networks with broad distribution of betweenness centralities, random failures will most often concern nodes with small centrality, whose removal has little impact. On the other hand, a random failure may sometimes (even if with small probability) occur on a large-betweenness node, having the same effect as a targeted attack. In such a case, a large cascading event and a global failure of the network would follow.

A crucial issue is how to improve the robustness of networks or avoid a cascade propagation in a network. It is intuitive that the addition of redundant edges on top of an existing structure may strongly decrease its vulnerability (da Fontoura Costa, 2004). For a given number of nodes and edges, optimizing the sum of thresholds for random and targeted attacks leads to a design in which all nodes have the same degree, except one which has a degree of order $N^{2/3}$ where N is the size of the network: such a design is, however, not very realistic (Paul *et al.*, 2004).

Motter (2004) has proposed an adaptive defense mechanism against cascading failures. It applies to models in which the load of a node is given by the number of shortest paths passing through it, with a capacity proportional to the initial load. It moreover relies on the idea that the initial failure can be detected *before it propagates* throughout the network. The action of immediately removing (or switching off) a certain number of nodes with the smallest loads then allows the cascade to be stopped (see Figure 11.9): these nodes contribute to the global traffic but since they are quite peripheral they do little for its handling. Cutting them out thus reduces the global traffic without causing too much damage in terms of efficiency. The

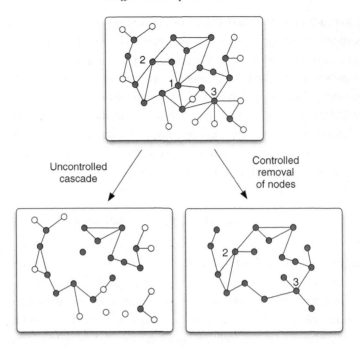

Fig. 11.9. Illustration of the defense mechanism proposed by Motter (2004). The cascade is provoked by the failure of node 1, which is removed from the network. Subsequently, nodes 2 and 3 become overloaded and fail as well, leading to a fragmentation of the network. If, on the other hand, a certain number of the most peripheral nodes, shown in white in the original network, are purposely removed (switched off) right after the failure of node 1, the traffic is alleviated, avoiding the failure of nodes 2 and 3, so that a large functioning component remains connected.

obtained size of the remaining giant component when using this strategy may then be much larger than for a uncontrolled cascade. The assumptions made here – such as the ability to detect the beginning of an important cascade quickly, to distinguish it from one of the many harmless random failures, and to remove nodes easily from the network *before* the cascade propagates – show the need for more investigations into the reinforcement issue.

11.4 Stylized models and real-world infrastructures

At the end of this chapter it is worth stressing again that while the approaches we have presented here are quite general and elegant, they do not account for many features of real-world networks. This has led to criticisms and a general skepticism from the computer science and engineering communities toward these approaches. Let us for instance consider the physical Internet. The routing policies of the

Internet are heavily affected by peering agreements between internet providers, commercial partnerships and other factors that make this system quite different from the stylized models used to study congestion phase transition. Another important issue is that even if the Internet and other information technology networks are self-organized complex systems, they possess, at the smaller scale, a high level of engineering design. In this sense the complex network topologies used in the models are not very good candidates for the modeling of small-scale properties of the network which on the other hand can also heavily affect the onset of congestions and traffic jams. This is also true for transportation systems. Analogously, avalanche processes are strongly dependent on the local nodes' properties and their reaction in case of failure and overload. In simple terms, it is hard to believe that a simple model could account for both a computer network and the railway system.

The value of the simplified models presented in this chapter stems from the fact that while the elements and technologies from which these networks are made are constantly changing, we can reasonably think that similar large-scale patterns and behaviors can be considered at the core of very different systems. This belief is not unmotivated but actually derives from the lesson we learned from statistical physics. The study of matter has taught us that although many materials or physical systems differ at the microscopic level, their large-scale behavior at critical points is essentially identical. For instance, a liquid–gas transition might appear very different from a change in the magnetic properties of ferromagnetic materials. However, both transitions are described by the same statistical laws and show the same phase diagram. More surprisingly, when emergent phenomena take over at the critical point, both systems exhibit the same scaling behavior even at a quantitative level. This feature, dubbed "universality," originates from the fact that when emergent cooperative phenomena set in, the large-scale behavior is essentially determined by the basic symmetry of the interactions and dynamics of the system's components, while other microscopic details become irrelevant (see also Chapter 5 and Chapter 13). In general, this is a common feature of complex systems with emergent properties. Naturally, the strict notion of universality must be relaxed when we leave the domain of phase transitions. However, the qualitative conservation with respect to changes of the very local details, of large-scale emerging properties such as heavy tailed-distributions or the absence of characteristic lengths, is a general property of cooperative phenomena. We can hope, therefore, that a large scale view of traffic on networks is no exception to this general scenario. On the contrary, the statistical physics approach might acquire a particular conceptual value, being focused on the properties that are likely to be preserved despite continuous technological changes.

It is important, however, that researchers interested in the large-scale view do not neglect the engineering and technology details. In particular, if the statistical approach is extremely useful in the quest to understand the basic mechanisms of a phenomenon, its predictive power can be limited for each particular case. Any real-world application and understanding is usually also based on these detailed features, and a complete view of networks can therefore be obtained only by combining the different perspectives in a unified approach.

12

Networks in biology: from the cell to ecosystems

Networks have long been recognized as having a central role in biological sciences. They are the natural underlying structures for the description of a wide array of biological processes across scales varying from molecular processes to species interactions. Especially at smaller scales, most genes and proteins do not have a function on their own; rather they acquire a specific role through the complex web of interactions with other proteins and genes. In recent years this perspective, largely fostered by the recent abundance of high-throughput experiments and the availability of entire genome sequences and gene co-expression patterns, has led to a stream of activities focusing on the architecture of biological networks.

The abundance of large-scale data sets on biological networks has revealed that their topological properties in many cases depart considerably from the random homogeneous paradigm. This evidence has spurred intense research activity aimed at understanding the origin of these properties as well as their biological relevance. The problem amounts to linking structure and function, in most cases, by understanding the interplay of topology and dynamical processes defined on the network. Empirical observations of heterogeneities have also revamped several areas and landmark problems such as Boolean network models and the issue of stability and complexity in ecosystems.

While concepts and methods of complex network analysis are nowadays standard tools in network biology, it is clear that a discussion of their relevance and roles has to be critically examined by taking into account the specific nature of the biological problem. The biological arena is incredibly vast (see the books of Palsson [2006] and Alon [2007a]), and this chapter is just a bird's eye view of the possible applications of dynamical processes on networks to relevant problems in biology. The material presented here should be considered an appetizer to possibly the most exciting and promising area for the application of network science methods and tools.

12.1 Cell biology and networks

Cell functioning is the consequence of a complex network of interactions between constituents such as DNA, RNA, proteins, and other molecules. A wealth of information on the cell is encoded in the DNA which includes both genes and non-coding sequences. The genes preside over the production of proteins and other molecules, which themselves interact with the genes on several levels. For instance, transcription factors, proteins that activate or inhibit the transcription of genes into mRNA, are the product of genes that may be active or not. Ultimately, genes regulate each other via a complex interaction pattern forming the genetic regulatory network. Proteins in their turn perform functions by forming interacting complexes. The map of physical interactions among proteins defines the protein–protein interaction network. Cell metabolism can also be thought of as a network whose fluxes are regulated by enzymes catalyzing the metabolic reactions. In general, all of these interaction networks are connected through a cascade of biochemical reactions when stimulated by a change in the cell environment, activating an appropriate transcriptional regulation that, in its turn, triggers the metabolic reaction enabling the cell to survive the change.

Researchers have long been constrained to focus on individual cellular constituents and their functions, but the development of high-throughput data collection techniques now provides the possibility of simultaneous investigation of many components and their interactions. Microarrays, protein chips, and yeast two-hybrid screens are some of the techniques that allow the gathering of data on gene, protein, and molecule interactions. These data can be used to obtain a system description usually mapped into interaction networks, which in turn form the network of networks that regulates cell life. These experiments are by construction error-prone, and many false positive and negative signals are usually found in the resulting data sets. This evidence has stimulated lively debate on how far these data sets can be considered reliable, and it is clear that continuous checking with specific high-confidence experiments is needed. On the other hand, for the first time it is possible to gather global information on cell functioning that challenges the research community to develop a systematic program to map and understand the many cell networks.

A first relevant result derived from the systematic analysis of a variety of cellular networks is that their topology is far from random, homogeneous graphs. On the contrary, the degree distribution of most cellular networks is skewed and heavy-tailed. In addition, it is generally accepted that these networks are fragmented into groups of diverse molecules or modules, each one being responsible for different cellular functions. One of the first pieces of evidence comes from the analysis of protein interaction networks (PIN) where nodes are single proteins and a link among them represents the possibility of binding interactions. PINs have been constructed for several organisms including viruses (McCraith *et al.*, 2000), prokaryotes (Rain *et al.*, 2001), eukaryotes such as the yeast *Saccharomyces cerevisiae* (Schwikowski, Uetz and Fields, 2000; Jeong *et al.*, 2001; Ito *et al.*, 2001; Gavin *et al.*, 2002; Ho *et al.*, 2002), and *Drosophila melanogaster* (Giot *et al.*, 2003). In all cases the topological analysis shows that the PINs have a giant connected component with small-world properties and a degree distribution which strongly differs from the Poisson distribution of homogeneous random graphs. In the case of yeast (see Figure 12.1), the network is characterized by hubs and presents non-trivial correlations, as measured by a slightly disassortative behavior of the degree–degree correlations, and by the clustering coefficient, which is around 10 times larger than that of an Erdős–Rényi random graph with

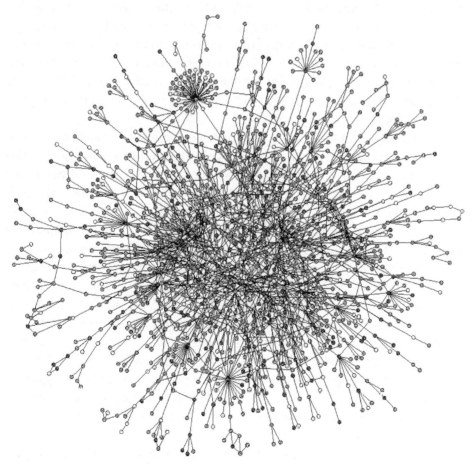

Fig. 12.1. Map of the protein–protein interaction network of yeast. Figure courtesy of H. Jeong.

the same number of nodes and links. Interestingly, in the case of PINs the network structure can be modeled by taking into account two mechanisms which are thought to be crucial for the understanding of the evolution of organisms, namely gene duplication and divergence (Ohono, 1970). Duplication means that, occasionally, an organism transmits two copies of one or more genes to an offspring. The divergence is then the result of possible mutations of one of the duplicated genes, which will survive and be further transmitted if the resulting organism is viable. This divergence process can lead to differences in the genes' functions and, at the proteome level, to different proteins with slightly different interaction patterns and functionalities. By using these ideas, a series of dynamical network models (see also Chapter 3) aimed at capturing the structure of PINs has been developed (Solé *et al.*, 2002; Vázquez *et al.*, 2003; Wagner, 2003). In addition to the issue of modeling, the representation of protein's interactions as a network provides interesting insights into the biological role and importance of the various proteins. For instance, it turns out that the lethality of a protein (i.e. the likelihood that the removal of a protein is lethal for the organism) is indeed correlated with its degree in the PIN (Jeong *et al.*, 2001). Hubs are about three times more likely to be essential than less connected proteins. This result can be related to the resilience of the PIN under damage and provides a vivid illustration of the relevance of targeted attacks in networks (see Chapter 6).

Further evidence for the complexity of biological networks is found in transcription regulatory networks, in which the nodes correspond to genes and transcription factors, and the directed links to transcriptional interactions. These networks are available for *E. coli* (Shen-Orr *et al.*, 2002) and *S. cerevisiae* (Guelzim *et al.*, 2002; Lee *et al.*, 2002). In this case the corresponding networks again exhibit heavy-tailed out-degree distributions while an exponential behavior is found for the in-degree distribution. These properties provide valuable information on the mechanisms presiding over gene regulation processes. The broad out-degree distribution signals that there is an appreciable probability that some transcription factors interact with a large number of genes. In contrast, each gene is regulated only by a small number of transcriptional factors as indicated by the exponential in-degree distribution. A detailed inspection of the basic transcriptional regulation patterns has been performed in Uri Alon's laboratory by looking at particular subgraphs, called motifs, which are much more recurrent than in the null hypothesis of a random graph (Milo *et al.*, 2002; Shen-Orr *et al.*, 2002; Alon, 2007b). Again, the relation between structure and function constrains the topology of the network and induces deviations from the simple random construction.

Like the networks previously described, the metabolic networks of various organisms display heterogeneous topologies and non-trivial correlations as revealed for instance by the clustering spectrum. Metabolic networks are directed and have various levels of representations that can consider up to three types of nodes (Jeong *et al.*, 2000; Wagner and Fell, 2001; Lemke *et al.*, 2004). Several simplified descriptions, however, have been proposed. In the most general representation, the metabolites (or substrates) constitute the nodes of the network and are connected through links which represent chemical reactions. Links are therefore directed, and for each metabolite the number of incoming and outgoing links corresponds to the number of reactions in which it participates as a product and ingredient, respectively. Along with the degree heterogeneity, the study of 43 different organisms (Jeong *et al.*, 2000) showed that the ranking (according to the total degree) of the most connected substrates is practically identical for all these organisms: species differences appear at the level of less connected substrates. The global picture which emerges from these results is the existence of a species-independent set of highly connected substrates which connect the other less connected nodes organized in modules serving species-specific enzymatic activities (Jeong *et al.*, 2000). Furthermore, the presence of modular and hierarchical network structures has been clearly documented in the case of metabolic networks (Ravasz *et al.*, 2002).

12.2 Flux-balance approaches and the metabolic activity

Although topological characterization of the cell functional networks can be extremely useful in achieving a system-level picture of the cell's machinery, it is clear that a complete understanding necessitates a dynamical description. In other words, the intensity and temporal aspects of the interactions have to be considered. The case for the importance of the dynamical description is illustrated by metabolic reactions. The chemical reactions defining metabolic networks may carry very heterogeneous fluxes: in fact, these reactions form *directed weighted networks*, and the characterization of the weights, given by the reaction fluxes, may provide crucial information in their study.

The metabolic flux-balance approaches can generate quantitative predictions for metabolic fluxes at equilibrium and provide testable hypotheses on the importance and weight of each reaction (Fell, 1996; Varma and Palsson, 1993; Segrè, Vitkup and Church, 2002). Let us consider a metabolite A_i; the variation in its concentration $[A_i]$ is given by the balance of the reactions in which it participates,

$$\frac{d}{dt}[A_i] = \sum_j S_{ij} v_j, \qquad (12.1)$$

where S_{ij} is the stoichiometric coefficient of metabolite A_i in the reaction j, and v_j is the flux of reaction j. The metabolite A_i is a product in this reaction if $S_{ij} > 0$ and an ingredient in the opposite case $S_{ij} < 0$. In this context, Almaas *et al.* (2004)

have analyzed on the global scale the metabolic fluxes of the bacterium *E. coli*, whose metabolic network contains 537 metabolites and 739 reactions, by assuming a steady state $(d[A_i]/dt = 0$ for all $A_i)$ and that the metabolism is optimized for a maximal growth rate. These assumptions make it possible to compute the flux of each reaction, leading to the conclusion that the distribution of fluxes for *E. coli* can be fitted by the form

$$P(\nu) \sim (\nu + \nu_0)^{-\alpha} \tag{12.2}$$

with an exponent of the order $\alpha \approx 1.5$, as shown in Figure 12.2. The experimental measure of metabolic fluxes is clearly very delicate, but the strong heterogeneity revealed by the numerical approach is confirmed by the experimental data (Emmerling *et al.*, 2002) which yield an exponent of order 1 (Figure 12.2). The metabolic network thus displays both topological heterogeneity and broadly distributed fluxes. Most metabolic reactions have a small flux, and the metabolism activity is dominated by a relatively small number of reactions with very high fluxes.

As presented in Chapter 1, the heterogeneity of weights at a global level may correspond to very different situations at the individual node level. The fluxes can be locally homogeneous or heterogeneous around a given metabolite: in the first case, the metabolite participates in similar proportions in the various reactions, while in the second case, it participates mostly in one or few reactions. The

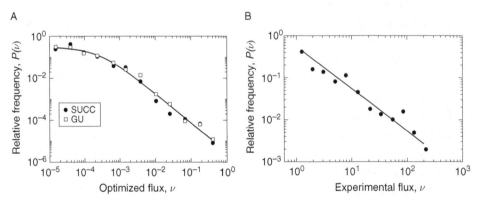

Fig. 12.2. Flux distribution for the metabolism of *E. coli*. A, Flux distribution for optimized biomass production when *E. coli* is put on succinate-rich (circles) and glutamate-rich (squares) uptake substrates. The solid line corresponds to the power law fit $(\nu + \nu_0)^{-\alpha}$ with $\nu_0 = 0.0003$ and $\alpha = 1.5$. B, The distribution of experimentally determined fluxes (Emmerling *et al.*, 2002) from the central metabolism of *E. coli* also displays a power-law behavior with an exponent value $\alpha \approx 1$. Data from Almaas *et al.* (2004).

disparity Y_2 distinguishes between these situations, and is defined for a metabolite i as (see also Chapter 1):

$$Y_2(i) = \sum_{j=1}^{k_i} \left(\frac{\hat{v}_{ij}}{\sum_{\ell=1}^{k_i} \hat{v}_{i\ell}} \right)^2 , \tag{12.3}$$

where k_i is the total number of reactions in which i participates, and $\hat{v}_{ij}$ is the mass carried by the reaction j which produces or consumes the metabolite i. If all the reactions producing (or consuming) the metabolite i have comparable fluxes, we will observe the scaling $Y_2(i) \sim 1/k_i$ (for large k_i). In contrast, Almaas *et al.* (2004) obtain for *E. coli* $Y_2(i) \sim k_i^{-\theta}$ with $\theta \approx 0.2$, showing that a relatively small number of reactions have dominant fluxes (see Figure 12.3).

The inhomogeneity of the local fluxes suggests that for most metabolites there is a single reaction dominating both its production and consumption. It is then possible to identify a high flux backbone (HFB) defined as the structure formed by metabolites linked by their dominant reactions. This backbone, which encompasses the subset of reactions that dominate the whole metabolism, forms a giant component including most of the metabolites. Surprisingly, only reactions with high fluxes are sensitive to changes in the environment: some are even completely inactive in one environment and extremely active in another. Changes

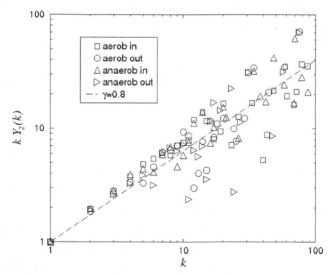

Fig. 12.3. Characterization of the local inhomogeneity of the metabolic flux distribution for the metabolic network of *E. coli*. The measured quantity $kY_2(k)$ calculated in various environments (with aerobic or anaerobic conditions) is averaged over all metabolites and displays a behavior characterized by an exponent $\gamma = 1 - \theta \approx 0.8$ (straight line). Similar behaviors are observed when k is the in- or the out-degree. Figure courtesy of E. Almaas.

in the external conditions therefore lead to (de)activation of certain metabolic reactions, without altering the fundamental identity and function of the major pathways that form the backbone (Almaas et al., 2004). Flux balance analysis combined with network theory thus provides new and interesting insights into the metabolic network. This type of analysis, if done on a growing number of metabolic networks of prokaryotic and eukaryotic organisms, will give access to more information about the interplay between the topology of the network and the reaction dynamics.

12.3 Boolean networks and gene regulation

Modeling the dynamical activity is also crucial in understanding the various genetic regulatory interactions and co-expression patterns. As we have previously explained, transcription factors are themselves produced by genes, and transcriptional interaction can effectively be described as an interaction between the genes. In addition, at a very coarse level, the gene can be represented as having just two possible states: active or inactive. For this reason, although detailed experimental data have only recently started to become available, the modeling of gene regulatory networks can be traced back to the pioneering works of Kauffman (1969), who put forward a simple model of interacting genes, each characterized by a binary on/off variable which describes their activity: an active gene is "on" while an inactive gene is "off". In the corresponding *random Boolean network model*, the genes are considered as the nodes of the network, and each gene receives inputs from a fixed number K of randomly selected genes: each node has in-degree K, its in-neighbors being chosen at random. For each gene i, the value of its Boolean variable σ_i is updated deterministically at each time step according to

$$\sigma_i(t+1) = F_i\left(\{\sigma_j(t)\}\right), \tag{12.4}$$

where $\{\sigma_j(t)\}$ is the set of K variables on the randomly chosen in-neighbors of the node i, and F_i is a Boolean function of its K inputs which is chosen at random for each node i. More precisely, for each set of values $\{\sigma_j\}$, $F_i(\{\sigma_j\})$ is equal to 1 with probability p and to zero with probability $1 - p$. Many studies have been devoted to this model and its variations, and we refer the reader to the book by Kauffman (1993) and to the recent reviews by Bornholdt (2001), Aldana, Coppersmith and Kadanoff (2003), Albert (2004) and Drossel (2007) for more details.

The random Boolean network model is of interest both as a pure dynamical system and as an applied model for biological processes. In both cases, a crucial point concerns the type of dynamical behavior these networks display. In a finite system of N genes (nodes), the configuration space of the set of binary variables $\{\sigma_i, i = 1, \ldots, N\}$ has a finite size 2^N. This ensures that, after a time of at most 2^N steps, the system reaches a state that it has already encountered. Since the dynamics is deterministic, the system will then repeatedly follow the same steps, which define an *attractor*. In fact, the system can either reach a fixed point (point attractor), or go successively and repeatedly through a certain set of configurations, which then form a cycle attractor. Different initial configurations lead to different attractors. In the biological context, the fact that the system goes repeatedly through the same set of configurations can be thought of as defining a set of functionalities. In this framework, each attractor would correspond to a different type of cell in an organism. Each cell contains the same genetic material, so that they may differ only by the pattern of gene activity. The computation of the number and lengths of the attractors is thus of particular interest (see for example Albert and Barabási [2000]; Kaufman, Mihaljev and Drossel [2005]; Correale *et al.* [2006]). Early numerical simulations by Kauffman with relatively small network sizes had led to the result that the mean number of attractors scales as $\sqrt{N}$ for $K = 2$ (and linearly with N for $K > 2$), in agreement with biological data available at the same time, under the assumption that the number of genes in a cell was proportional to the DNA mass. This apparent agreement has spurred much excitement since it seemed to give support to the existence of universal laws, which could be reproduced by very simple models. More recent results, however, have shown that both early empirical data and computer simulations have led to incorrect conclusions. On the one hand, the number of attractors in random Boolean networks with $K = 2$ increases faster than any power law as a function of the number of genes, as uncovered by numerical simulations with large enough sizes and analytical calculations (Drossel, 2007; Kaufman *et al.*, 2005). On the other hand, the sequencing of genomes has revealed that the number of genes is not proportional to the mass of the DNA. The existence of universal laws concerning the number of attractors therefore remains an open issue.

Another, and maybe more essential, characteristic of a Boolean network is that such a system can be either ordered or chaotic: in the ordered phase, most nodes are frozen, i.e. their Boolean internal variable does not change value, while most nodes are continuously changing state in the chaotic phase. The transition from order to chaos ("the edge of chaos") is characterized by the percolation of clusters of non-frozen nodes. The difference between ordered and chaotic states lies also in the response of the system to a perturbation: in the chaotic phase, a perturbation of a single element can propagate to the whole system and make it jump from one cycle attractor to another, while perturbations die out in the ordered or "robust" phase. This difference is crucial as robust features are clearly needed for a biological system to perform its tasks.

The original model considers that each gene receives inputs from exactly K in-neighbors. If this assumption is relaxed, and the genes form an Erdős–Rényi graph with average in-degree K, it can be shown (Derrida and Pomeau, 1986) that there exists a critical value K_c such that for $K > K_c$ the system is in the chaotic regime:

$$K_c = \frac{1}{2p(1-p)},$$

(12.5)

where p is the probability that the output of a Boolean function is 1 in Equation (12.4). In particular, for $K \leq 2$ the network is always in the ordered phase, while for larger average degrees the ordered phase is obtained only for $p < p_c$ or $p > 1 - p_c$, with $p_c = 1/2(1 - \sqrt{1 - 2/K})$. This implies a narrow range of p values as soon as K is not small. For example, for $K = 3$, the system is robust with respect to perturbations only for $p \geq 0.788$ or $p \leq 0.21$. In other words, a fine-tuning of the model is needed in order to build ordered systems, robust with respect to perturbations: the number of inputs has to be small or the bias in the choice of the Boolean functions has to be large.

These results, however, lead to a contradiction with recent experimental evidence. On the one hand, indeed, real cell cycle networks appear to be robust (Fox and Hill, 2001; Li *et al.*, 2004a; Shmulevich, Kauffman and Aldana, 2005), with either fixed points or periodic behaviors. On the other hand, the average number of inputs per element may be larger than 2, and some elements in particular may present a very large in-degree. In fact, broad degree distributions are observed (Guelzim *et al.*, 2002; Tong *et al.*, 2004), and the impact of such topological heterogeneities has to be taken into account. Fox and Hill (2001) therefore consider Boolean networks with various distributions of in-degrees $P(k_{in})$: the initial model corresponds to $P(k_{in}) = \delta(k_{in} - K)$, but Poisson or even scale-free distributions $P(k_{in}) \sim k_{in}^{-\gamma}$ may lead to different behaviors. Interestingly, the stability criterion turns out to involve only the first moment of the distribution. More precisely, in the thermodynamic limit, the critical point for the transition from order to chaos depends only on the average degree of the network, and not on the degree fluctuations (Fox and Hill, 2001; Aldana, 2003). Despite this result, more detailed investigations show that scale-free topology leads to more stable systems, with larger fractions of frozen nodes (Fox and Hill, 2001). This enhanced stability stems from the fact that scale-free networks contain many nodes of small degree, which may easily become frozen. Moreover, hubs have many inputs of small degree, and therefore can themselves become frozen. In other words, the order introduced by the large number of nodes with small in-degree may outweigh the disorder effect due to nodes with a large number of inputs.[1]

Aldana (2003) has built on this result by noting that the exponent of the degree distribution is often a more relevant quantity than the average degree in scale-free topologies. In particular, if the in-degree follows the distribution $P(k_{in}) \sim k_{in}^{-\gamma}$ with $1 < \gamma < 3$, for $k_{in} \geq 1$, the average degree is given by $\zeta(\gamma - 1)/\zeta(\gamma)$, where $\zeta(x) = \sum_{k=1}^{\infty} k^{-x}$ is the Riemann zeta function. The transition to chaos therefore takes place at a critical value γ_c given by

$$\frac{\zeta(\gamma_c - 1)}{\zeta(\gamma_c)} = \frac{1}{2p(1 - p)}. \qquad (12.6)$$

This equation can be solved numerically and the result, displayed in Figure 12.4, shows that scale-free topologies lead to a wide range of parameter values for which the network has a robust behavior. In particular, one observes that the maximum value reached is $\gamma_c \approx 2.48$ for $p = 0.5$. This is an interesting result since many networks seem to present an exponent in the range $[2, 2.5]$, meaning that they are close to the edge of chaos. To reconcile this result with the fact that the critical

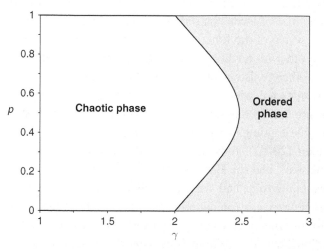

Fig. 12.4. Phase diagram for scale-free networks with power-law in-degree distribution $P(k_{in}) \propto k_{in}^{-\gamma}$ with exponent γ and minimum in-degree $k_{in,min} = 1$. The quantity p describes the probability that each Boolean function has an output equal to one for a given input. From Aldana (2003).

---

[1] Oosawa and Savageau (2002) also consider various distributions for the out-degree, for a network with uniform in-degree, and show that the percentage of active sites is smaller for a broad out-degree distribution. On the other hand, the system exhibits a high degree of dependence upon the few nodes with large out-degree.

point is independent of the topology (Fox and Hill, 2001), it is important to emphasize that Equation (12.6) is valid under the condition that the minimum in-degree is $k_{in,min} = 1$: when γ increases, an increasing number of nodes have degree 1, leading to increased order (as $\gamma \to \infty$ for instance, all nodes have degree 1). For networks with $k_{in,min} > 1$ the range of stability is strongly reduced.[2] When perturbations are applied, the hubs turn out to play a crucial role, as usual in complex networks: the network is sensitive to perturbations applied to the most connected elements even in the ordered phase (Aldana and Cluzel, 2003).[3]

While taking into account the heterogeneous topology of real regulation networks is certainly an important step in their understanding, the interplay between dynamics and topology also depends on the fact that neither the biological networks nor the dynamics of single nodes are random (for example, some particular subgraphs are over-represented with respect to random networks, see Milo *et al.* [2002]). The validation of Boolean network models against empirical data therefore requires these aspects to be taken into account. Kauffman *et al.* (2003) consider, for example, the known architecture of the yeast transcriptional network, and show that the system is marginally stable for randomly chosen Boolean functions, but that the robustness is greatly enhanced if the Boolean functions are *nested canalizing* (Kauffman, 1993). A Boolean function F_i is said to be canalizing if there exists one input j_0 such that fixing σ_{j_0} to a certain value a (0 or 1) completely determines the value of the function: the Boolean variable σ_i becomes then independent of i's other in-neighbors. A function is nested canalizing if, when σ_{j_0} is not equal to a, there exists another input which canalizes F_i, and so on. It turns out that such rules are widely present in empirically observed networks (Harris *et al.*, 2002), and that such nested canalizing rules lead to stable systems for a wide range of possible topologies (Kauffman *et al.*, 2004). Albert and Othmer (2003) (see also Albert [2004]) study in detail a Boolean model where both the structure of the network and the Boolean rules are taken from empirical data for the fruit fly *D. melanogaster*, and show that this model is able to reproduce the various observed gene expression patterns, even if it neglects the existence of various time scales for the chemical reactions, and abstracts the genes and their products as being either "on" or "off" while their concentration can instead take a continuum of values (see also Shmulevich *et al.* [2005] and Li *et al.* [2004a] for the study of the robustness of other regulatory networks).

---

[2] Few studies also take into account that nodes with no input, i.e. 0 in-degree, may exist. These nodes then correspond to the propagation of external variables (Mahmoudi *et al.*, 2007).

[3] In this study, a system is perturbed by setting one node's Boolean variable to randomly chosen values instead of the one it should take according to its Boolean function F_i. The overlap between the perturbed and an unperturbed system is then computed and averaged over time.

Such results may seem puzzling, since the modeling of regulatory networks by simple on/off "switches" can appear to be too crude a representation of the biological complexity. Moreover, the evolution of Boolean networks assumes a synchronous updating: in other words, all units are updated together in discrete time steps. In this respect, Klemm and Bornholdt (2005) have shown that, if this particular updating rule is slightly modified by introducing small shifts of update events forward or backward in time, many attractors become unstable. Since a certain amount of noise is unavoidable in real biological systems, this seems to contradict the reproduction of biological results by the Boolean network modeling approach. Zhang *et al.* (2006) and Braunewell and Bornholdt (2007) have therefore studied how the introduction of noise affects the update dynamics of the Boolean network of yeast genes. Very interestingly, both studies (which introduce two different kinds of noise) reach the conclusion that both the biologically relevant attractor and its basin of attraction (the set of initial configurations which lead the system to this attractor) are stable against such noise in the updating rule. All these results suggest that the precise kinetic details of the activation or inhibition of genes might not be essential to understand the system level functioning of regulatory networks (Bornholdt, 2005).

12.4 The brain as a network

While protein interaction networks and metabolic networks operate inside cells, and the food-webs described in the next section concern large-scale phenomena, biological networks can also be encountered at intermediate scales. A prime example is one of the most complex objects in existence: the human brain. Trying to understand its functioning, and how the microscopic biochemical and biophysical processes at work inside and between neurons can give rise to high level phenomena such as perception, memory, or language, is perhaps the most challenging open problem of modern science. Despite much progress made in neuroscience, in particular at the level of molecular and genetic mechanisms, most questions remain largely open.

It is no surprise that the set of neurons and their interconnections through synapses can be considered as a network. In fact, the networked structure of the brain has inspired a whole field of research about *artificial neural networks*. In such systems, the nodes are elements connected together in order to perform certain tasks, such as pattern recognition, function approximation, or data processing. These networks, however, are not "complex" in the sense that they are engineered and put together according to definite patterns. We will therefore not deal with this huge research area and refer the interested reader to the numerous recent books, for example by Dreyfus (2005) and Haykin (2007). On the other hand, the recent advances and tools developed in the area of complex networks have also been applied to the field of neuroscience (Sporns, 2003; Sporns *et al.*, 2004; Stam and Reijneveld, 2007).

The human cortex contains about 10^{10} neurons with about 10^{14} connections, whose topography, forming the human "connectome," remains largely unknown (Bota, Dong and Swanson, 2003; Sporns, Tononi and Kötter, 2005). More information is available on some animal brains, and various projects have started to collect data and create repositories for neuroscience researchers in order to foster the efforts of this community. For instance, the Brain Architecture Management System (http://brancusi.usc.edu/bkms/) gathers data on brain connectivity structures from different species, collated from the literature. A graph theory toolbox can be found on the website http://www.indiana.edu/~cortex/connectivity.html to analyze neural connectivity patterns. The Blue Brain project (see Markram [2006] and http://bluebrain.epfl.ch/) aims at creating a biologically accurate, functional model of the brain in a supercomputer, starting from realistic models of neurons connected in a biologically accurate manner.

When considering the brain as a network, a first aspect concerns the *anatomical* connectivity, which describes the biological connections between neurons. Data on this network can unfortunately be obtained only through invasive techniques (injection of markers), so that they are available only for certain animals (mostly the cat or primates such as the macaque). The connectivity patterns reveal a structure formed of densely connected clusters linked together in a globally small-world network (Sporns and Zwi, 2004).[4] Such a structure turns out to be essential for the brain since it combines two fundamental aspects of its functioning. The first one is *segregation*: similar specialized neuronal units are organized in densely connected groups called "columns". In order to obtain globally coherent states, these groups need to be interconnected, leading to the second necessary aspect: the functional *integration* which allows the collaboration of a large number of neurons to build particular cognitive states (Sporns, 2003).

A fundamental question concerns the impact of the particular observed structure on the dynamics of the brain. A natural way to tackle this problem consists of simulating the dynamics of neurons interacting on various topologies (see Stam and Reijneveld [2007] for a recent review). For example, Kötter and Sommer (2000) have used the real (experimentally known) anatomical connectivity of the cat cortex to simulate the propagation of neuronal activity and compare it with a randomly connected network in order to demonstrate the existence of a relationship between

---

[4] Interestingly, the seminal paper of Watts and Strogatz (1998) which introduced the Watts–Strogatz model showed that the neural network of *Caenorhabditis elegans* is a small world.

structure and function. Other studies have compared various network structures and shown that the combined strong local clustering and short topological distances favor the interplay between segregation and integration (Sporns and Tononi, 2002), and allow faster learning or information transmission than regular or fully random topologies (Simard, Nadeau and Kröger, 2005). Many works have also focused on the effect of particular model topologies (small-world, scale-free...) on synchronization phenomena on complex networks, as described in Chapter 7, showing for example how predefined dynamical patterns of firing neurons can be obtained by specifically designing the corresponding network of connections between neurons (Memmesheimer and Timme, 2006). In this framework, Zhou *et al.* (2006) study the synchronization dynamics in a realistic anatomical network of the cat cortex (see also Zhou *et al.* [2007]). This approach considers a multilevel network in which each node is a region of the cortex. Since detailed anatomical information about the connectivity at the neuronal level is lacking, these regions are modeled by small-world subnetworks of interacting neurons. The regions (nodes) are themselves connected using the anatomically known connections. Biologically plausible dynamics emerge as a result of this architecture, and the synchronization patterns can be analyzed in detail, providing insights into the relationship between the hierarchical organization of the network structure and its dynamics.

Interestingly, complex network approaches have recently been used to study a different kind of "brain network." The dynamics of the neurons give rise to a (virtual) network, which describes the *functional* connectivity of the brain. More precisely, in this representation the nodes of the functional network are millimetric regions of the brain, and a link is considered to exist between two nodes if the corresponding regions display a correlated activity. In this construction, the links are therefore not necessarily physically present, but they describe the existence of correlations in the dynamics. Data on these correlations can be obtained through various non-invasive techniques such as electroencephalography (EEG), magnetoencephalography (MEG) or functional magnetic resonance imaging (fMRI), and therefore even the human brain can be examined in this way (see Bassett and Bullmore [2006]; Stam and Reijneveld [2007] for reviews on recent experiments). The activity of brain areas of millimetric size (called "voxels") is measured either by the electromagnetic field generated locally by neuronal activity (EEG or MEG) or by the level of oxygen in the blood (fMRI), which also depends on the local activity of neurons. Time-series of the activity of the voxels are recorded, and a link between two voxels is drawn whenever the linear correlation coefficient of their time-series exceeds a given threshold. The obtained network, which defines the functional connectivity, is therefore the result of the activity of the brain itself, and different states of the brain (for instance, rest, or certain predefined tasks such as finger tapping) give rise to different networks. Various questions then naturally arise about the structure of these networks, and the relationship between anatomical and functional connectivity (Sporns, 2004). The application of graph theoretical techniques to the analysis of functional connectivity is in fact quite recent (Dodel, Herrmann and Geisel, 2002), and has allowed interesting properties to be uncovered, such as small-world features with large clustering coefficient and small average shortest path length (Stam, 2004), and scale-free degree distributions (Eguíluz *et al.*, 2005a), possibly with exponential cut-offs (Achard *et al.*, 2006). Strikingly, recent studies of the functional connectivity of patients suffering various types of brain diseases (Alzheimer's disease, schizophrenia, epilepsy) have provided evidence of an association of these diseases with deviations of the functional network topology from the small-world pattern, with an increase of the average shortest path length (see e.g. Stam *et al.* [2007] and the review by Stam and Reijneveld [2007]).

All these experimental approaches are still in their initial stages, and show how the integration of graph theoretical analysis and the use of complex network theory opens the door to many further investigations on the links between anatomical and functional connectivity patterns in the brain. For instance, one can wonder how the anatomical structure evolves during growth and development, possibly through feedback effects in which function shapes the anatomical structure. Finally, the impact of genetic or environmental factors on both anatomical and functional properties remains to be asserted, and the possible relationship between pathologies and particular network properties appears to be a very exciting area of research (Stam and Reijneveld, 2007).

12.5 Ecosystems and food webs

The stability of ecosystems and the preservation of biodiversity are very real and critical topics. Biodiversity reflects the number, variety, and variability of living organisms which is, at present, strongly decreasing. One of many possible examples is the 30% decrease in the number of vertebrate species in the last 30 years. In this context, the theoretical approach to model ecosystems provides valuable insights into issues such as the fragility and stability of populations and species interactions.

An ecosystem can be abstracted as a set of species interacting with the environment. From a global point of view, it is a system that, through the food chain, allows the flow of energy across various levels of the ecosystem, from the basic nutrients to higher order vertebrates. This energy flow is encoded in the ecosystem food web which is specified by the set of relations identifying, at the level of species, "who eats whom". This set of relations can be naturally represented as a directed network where the nodes are the species, and any directed link from a species A to another species B implies that the species B feeds on the species A. More precisely, a directed link is present between the "prey" or "resource" and its "predator" or "consumer". Links are therefore in the direction of the energy flow as the prey is the energy source for the predator.[5] While the nodes represent species in most studies, they can also represent a whole group of taxonomically related species or another type of grouping (in some studies, the term "taxa" is used instead of "species"). Food webs are also organized in *trophic levels* which are sets of species sharing the same set of predators (out-neighbors in the directed network) and prey (in-neighbors) and therefore play equivalent roles in the ecosystem. Food webs usually have a small number (typically about 3) of trophic levels. Figure 12.5 displays a typical example of such a chain of trophic levels: green plants form the first trophic level, herbivores the second, while carnivores form the third and usually the last trophic level. The flow of energy in the system goes from the lowest trophic level to the highest. Several catalogs of food webs have been proposed in the last three decades, and we refer the interested reader to the book of Pimm (2002) for an updated review. The recent availability of data sets with increasing accuracy and the interest in complex networks have also stimulated the activity in the field, as testified in the reviews of Drossel and McKane (2003) and Montoya *et al.* (2006).

Food webs, as other networks, can be characterized by simple global parameters such as the number of links E, the number of species (nodes) N, and average clustering coefficient. More local information can be gained from quantities such as the degree distribution $P(k)$. Since the network is directed (see Chapter 1), the number of links per species E/N is equal to the average in-degree – and also to the average out-degree –, i.e., to the average number of prey or predators. The connectance in its turn measures the "density" of links and is given for a directed network by[6]

$$\mathcal{D} = \frac{E}{N(N-1)}, \tag{12.7}$$

which is the ratio between the actual number of directed links and the maximum possible number of links, self-loops being excluded. As reviewed by Dunne (2006), there is a lively debate in the research community about the values of E/N and $\mathcal{D}$ and their dependence with the food-web size N. In particular, the average number of prey or predators of a species, E/N, has been considered for a long time to be almost constant, with values in the interval $[1, 2]$. This conjecture, known as the "link-species relation," means that species have the same constant number of prey regardless of the global size of the food web. It has, however, been argued

[5] In this representation omnivorous species will have a large in-degree and cannibal species present self-loops.
[6] Note that, in undirected networks, the connectance is $2E/N(N-1)$ (see Chapter 1).

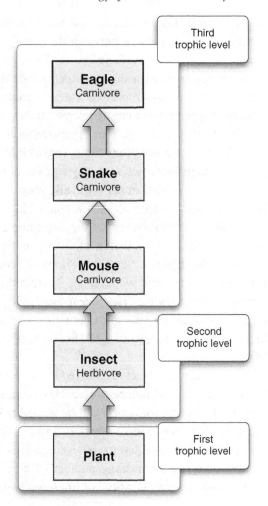

Fig. 12.5. Example of a terrestrial food chain illustrating the concept of trophic levels. The energy flow goes from the lower level usually constituted by plants to the higher order levels such as carnivores.

that the number of different prey does not in general increase with N, in contrast to the number of predators against which a species has to protect itself. The recent availability of larger data sets supports the idea that E/N does depend on N (Dunne, 2006), with empirical studies showing that in general

$$E \sim N^b \qquad (12.8)$$

where the exponent b can vary from $b = 1$ (which corresponds to the link-species relation) to $b = 2$. The case $b = 1$ corresponds to a sparse graph with a finite average degree and with a small connectance. In contrast, the case $b = 2$ corresponds to a very dense graph whose average degree increases with size and where the connectance is constant. The empirical results demonstrate that the connectance can reach values as large as 10% and that the value of b is not universal, varying from one ecosystem to the other.

Various empirical studies have allowed the characterization of other topological properties of food webs. Montoya and Solé (2002) and Williams *et al.* (2002) have shown on several examples of food webs that the diameter is small (with typically "two degrees of separation") with a large clustering coefficient. A more extensive analysis has been performed by Dunne *et al.* (2002a) who analyze 16 different ecosystems with different sizes (the number of species ranging from 25 to 172) but also different connectances (from 0.026 to 0.315) and average degrees (from about 3 to 50). Larger connectance corresponds to larger average degree, larger clustering coefficient and smaller diameter, all simple consequences of a larger density of links. Interestingly, some particular subgraphs (motifs) have also been shown to be over-represented with respect to a random reshuffling of links (Milo *et al.* 2002). Figure 12.6 shows two particular regular patterns which are often found in food webs. For instance, the abundance of the bi-parallel motif shown in Figure 12.6B suggests that species that share the same predator also tend to share the same prey.

A number of empirical studies have tackled the analysis of degree distributions and other statistical properties. It is worth remarking that the limited size of the available data sets is an obvious limitation to this type of analysis. In general, however, the degree distribution does not exhibit a remarkable level of universality (see also Figure 12.7). While the degree distribution is in some cases approximated by homogeneous and Poisson-like behavior, it is skewed and possibly heavy-tailed in other instances (Camacho, Guimerà and Amaral, 2002b; Montoya and Solé, 2002; Dunne *et al.*, 2002a). In fact, the study of Dunne *et al.* (2002a) shows that $P(k)$ evolves with the connectance of the food web; the larger the connectance, the more homogeneous the degree distribution.

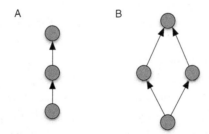

Fig. 12.6. Motifs over-represented in seven food webs (Milo *et al.*, 2002). A, Observed in five food webs. B, Observed in all seven food webs.

A   B

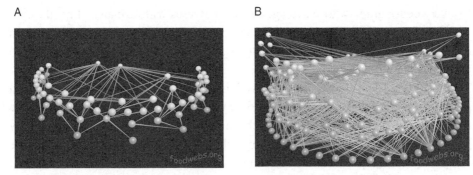

Fig. 12.7. Two examples of food webs (considered here as undirected). A, Example of a common food web with a skewed degree distribution (grassland). B,
A food web with an exponentially decaying degree distribution (El Verde rain-
forest). Figures courtesy of Pacific Ecoinformatics and Computational Ecology
Laboratory, www.foodwebs.org.

A review of topological models for food webs is beyond the scope of this book,
and we mention only a few of the interesting models put forward by the research
community. The most basic model simply considers that a directed link is drawn
from one species to another with uniform probability p. It is therefore simply a
directed version of the Erdős–Rényi model that does not take into account biological constraints, and reproduces simply the connectance (and the average degree),
the average number of links being $p \times N(N - 1)$ (Cohen, Briand and Newman, 1990). Despite its simplicity, it can be used as a benchmark in order to observe
and identify deviations from a purely random model. To restore some structure
between the trophic levels, the cascade model (Cohen *et al.*, 1990) assigns to each
species i a random number n_i drawn uniformly from the interval [0, 1]. A species
i can prey on another species j only if $n_i > n_j$ and with probability p. For two
randomly chosen species i and j, the probability that $n_i > n_j$ is simply $1/2$, so the
probability that there is a link from i to j is $p/2$, and the average number of links
is therefore $p \times N(N - 1)/2$. Taking $p = 2\mathcal{D}N/(N - 1) \simeq 2\mathcal{D}$ (for large $N \gg 1$),
where $\mathcal{D}$ is the desired connectance, one obtains the correct number of links on
average. This model, however, has a certain number of drawbacks, including the
overestimation of the food-chain length.

A more elaborate framework is the niche model proposed by Williams and Martinez (2000). In this model, analogously to the cascade model, species are described
by a "niche parameter" n_i but a species i can prey on all species contained in an
interval of width r_i centered at a random position c_i drawn uniformly in the interval $[r_i/2, n_i]$ (see Figure 12.8). Species may remain isolated and are thus removed
from the web. Species with the same list of prey and predators are instead merged.

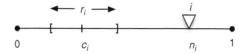

Fig. 12.8. Niche model (Williams and Martinez, 2000): Each species i preys on all species in the interval of length r_i centered on $c_i \in [r_i/2, n_i]$.

The length of the niche interval is chosen as $r_i = xn_i$ where x is a random variable (independent from n_i) drawn in the interval $[0, 1]$ according to the distribution $p(x) = b(1 - x)^{b-1}$, this functional form being chosen for the sake of simplicity. In contrast with the cascade model, a species can prey here also on species with larger values of n_i (up to $n_i + r_i/2$ if $c_i = n_i$). The average of n is $1/2$ and thus $\bar{r} = \bar{x}/2 = 1/[2(1 + b)]$. The values of the parameters b and N then determine the total average degree and the directed connectance: $\langle k \rangle = 2N\bar{r}$ and $\mathcal{D} = \bar{r}$. The niche model yields a good agreement with empirical data for various food-web characteristic quantities such as the food-chain length, the fraction of top species (having no predators), and the fraction of basal species (having no prey). They therefore provide an interesting starting point for a realistic modeling of food webs (Williams and Martinez, 2000). Interestingly, Camacho, Guimerà and Amaral (2002a) have obtained analytically the degree distributions characterizing the niche model. The results show that both the in-degree and the out-degree of a species i decay very rapidly (at least as an exponential) with a typical cutoff of the order of the average degree. In addition there is only one characteristic parameter $\langle k_{\text{in}} \rangle = \langle k_{\text{out}} \rangle$ governing these distributions. Even if these results are in relative good agreement with empirical data for some ecosystems, they do not explain why some ecosystems with a small connectance have a distribution decaying much slower than an exponential. On the other hand, many factors are not taken into account in the niche model, including the predator–prey dynamics occurring on the network.

12.5.1 Dynamics and stability of ecosystems

Although much effort has been devoted to the study of the static and topological properties of food webs, it is clear that their descriptions must include the dynamics of species: in particular, the number of individuals per species varies, and the number of species itself evolves, on a larger time scale, through extinction, mutation, and invasions. The interaction strength between species in its turn may depend upon the species populations and on external resources. The basic building block of dynamical ecosystem models is the prey–predator interaction. Population dynamics models for general webs are usually considering the Lotka–Volterra scheme. If we define by N_i the population size of the species i, the Lotka–Volterra equations for a web of N interacting species can be written as

$$\frac{dN_i}{dt} = N_i \left(b_i + \sum_j a_{ij} N_j \right). \tag{12.9}$$

In this set of equations the parameters b_i and a_{ij} are usually independent of the population size and encode the birth/death rates of species and their interactions. In general, b_i is a positive birth rate for the basal species, and a negative death rate for the other species. The parameters a_{ij} describe the rate at which predators convert prey (food) into predator births and conversely the rate of prey consumed by each predator. Therefore, if i is a predator and j is prey, a_{ij} is positive. If i is prey and j is a predator a_{ij} is instead negative. In general, given the predator–prey relation (i, j), the two interactions are related as $a_{ij} = -\lambda a_{ji}$. The factor λ, called the ecological efficiency, expresses the fraction of consumed prey that is actually transformed into the birth of predators. Some general variations of the model consist of including a saturation term by allowing $a_{ii} < 0$ for basal species and a competition term $a_{ji} = a_{ij} < 0$ if two predators i and j feed on the same species. The food web structure and topology enters the dynamics of the ecosystem by imposing the structure of the interaction matrix a_{ij} and therefore defining the set of Lotka–Volterra equations. The general study of this form of equations has, however, led to models where random numbers are used for the interactions a_{ij} and in some cases the constraints on the opposite sign of a_{ij} and a_{ji} are relaxed.

One of the most important issues in the study of food webs concerns their stability with respect to perturbations, and the effect of their structure on this stability (recent reviews can be found for example in Pimm [2002] and Dunne [2006]). Early considerations lead to the argument that a large number of paths through each species is necessary in order to limit overpopulation effects, since each species is regulated by many others (MacArthur, 1955). More complex food webs, i.e. food webs with more species and more links, would therefore be more stable.[7] In this context, May (1973), following the work of Gardner and Ashby (1970), produced an argument in the opposite direction, showing that the stability is *decreased* by increasing the number of species or the connectance of the network. May used a linear stability analysis by considering small perturbations around a supposed equilibrium point N_i^* of the Lotka–Volterra equations so that $N_i = N_i^* + \delta N_i$. In this case it is possible to

---

[7] We note here that the definition of complexity in terms of connectance is not analogous to the modern definition of complex systems we have provided in this book. In this context we are more inclined to talk about the intricacy of the web.

write the linearized Lotka–Volterra equations for the deviations of the population size as

$$\frac{d}{dt}(\delta N_i) = \sum_j w_{ij}\delta N_j, \tag{12.10}$$

where w_{ij} is the community matrix obtained by the Taylor expansion of each basic Equation (12.9) whose elements characterize the interaction among species near the equilibrium point. The diagonal elements w_{ii} represent the relaxation time to equilibrium in the absence of any interactions and for the sake of simplicity are all set to be $w_{ii} = -1$ by a proper rescaling on the time scale. The choice of the w_{ij} corresponds to the choice of the food-web structure and in order to be as general as possible May (1973) considers that the non-diagonal matrix elements are randomly set equal to zero with probability $1 - \mathcal{D}$ (the connectance), and chosen from a distribution of average 0 and width w with probability $\mathcal{D}$. The issue of stability is then studied by looking at the behavior of the set of differential equations. Stability implies that the real part of all eigenvectors of w_{ij} are negative so that the system goes back to equilibrium after a perturbation. By using general results from random matrix theory (Mehta, 1991), it is possible to reach the general conclusion that in the space of all possible random community matrices the system will be almost certainly unstable (with probability 1) if

$$w\sqrt{N\mathcal{D}} > 1, \tag{12.11}$$

and almost certainly stable otherwise. In other words, the stability decreases for increasing values of $w\sqrt{N\mathcal{D}}$ as the negative eigenvalues move to zero. According to this result, food webs should become unstable as the complexity (measured by the average strength, the number of species, or the connectance) increases. While this could be accepted in the framework of constant connectivity networks ($E \propto N$, i.e. finite $N\mathcal{D}$ as $N \to \infty$), it is, however, hardly compatible with the evidence of superlinear behavior $E \propto N^b$ with $b > 1$, which implies that many food webs should be unstable.

Although the seminal work of May sets the framework for the discussion of ecosystem stability, the general results concerning random community matrices are obtained at the price of introducing several modeling assumptions. While one issue is the structure of the Lotka Volterra equations which can be certainly generalized to more complicate functional form, possibly the weakest point is the assumption of random matrix elements. Real food-web networks are far from random graphs and the structure used for the community matrix should contain elements of biological realism. Since the 1970s, work has been done in constructing more realistic topologies and studying their stability (De Angelis, 1975; Lawlor, 1978; Yodzis, 1981).

Another important issue is the definition of stability itself, which can be generalized to include processes that go beyond small perturbation to the equilibrium state, such as the complete deletion of a species (Pimm, 1979). In this context, various authors investigate the link between the topology of a food web and its stability (Jordán and Molnar, 1999; Solé and Montoya, 2001; Dunne, Williams and Martinez, 2002b; Memmott, Waser and Price, 2004; Allesina, Bodini and Bondavalli, 2005). In particular, considering the results concerning the resilience and robustness of complex heterogeneous networks presented in Chapter 6, several studies deal with the effect on the food-web topological structure of the removal of a certain fraction of species (Solé and Montoya, 2001; Dunne *et al.*, 2002b; Memmott *et al.*, 2004) and the corresponding interactions. Species (nodes) can be removed either at random or by following selective procedures such as targeting species with largest degree. Since food webs are directed, the removal of species with largest out-degree, i.e. those with most predators, has been studied, to model for instance the extinction of large herbivores (Solé and Montoya, 2001). The robustness with respect to the removal of the most connected nodes can also be investigated (Dunne *et al.*, 2002b), with the exception of basal species whose disappearance generally leads to the immediate collapse of the whole ecosystem. As customary in the study of complex networks (see Chapter 6), the robustness of a food web can be measured by the size of the largest connected component after a fraction f of nodes has been removed. In the ecological context, however, another key assessment of the importance of species removal is the possibility of *secondary extinctions*, i.e. the disappearance of species caused by the removal of others. In particular, species that lose all their prey become extinct: a node whose in-degree vanishes because of the removal of other nodes is therefore removed from the network. The studies of Solé and Montoya (2001) and Dunne *et al.* (2002b) show that food webs are robust with respect to random removals but highly fragile and vulnerable if species with large degree are removed, with a large number of secondary extinctions in the latter case.[8] Notably, both in- and out-degree play an important role: a high degree corresponds to a high potential to affect the community, whether this degree is due to a large number of prey or predators.[9] Food webs with heterogeneous topologies appear to be the most vulnerable, but the difference between the various removal procedures is evident also in food webs with exponentially decaying degree distributions. A slightly different definition of robustness considers the fraction of removed species that triggers a total loss of at least 50% of the

---

[8] The protection of basal species leads in certain cases to a vulnerability decrease (Dunne *et al.*, 2002b).
[9] The extinction of a particular species with relatively few links may also lead to important secondary extinctions, in a way linked to detailed, precise and non-general connectivity patterns.

species. By using this definition it is found that the robustness of the web increases with its connectance (Dunne *et al.*, 2002b).

Purely structural studies of robustness completely ignore the population dynamics, while the linear stability analysis neglects the importance of network structure. The integration of both topological and dynamical aspects therefore appears necessary in order to provide a more complete picture. Recent attempts in this direction include the work of Borrvall, Ebenman and Jonsson (2000), who consider various fixed structures in which species are divided into functional groups, all species inside a functional group having the same prey and predators, and evolve according to Lotka–Volterra dynamics. The risk of cascading extinction events is then shown to be smaller when the number of species per functional group increases. Chen and Cohen (2001) also consider web structures obtained from the cascade model together with Lotka–Volterra dynamics, but on small number of species. Quince, Higgs and McKane (2005) finally build on the model of Caldarelli, Higgs and MacKane (1998) and Drossel, Higgs and MacKane (2001) to study community responses to species deletions using realistic global dynamics with realistic network structures. In this last study, no evidence was found that complexity destabilizes food webs. We finally refer the reader to Jordán and Scheuring (2004), Martinez, Williams and Dunne (2006), Dunne (2006) and Montoya *et al.* (2006) for reviews on these promising new research directions.

12.5.2 Coupling topology and dynamics

The investigation of stability and dynamical features of food webs naturally raises the question of the effect of population dynamics in the shaping of food-web topology. In other words, a more fundamental approach to the modeling of the formation and structure of food webs should consider the coevolution of the populations and their interaction patterns. Generally, evolutionary models separate the dynamics on short and long time scales. On the short time scale, species are fixed and their populations evolve according to the predator–prey dynamics. On long time scales species mutate or change their predation links because of competition effects. For instance, Caldarelli *et al.* (1998) consider a short time scale evolution where a certain percentage of each species is eaten by its predators, while Drossel *et al.* (2001) and Lässig *et al.* (2001) use more detailed Lotka–Volterra like dynamics. These models also allow for mutations and changes of predator–prey relations at large time scales. The resulting network structure and the strength of the interactions are therefore the outcome of the dynamical evolution at the various time scales (Caldarelli *et al.*, 1998; Drossel *et al.*, 2001; Lässig *et al.*, 2001; Quince *et al.*, 2005).

Evolutionary models are capable of generating artificial food webs whose structures are compatible with real data. For instance, a quantity of interest is the number of species present in different trophic levels. This function – the shape of the food web – is represented in Figure 12.9 for different real food webs, and the model of Lässig *et al.* (2001) is able to reproduce such a shape and to explain the existence of a maximum. At low trophic levels, species feed on, and compete for, limited external resources (large open symbols in Figure 12.9A), while resisting predation. The model of Lässig *et al.* (2001) predicts that, as the trophic level *l* increases, the number of species increases thanks to the increasing prey diversity, but that the population of each species decreases. For still increasing trophic level, therefore, more and more species have populations too small to support predation, and the number of species starts to decrease.

Other ecological ingredients such as predator–predator competition or the limited number of resources are obviously crucial ingredients for a realistic description of food webs (Chen and Cohen, 2001; Jordán and Scheuring, 2004; Martinez *et al.*, 2006; Dunne, 2006). Although no definitive model exists at this time, the recent modeling approaches suggest that it is crucial to allow the population dynamics to affect and coevolve with the food-web structure.

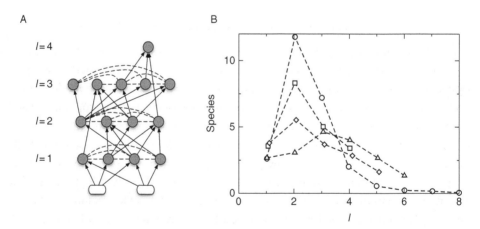

Fig. 12.9. A, Example of a food web (Pamlico estuary food web with 14 species represented by filled circles) with four trophic levels starting from detritus ($l = 1$) which feed on external resources (large open symbols) and finishing with predatory fishes ($l = 4$). Arrows point from prey to predator and the dashed lines connect species with some common prey or predators. B, Average number of species versus the trophic rank for a set of natural ecosystems (different symbols correspond to different data sets). Data from Lässig *et al.* (2001).

12.6 Future directions

Without any doubt, biological research is producing terrific advances at an unprecedented pace. At the same time, however, many areas are still in their infancy as the research community is facing for the first time such an abundance and continuous stream of data. While data reliability has to be assessed and its precision improved, a large part of our understanding depends on the possibility of organizing these data in a meaningful way. Multiscale modeling of the dynamics occurring on networks unavoidably forms one of the key issues in our understanding of biological systems. Although in the biological realm a certain degree of universality appears to be present in the large-scale organization of networks and the ensuing dynamics, it is also clear that it is important to introduce specific features characterizing the different scales and phenomena of the system at hand. At the same time we can see that new challenges lie ahead, in which networks and the dynamics of networks are important. Applications ranging from drug design and pathogen–host interactions to disease networks (Uetz *et al.*, 2006; Goh *et al.*, 2007) are just starting to be explored.

13

Postface: critically examining complex networks science

Although we have reviewed or mentioned more than 600 scientific works on equilibrium and non-equilibrium processes in complex networks, the present book is by no means intended as an exhaustive account of all research activities in network science. It would have been extremely easy to list more than twice as many scientific papers and, as we have stressed in several passages of this book, we have decided to focus our attention on that subset of network research that deals with dynamical processes in large-scale networks characterized by complex features such as large-scale fluctuations and heterogeneities. This implies the selection of studies and approaches tailored to tackle the large-scale properties and asymptotic behavior of phenomena and, as a consequence, models that usually trade details and realism for generality and a high level of abstraction. This corresponds to a methodological approach that has its roots in statistical physics and has greatly contributed to generate an entire research area labeled in recent years as the "new science of networks" (Watts, 2004). As network science is an interdisciplinary endeavor, however, we do not want to overlook the criticisms raised by several authors to this "new science of networks" and to the statistical physics and complex systems approaches. These criticisms are articulated around several key points that are echoed in several commentaries and essays (see for instance Urry [2004], Keller [2005] and Mitzenmacher [2006]), and at the end of 12 long chapters we believe we are obliged to provide a discussion of some of these key points.

One of the criticisms often raised is that the "new science of networks" is not new at all; that is, most of the papers considering complex networks are just re-inventing the wheel. This criticism is rooted in the fact that the results produced by complex network science borrow heavily from previous works and sometimes recover results implicitly contained in those works. On the other hand the new generation research is based on a truly interdisciplinary approach that benefits from unprecedented computing power and data and allows the use of computational and analytical techniques which were not available in the past. For instance, the intrusion of physics techniques and methodology, as with all cross-fertilization of research areas, is producing truly novel approaches to old problems. In some cases this corresponds to conceptually new results and understanding. In many other cases, the innovation lies in the approach itself. Statistical physics has long dealt with the mathematical modeling of very complicated systems, acquiring vast experience in isolating the main ingredients and performing approximations to allow the understanding of complex problems without resorting to rigorous mathematical proofs. Such proofs may be very elegant but are often out of reach of all but a few mathematically gifted minds. Physics is in many cases the art of approximation and often uses the quick and dirty approach which has the advantage of lying within reach of a much larger audience. In recent years the use of numerical simulations and agent-based modeling, in which physicists are at the forefront of research and implementation, is also helping to elucidate concepts and phenomena which sometimes hide behind the veil of formal mathematics. The immediacy of concepts and the possibility of easy manipulation of models are extremely important.

The critics of complex networks science also address a number of technical points that should be carefully scrutinized. A recurrent criticism found in the literature focuses on the poor evidence for power-law behavior and the fact that most of the so-called "power laws" would fail more rigorous statistical tests. As we discussed in Chapter 2, this is technically true in many cases, since most of the real world networks exhibit upper and lower cut-offs, offsets, different functional regimes and all the fluctuations due to the dirt and dust of real-world phenomena. In strict mathematical and statistical terms it is therefore correct to state that most of the claimed power-laws are actually just an approximation of the actual data. In addition, careful statistical tests may be relevant in order to identify and discriminate other possible functional forms. On the other hand, an over-zealous approach to "fitting" loses the focus on the main conceptual issue. Most real-world networks differ from the Poissonian graphs we are used to considering as zeroth order approximations. They are skewed, heterogeneous, heavy-tailed networks with impressive levels of statistical fluctuations. Their modeling with power laws is a better approximation than a Poissonian random graph but nonetheless an approximation. Lengthy debates on how accurate an exponent is, if at all measurable, are therefore a kind of academic exercise that misses the crucial point: the real world is in many cases better approximated by heavy-tailed graphs. Furthermore, many of the small and large-scale deviations from the power-law behavior are in fact predicted in the more detailed models, thus becoming a confirmation of the overall heavy-tailed nature of the system. The power law should then be viewed simply as a conceptual modeling framework that allows the analytical estimation of the effect of fluctuations on dynamical processes. We should always keep in mind that this is an approximation, but also that, if physicists had started to criticize the theory of electronic bands in crystals by saying that actually most real materials are much more complicated, we would still have computers working with valves instead of microchips.

Another major criticism is that the scale-free ideas and the quest for universal laws are a physicist's obsession that cannot apply to network science which deals with a variety of systems governed by very different dynamical processes. Well, that is true, and all knowledgeable physicists would agree on that. This criticism indeed comes from a general misconception, and also often from a superficial study of the field of phase transition by scientists with non-physics training. Physicists know extremely well that the universality of some statistical laws is not to be confused with the "equivalence of systems", and we have already warned against this confusion in Chapters 3 and 5. Systems sharing the same universal critical properties have at the same time enormous macroscopic differences. The scaling exponents of a ferromagnetic system and a lattice gas at the critical point may be the same, but any other quantities such as density or temperature at the transition are completely different and their calculation requires the inclusion of the system details. Even the evaluation of scaling and critical exponents generally needs to take into account some specific features of the system. A non-physicist will find it hard to believe, but the Ising model so widely used as the prototype for phase transitions does not work for real-world ferromagnets. The quantum nature of spins and other properties that alter the basic symmetries of real systems have demanded the development of more refined models such as the Heisenberg model, the XY model and many others, all of them having different critical behavior. What, then, is the origin of the success of the Ising model and universal phase transitions? It lies in its conceptual power. To calculate the critical behavior and the scaling of a system we need to focus on basic symmetries and interactions, which are responsible for the large-scale statistical behavior. The concept of universality allows the discrimination of what is relevant and what is superfluous in the determination of a specific set of statistical properties, defining general classes of the system's large-scale behavior. This does not imply that these are the sole relevant properties of the system. The system diversity indeed may be contained in features that are not relevant for the large-scale scaling properties but are crucial for the other properties. In many cases, outside the physics community, the ideas of scaling behavior and the mechanisms ruling its emergence in complex networks are equivocated with the claim that "all networks are equal in the eyes of the physicist." This is far from reality, as testified by the fact that while other scientific communities are writing dozens of papers to discredit the BA model as a realistic model for a specific system, unknown to them physicists have already written hundreds of papers in which different ingredients and dynamics more specifically devised for different systems are developed. The value of the BA model, like that of the Ising model, is purely conceptual. No one considers the Ising model as a realistic candidate for most physical systems, and so it should be with the BA and the Watts–Strogatz models. For instance, the various extensions of the BA model developed for specific systems are surely more appropriate than the original BA model itself. At the same time, the study of conceptual models is well worth while, since they represent the guinea pigs on which the theoretician is supposed to test ideas and achieve understanding before moving to more realistic situations.

The issue of engineering versus self-organization represents still another point of discussion. As we discussed in some of the previous chapters, engineering and function play a crucial role in the shaping of networks and dynamical processes. Notwithstanding engineering, even networks which are human artifacts do not have a blueprint above a certain scale. The Internet, for instance, is clearly blueprinted at the level of Local Area Networks and large Autonomous Systems. It is also constrained by technical, economic, and geographical features which are very specific and to some extent time-dependent. The large-scale structure, however, is not planned and is surely the outcome of a self-organization process. The large-scale statistical approach is therefore not in conflict with the detailed engineering approach. On the contrary, the two approaches deal with different scales and features of the problem, and a final understanding of the internet structure will result only from the combination of the two. This combination is crucial but unlikely to happen before the two approaches have been extensively exploited at their respective scales and all their facets mastered.

In this book we have focused mostly on approaches dealing with the large-scale and general properties of systems by using stylized models. There is no doubt that a complete understanding of networked structure requires diving into the specifics of each system by adopting a domain specific perspective. We believe, however, that domain specialists have a lot to gain from the general approach and the "new science of complex networks." The models presented are amenable to many refinements and the analytical techniques used can be easily exported. New observables and measurements can be suggested. The interplay of the micro-level with the macro-level is a major question in most of the systems we are facing nowadays and the approach to complex networks presented here can provide basic insights into this issue. We hope that this book will contribute to a bidirectional exchange of ideas, eventually leading to a truly interdisciplinary approach to network science.

Appendix 1

Random graphs

A network is said to be random when the probability that an edge exists between two nodes is completely independent from the nodes' attributes. In other words, the only relevant function is the degree distribution $P(k)$. In the case that we are mainly interested in the node degree, by using the correlation function language used in Chapter 1 this implies that all degree correlation functions must be trivial. Even uncorrelated networks, however, must satisfy some basic constraints. The first one is the normalization relation

$$\sum_k P(k) = 1. \tag{A1.1}$$

A second constraint is imposed by the fact that all edges must point from one vertex to another, so that no edges with dangling ends exist in the network. This amounts to a degree detailed balance condition imposing that the total number of edges pointing from vertices of degree k to vertices of degree k' must be equal to the number of edges that point from vertices of degree k' to vertices of degree k. In order to state this condition mathematically let us denote by N_k the number of vertices with degree k. Since $\sum_k N_k = N$, we can define the degree distribution as

$$P(k) \equiv \lim_{N \to \infty} \frac{N_k}{N}. \tag{A1.2}$$

To define the network completely, apart from the relative number of vertices of a given degree, we need to specify how the vertices are connected through the symmetric matrix $E_{kk'}$ of the total number of edges between vertices of degree k and vertices of degree k' for $k \neq k'$. The diagonal values E_{kk} are equal to twice the number of connections between vertices in the same degree class, $k = k'$, yielding the following identities:

$$\sum_{k'} E_{kk'} = k N_k, \tag{A1.3}$$

$$\sum_{k,k'} E_{kk'} = \langle k \rangle N = 2E, \tag{A1.4}$$

where $\langle k \rangle$ is the average degree and E is the total number of edges in the network. In fact, the first relation states that the number of edges emanating from all the vertices of degree k is simply $k N_k$, while the second indicates that the sum of all the vertices' degrees is equal to twice the number of edges. The first identity allows us to write the conditional probability $P(k'|k)$, defined as the probability that an edge from a k vertex points to a k' vertex, as

$$P(k'|k) = \frac{E_{kk'}}{kN_k}. \tag{A1.5}$$

The second identity allows the definition of the joint degree distribution as

$$P(k,k') = \frac{E_{kk'}}{\langle k \rangle N}, \tag{A1.6}$$

with the symmetric function $(2 - \delta_{k,k'})P(k,k')$ being the probability that a randomly chosen edge connects two vertices of degrees k and k'. This yields

$$P(k'|k) = \frac{E_{kk'}}{kN_k} \equiv \frac{\langle k \rangle P(k,k')}{kP(k)}, \tag{A1.7}$$

from where the detailed balance condition

$$kP(k'|k)P(k) = k'P(k|k')P(k'), \tag{A1.8}$$

follows immediately as a consequence of the symmetry of $P(k,k')$.

For uncorrelated networks, in which $P(k'|k)$ does not depend on k, application of the normalization condition $\sum_k P(k|k') = 1$ into Equation (A1.8) yields the form

$$P(k'|k) = \frac{1}{\langle k \rangle}k'P(k'), \tag{A1.9}$$

that has been used throughout this book. This recovers that for generalized random networks the conditional probability is just proportional to the relative proportion of edges departing from nodes k' and it is the same for all edges whatever the degree k of the emitting node. Equation (A1.9) allows the explicit calculation of some basic quantities of random networks with arbitrary degree distribution $P(k)$.

Following Newman (2003b), let us consider a generalized random graph with arbitrary degree distribution $P(k)$. Since edges are assigned at random between pairs of vertices, Equation (A1.9) implies that the probability that any edge points to a vertex of degree k is given by $kP(k)/\langle k \rangle$. Consider now a vertex i; following the edges emanating from it, we can arrive at k_i other vertices. The probability distribution that any of the neighboring vertices has k edges pointing to other vertices different from i (plus the edge from which we arrived) is therefore given by the function

$$q(k) = \frac{(k+1)P(k+1)}{\langle k \rangle}. \tag{A1.10}$$

In other words, $q(k)$ gives the probability distribution of the *number of second nearest neighbors* that can be reached following a given edge in a vertex. The average number of these second nearest neighbors is then given by

$$\sum_k kq(k) = \frac{1}{\langle k \rangle}\sum_k k(k+1)P(k+1) = \frac{\langle k^2 \rangle - \langle k \rangle}{\langle k \rangle}. \tag{A1.11}$$

The absence of correlations also yields that any vertex j, neighbor of the vertex i, is connected to another vertex l, which is at the same time a neighbor of i, with probability

$k_j k_l/\langle k \rangle N$. Thus, the clustering coefficient is simply defined as the average of this quantity over the distribution of all possible neighbors of i, i.e.

$$\langle C \rangle = \frac{1}{N\langle k \rangle} \sum_{k_j} \sum_{k_l} k_j k_l q(k_j) q(k_l) = \frac{1}{N} \frac{(\langle k^2 \rangle - \langle k \rangle)^2}{\langle k \rangle^3}. \tag{A1.12}$$

In order to provide an approximate expression for the scaling of the diameter of random networks[1] we can compute iteratively the average number of neighbors z_n at a distance n away from a given vertex as

$$z_n = \frac{\langle k^2 \rangle - \langle k \rangle}{\langle k \rangle} z_{n-1}. \tag{A1.13}$$

Finally, by considering that $z_1 = \langle k \rangle$, it is possible to obtain the explicit expression

$$z_n = \left(\frac{\langle k^2 \rangle - \langle k \rangle}{\langle k \rangle} \right)^{n-1} \langle k \rangle. \tag{A1.14}$$

If the average shortest path length is $\langle \ell \rangle$, then the number of neighbors at this distance must be approximately equal to the size of the graph N, thus obtaining $z_{\langle \ell \rangle} = N$. Using the same argument as in the Erdős–Rényi model, we readily obtain

$$\langle \ell \rangle \approx 1 + \frac{\log[N/\langle k \rangle]}{\log[(\langle k^2 \rangle - \langle k \rangle)/\langle k \rangle]}. \tag{A1.15}$$

The small-world properties are thus evident also for generalized random graphs and it is easy to check that for the case of a Poisson distribution, with second moment $\langle k^2 \rangle = \langle k \rangle + \langle k \rangle^2$, one recovers the results for the Erdős–Rényi model. However, we must keep in mind that the previous expression is a rather crude approximation, which might fail for more complex degree distributions (especially in the presence of heavy tails).

The absence of degree correlations allows also the calculation of the dependence of the degree cut-off $k_c(N)$ induced by the finite size N of the network in scale-free networks.[2] The presence of the degree cut-off translates in the degree distribution into an explicit dependence on the network size (or time) that we can write in the scaling form

$$P(k, N) = k^{-\gamma} f\left[\frac{k}{k_c(N)} \right], \tag{A1.16}$$

where $f(x)$ is constant for $x \ll 1$ and decreases very quickly for $x \gg 1$. It is possible to obtain an upper bound for the functional dependence on N of the degree cut-off for generalized uncorrelated random graphs using an extremal theory argument. In the continuous k approximation, consider a random graph with normalized degree distribution in the infinite size limit $P(k) = (\gamma - 1)m^{\gamma-1}k^{-\gamma}$, with $\gamma > 2$ and $k \in [m, \infty]$, where m is the minimum degree of the graph. Consider now that we generate a graph of size N by sorting N independent random variables according to the distribution $P(k)$, obtaining

---

[1] This calculation is analogous to that used for the Erdős–Rényi model in Section 3.1.

[2] The result obtained here is valid only if the only origin of the degree cut-off resides in the finite number of vertices forming the network. In other situations, networks may exhibit a degree cut-off due to external constraints and finite connectivity resources (Amaral *et al.*, 2000; Mossa *et al.*, 2002). In this case k_c is not related to the network size and has to be considered as an external parameter.

the sample $\{k_1, \ldots, k_N\}$. Let us define K the maximum value of this particular sample, $K = \max\{k_1, \ldots, k_N\}$. When generating an ensemble of graphs, we will obtain in each case a different value of the maximum degree K. Thus, we can define the cut-off $k_c(N)$ as the average value of K, weighted by the distribution $P(k)$. It is easy to see that the probability of this maximum being less than or equal to K is equal to the probability of all the individual values k_i being in their turn less than or equal to K. This means that the distribution function of the maximum value K is just

$$\Pi(K, N) = \left[\Psi(K)\right]^N, \tag{A1.17}$$

where $\Psi(K)$ is the distribution function of the probability $P(k)$, i.e. $\Psi(K) = \int_m^K P(k)\,dk = 1 - (K/m)^{-\gamma+1}$. By differentiating Equation (A1.17), we obtain the probability distribution of maximum values, namely

$$\pi(K, N) = \frac{d\Pi(K, N)}{dK} = \frac{N(\gamma - 1)}{m}\left(\frac{K}{m}\right)^{-\gamma}\left[1 - \left(\frac{K}{m}\right)^{-\gamma+1}\right]^{N-1}. \tag{A1.18}$$

If the degree cut-off is defined as the average value of the maximum K, then we have that

$$k_c(N) = \int_m^\infty K\pi(K, N)\,dK = \frac{Nm}{\lambda}\frac{\Gamma(1 + \lambda)\Gamma(N)}{\Gamma(N + \lambda)}, \tag{A1.19}$$

where $\Gamma(x)$ is the Gamma function and we have defined the constant $\lambda = (\gamma - 2)/(\gamma - 1)$. Using the asymptotic relation $\lim_{N\to\infty} \Gamma(N + a)/\Gamma(N + b) \simeq N^{a-b}$ (Abramowitz and Stegun, 1972), we obtain the leading behavior for large N

$$k_c(N) \simeq m\frac{\Gamma(1 + \lambda)}{\lambda}N^{1-\lambda} \sim mN^{1/(\gamma-1)}. \tag{A1.20}$$

The previous equation is in fact an upper bound for $k_c(N)$, since we have only considered the possible values that the random variables k_i can take, according to the probability distribution $P(k)$. If those values must represent the degree sequence of an actual graph, some constraints would then apply, especially if we want to avoid the presence of loops or multiple edges.

The previous scaling relation with the size of the graph allows, finally, an important consideration on the scaling of $\langle C \rangle$ and $\langle \ell \rangle$. In generalized random graphs these quantities depend essentially on the first and second moments of $P(k)$. For a bounded degree distribution, in which the degree fluctuations $\langle k^2 \rangle$ are finite, we observe that $\langle C \rangle \sim 1/N$ and $\langle \ell \rangle \sim \log N$, in agreement with the results found for the Erdős–Rényi model. On the other hand, for degree distributions with a fat tail, such as a power law, the second moment $\langle k^2 \rangle$ can be very large, and even diverge with N. In this case, the prefactor in Equation (A1.12) can be noticeably large, and yield a clustering coefficient that might be higher than the one corresponding to an Erdős–Rényi graph with the same size and average degree. Indeed, if we consider a scale-free graph with degree distribution $P(k) \sim k^{-\gamma}$, we have that $\langle k \rangle$ is finite, while the degree fluctuations scale as $\langle k^2 \rangle \simeq k_c^{3-\gamma}$, where k_c is the maximum degree present in the graph. By plugging the behavior of k_c given by Equation (A1.20) into Equation (A1.12), we obtain an average clustering coefficient depending on the network size as

$$\langle C \rangle_N \simeq N^{(7-3\gamma)/(\gamma-1)}. \tag{A1.21}$$

Since the clustering coefficient cannot be larger than 1, Equation (A1.12) must be restricted to values of the degree exponent $\gamma > 7/3$. The last consideration points out that correlations must arise naturally to allow the connectivity pattern observed in some real-world networks. The random approximation is, however, a basic model often used in calculations as it generally allows an explicit analytic expression to be obtained.

Appendix 2

Generating functions formalism

The percolation theory in networks with arbitrary degree distribution developed in Chapter 6 can be treated using the generating function formalism, as developed by Callaway *et al.* (2000) and Newman *et al.* (2001).

Let us note for simplicity $P_k = P(k)$, the probability that a randomly chosen node has degree k. The *generating function* of the distribution P_k is

$$G_0(x) = \sum_k P_k x^k. \tag{A2.1}$$

The name "generating function" comes from the fact that each P_k can be recovered from the knowledge of $G_0(x)$ through the following formula:

$$P_k = \frac{1}{k!} \frac{d^k}{dx^k} G_0 \bigg|_{x=0}, \tag{A2.2}$$

i.e. by taking the kth derivative of G_0, while the moments of the distribution are given by

$$\langle k^n \rangle = \left(x \frac{d}{dx} \right)^n G_0 \bigg|_{x=1}. \tag{A2.3}$$

For example, it is easy to see that the average degree $\langle k \rangle = \sum_k k P_k$ is equal to $G_0'(1)$.

Another important property of G_0 is the following: the distribution of the sum of the degrees of m vertices taken at random is generated by the mth power of G_0, G_0^m. This is easy to see in the case $m = 2$:

$$G_0^2(x) = \left(\sum_k P_k x^k \right)^2 = \sum_k \left(\sum_{k_1, k_2} P_{k_1} P_{k_2} \delta_{k_1+k_2, k} \right) x^k, \tag{A2.4}$$

where the Kronecker symbol is such that $\delta_{i,j} = 1$ if $i = j$ and $\delta_{i,j} = 0$ if $i \neq j$. The coefficient of x^k thus sum of all terms $P_{k_1} P_{k_2}$ such that the sum of k_1 and k_2 is k, i.e. exactly the probability that the sum of the degrees of two independent vertices is k.

While P_k is the probability that a randomly chosen node has degree k, the probability that a node reached *by following a randomly chosen edge* has degree k is $k P_k / \langle k \rangle$. The probability that such a node has k "outgoing" links, i.e. k other than the edge followed to reach it, is thus $q_k = (k+1)P_{k+1}/\langle k \rangle$, and the corresponding generating function therefore reads

$$G_1(x) = \sum_k \frac{(k+1)P_{k+1}}{\langle k \rangle} x^k = \frac{1}{\langle k \rangle} G_0'(x). \tag{A2.5}$$

The definition and properties of generating functions allow us now to deal with the problem of percolation in a random network: let us call $H_1(x)$ the generating function for the distribution of the sizes of the connected components reached by following a randomly chosen edge. Note that $H_1(x)$ considers only *finite* components and therefore excludes the possible giant cluster. We neglect the existence of loops, which is indeed legitimate for such finite components. The distribution of sizes of such components can be visualized by a diagrammatic expansion as shown in Figure A2.1: each (tree-like) component is composed by the node initially reached, plus k other tree-like components, *which have the same size distribution*, where k is the number of outgoing links of the node, whose distribution is q_k. The probability that the global component Q_S has size S is thus

$$Q_S = \sum_k q_k \, \text{Prob(union of } k \text{ components has size } S - 1) \tag{A2.6}$$

(counting the initially reached node in S). The generating function H_1 is by definition

$$H_1(x) = \sum_S Q_S x^S, \tag{A2.7}$$

and the distribution of the sum of the sizes of the k components is generated by H_1^k (as previously explained for the sum of degrees), i.e.

$$\sum_S \text{Prob(union of } k \text{ components has size } S) \cdot x^S = (H_1(x))^k. \tag{A2.8}$$

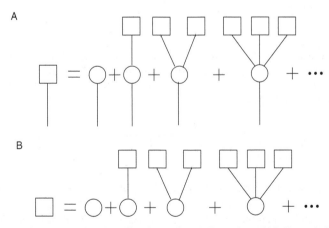

Fig. A2.1. Diagrammatic visualization of (A) Equation (A2.9) and (B) Equation (A2.10). Each square corresponds to an arbitrary tree-like cluster, while the circle is a node of the network.

Putting together equations (A2.6), (A2.7), and (A2.8), one obtains the following self-consistency equation for H_1:

$$
\begin{aligned}
H_1(x) &= xq_0 + xq_1 H_1(x) + xq_2(H_1(x))^2 + \cdots + xq_k(H_1(x))^k + \cdots \\
&= xG_1(H_1(x)).
\end{aligned} \tag{A2.9}
$$

We now define $H_0(x)$ as the generating function of the sizes of connected components obtained starting from a randomly chosen vertex. A similar diagrammatic expansion can be written (Figure A2.1): a component either is restricted to the vertex if it has degree 0, or is the union of k components reached through the k links of the vertex. As previously the size of this union is generated by H_1^k, so that:

$$
\begin{aligned}
H_0(x) &= xP_0 + xP_1 H_1(x) + xP_2(H_1(x))^2 + \cdots + xP_k(H_1(x))^k + \cdots \\
&= xG_0(H_1(x)).
\end{aligned} \tag{A2.10}
$$

Equation (A2.9) is in general complicated and cannot be easily solved. Together with Equation (A2.10), however, it allows us to obtain the average size of the connected clusters. This size is given by $\langle s \rangle = H_0'(1)$, i.e. by deriving Equation (A2.10) in $x = 1$:

$$
H_0'(1) = G_0(H_1(1)) + 1 \cdot H_1'(1) \cdot G_0'(H_1(1)). \tag{A2.11}
$$

If no giant component is present, normalization of the size distribution ensures that $H_1(1) = 1$, and $G_0(1) = 1$ as well by normalization of P_k. The derivative $H_1'(1)$ can also be computed by deriving Equation (A2.9) in $x = 1$, yielding

$$
H_1'(1) = \frac{1}{1 - G_1'(1)}. \tag{A2.12}
$$

Moreover, $G_0'(1) = \langle k \rangle$, and $G_1'(1) = G_0''(1)/\langle k \rangle = \kappa - 1$ where $\kappa = \langle k^2 \rangle / \langle k \rangle$ is the heterogeneity parameter of the network, so that one finally obtains

$$
\langle s \rangle = 1 + \frac{\langle k \rangle}{2 - \kappa} \tag{A2.13}
$$

which diverges when $\langle k^2 \rangle = 2\langle k \rangle$. We recover the Molloy–Reed criterion for the transition percolation, Equation (6.13). When a giant component is present, i.e. for $\langle k^2 \rangle > 2\langle k \rangle$, the computations can still be carried out, with slight modifications since the size distribution is no longer normalized (Newman *et al.*, 2001).

Appendix 3

Percolation in directed networks

In Chapter 6, the case of percolation and resilience of undirected graphs has been considered. Various authors have tackled the problem of its generalization to directed networks (Newman *et al.*, 2001; Schwartz *et al.*, 2002; Dorogovtsev *et al.*, 2001a; Boguñá and Serrano 2005).

As seen in Chapter 1, directed networks are topologically richer and more complex than their undirected counterparts. The links attached to a node can be either incoming or outgoing. The number of in-links is the in-degree k_{in}, and the number of out-links is the out-degree k_{out}. Since some pairs of nodes can be linked by two directed links pointing in the opposite directions, it is also possible to distinguish between strictly incoming, strictly outgoing or bidirectional links, the total degree being the sum of their respective numbers k_i, k_o, k_b, with $k_{in} = k_i + k_b$ and $k_{out} = k_o + k_b$. The degree distribution is thus the joint distribution of these degrees. Most often, strictly directed networks are considered, with no bidirectional links: $k_b = 0$. Note that by a simple conservation rule, the averages of k_i and k_o are then equal: $\langle k_i \rangle = \langle k_o \rangle$. Each (weakly) connected component (WCC) of a directed graph has also an internal structure as described in Chapter 1, and is composed of a strongly connected component (SCC), an in and an out components, tendrils and tubes. The giant WCC (GWCC) appears at the percolation threshold for the undirected version of the graph. On the other hand, giant SCC, in and out components as sketched in Figure 1.2 appear at another phase transition, as explained in the following.

A3.1 Purely directed networks

Let us first consider the case of purely directed networks: all links are directed, and no bidirectional links exist. The formalism of generating functions (see Appendix 2) can be adapted (Newman *et al.*, 2001) to this case by defining $p_{k_i k_o}$ as the probability for a node to have k_i incoming links and k_o outgoing links, and its generating function of two variables $\sum_{k_i k_o} p_{k_i k_o} x^{k_i} y^{k_o}$. We follow here instead the simple argument given by Schwartz *et al.* (2002) for the existence of a giant connected component. Let us consider a site b, reached from another site a. In order for it to be part of the giant component, such a site must have on average at least one outgoing link:

$$\langle k_o(b) | a \rightarrow b \rangle \geq 1, \tag{A3.1}$$

and, if $P(k_i, k_o | a \rightarrow b)$ is the probability for b to have in-degree k_i and out-degree k_o *knowing that* there is a link from a to b, we obtain

$$\sum_{k_i, k_o} k_o P(k_i, k_o | a \rightarrow b) \geq 1. \qquad (A3.2)$$

We can now use Bayes rule to write, with obvious notations:

$$
\begin{aligned}
P(k_i, k_o | a \rightarrow b) &= P(k_i, k_o, a \rightarrow b)/P(a \rightarrow b) \\
&= P(a \rightarrow b | k_i, k_o) P(k_i, k_o)/P(a \rightarrow b). \qquad (A3.3)
\end{aligned}
$$

In a random network of N nodes, $P(a \rightarrow b) = \langle k_i \rangle/(N-1)$ and $P(a \rightarrow b | k_i, k_o) = k_i/(N-1)$, so that the condition which generalizes the Molloy–Reed criterion finally reads

$$\sum_{k_i, k_o} \frac{k_i k_o}{\langle k_i \rangle} P(k_i, k_o) \geq 1 \qquad (A3.4)$$

i.e.

$$\langle k_i k_o \rangle \geq \langle k_i \rangle. \qquad (A3.5)$$

Let us now consider a directed network from which a fraction p of nodes has been randomly deleted. A node has then in-degree k_i and out-degree k_o if it has in-degree k_i^0 and out-degree k_o^0 in the undamaged network, and if $k_i^0 - k_i$ of its in-neighbours have been removed, together with $k_o^0 - k_o$ of its out-neighbours. The joint probability distribution of k_i and k_o can thus be written as

$$P_p(k_i, k_o) = \sum_{k_i^0, k_o^0} P_0(k_i^0, k_o^0) \binom{k_i^0}{k_i} (1-p)^{k_i} p^{k_i^0 - k_i} \binom{k_o^0}{k_o} (1-p)^{k_o} p^{k_o^0 - k_o}, \qquad (A3.6)$$

where P_0 is the degree distribution of the undamaged network. According to the condition (A3.5), a giant component is present if and only if

$$\sum_{k_i, k_o} k_i k_o P_p(k_i, k_o) \geq \sum_{k_i, k_o} k_o P_p(k_i, k_o). \qquad (A3.7)$$

We use on each side of the inequality the relation

$$\sum_{k=1}^{k^0} k \binom{k^0}{k} x^k y^{k^0 - k} = x \frac{d}{dx}(x+y)^{k^0} = x k^0 (x+y)^{k^0 - 1} \qquad (A3.8)$$

with $x = 1 - p$ and $y = p$, to obtain

$$(1-p)^2 \langle k_i^0 k_o^0 \rangle \geq (1-p) \langle k_o^0 \rangle. \qquad (A3.9)$$

The critical value for the existence of a giant component is therefore given by

$$p_c = 1 - \frac{\langle k_o^0 \rangle}{\langle k_i^0 k_o^0 \rangle}. \qquad (A3.10)$$

Such a result shows that directed networks are resilient if and only if $\langle k_i^0 k_o^0 \rangle$ diverges in the infinite-size limit.

The case of uncorrelated in- and out-degree corresponds to $\langle k_i^0 k_o^0 \rangle = \langle k_i^0 \rangle \langle k_o^0 \rangle$, i.e. to a critical value

$$p_c = 1 - \frac{1}{\langle k_o^0 \rangle} \tag{A3.11}$$

strictly lower than 1: contrarily to undirected networks, uncorrelated directed scale-free networks are not particularly resilient to random failures. On the other hand, correlations between in- and out-degrees are often observed and can modify this property. For example, Schwartz *et al.* (2002) study the case of scale-free distributions of in- and out-degrees, given by

$$P_i(k_i) = (1 - B)\delta_{k_i,0} + Bc_i k_i^{-\lambda_i}(1 - \delta_{k_i,0}); \quad P_o(k_o) = c_o k_o^{-\lambda_o}. \tag{A3.12}$$

where the additional parameter B is needed to have the possibility to ensure $\langle k_i \rangle = \langle k_o \rangle$. Moreover, for each site, k_o is either completely uncorrelated from k_i (with probability $1 - A$) or fully determined by k_i (with probability A) through a deterministic function $k_o = f(k_i)$. The condition of scale-free distributions implies that $f(k_i) \sim k_i^{\lambda_i - 1/\lambda_o - 1}$, and the joint distribution of in- and out-degrees thus reads

$$
\begin{aligned}
P(k_i, k_o) &= (1 - A)Bc_i c_o k_i^{-\lambda_i} k_o^{-\lambda_o} + ABc_i k_i^{-\lambda_i} \delta_{k_o, f(k_i)} \quad \text{if } k_i \neq 0 \\
&= (1 - B)c_o k_o^{-\lambda_o} \quad \text{if } k_i = 0.
\end{aligned}
\tag{A3.13}
$$

It is then easy to see that, for any finite fraction of fully correlated sites, the average $\langle k_i k_o \rangle$ diverges if and only if

$$(\lambda_i - 2)(\lambda_o - 2) \leq 1, \tag{A3.14}$$

leading to a very resilient network with $p_c \to 1$, while full uncorrelation ($A = 0$) gives $p_c = 1 - 1/\langle k_o \rangle < 1$ even for scale-free distributions.

A3.2 General case

Boguñá and Serrano (2005) have treated the more general case in which a network contains both directed and undirected (bidirectional) links. The degree of a node is then a three-component vector $\mathbf{k} = (k_i, k_o, k_b)$ where k_i, k_o, k_b are respectively the number of incoming, outgoing and bidirectional links. The degree distribution is then denoted $P(\mathbf{k})$, and correlations are encoded in the conditional probabilities $P_a(\mathbf{k}'|\mathbf{k})$ with $a = \text{i}$ or $a = \text{o}$ or $a = \text{b}$ for the probability of reaching a vertex of degree $\mathbf{k}'$ when leaving a vertex of degree $\mathbf{k}$ through respectively an incoming, out-going or bidirectional link. It can then be shown through the generating function formalism that giant in- and out-components appear when $\Lambda_m > 1$, where Λ_m is the largest eigenvalue of the correlation matrix (Boguñá and Serrano, 2005).

$$
C_{\mathbf{kk}'}^o = \begin{pmatrix} k_o' P_o(\mathbf{k}'|\mathbf{k}) & k_b' P_o(\mathbf{k}'|\mathbf{k}) \\ k_o' P_b(\mathbf{k}'|\mathbf{k}) & (k_b' - 1) P_b(\mathbf{k}'|\mathbf{k}) \end{pmatrix} \tag{A3.15}
$$

(the matrix $C_{\mathbf{kk}'}^i$ obtained by replacing the indices o by i in the expression of $C_{\mathbf{kk}'}^o$ has the same eigenspectrum, so that the condition for the presence of giant in- and out-components are the same).

Note that both cases of purely directed and purely undirected networks can be recovered from the condition $\Lambda_m > 1$. Indeed, for purely undirected networks only k_b is defined, and the correlation matrix becomes $(k' - 1)P(k'|k)$, giving back the results of Moreno and Vázquez (2003). For purely directed networks, on the contrary, $k_b = 0$ and $C_{\mathbf{kk'}}^o = k_o' P_o(\mathbf{k'}|\mathbf{k})$; if the degrees of neighbouring nodes are uncorrelated, moreover, $P_o(\mathbf{k'}|\mathbf{k}) = k_i' P(\mathbf{k'})/\langle k_i \rangle$ and one recovers the condition $\langle k_i k_o \rangle > \langle k_i \rangle$ for the existence of a giant component.

Appendix 4

Laplacian matrix of a graph

The discrete Laplacian on a network and acting on a function ϕ is given by

$$\Delta\phi(v) = \sum_{w \in V_{(v)}} (\phi(w) - \phi(v)). \qquad (A4.1)$$

From this definition, one introduces the Laplacian matrix of a graph given by $\mathbf{L} = -\mathbf{\Delta}$, which can be rewritten as

$$\mathbf{L} = \mathbf{D} - \mathbf{X} \qquad (A4.2)$$

where $\mathbf{D}$ is the diagonal degree matrix with elements $D_{ij} = \delta_{ij}k_i$ and $\mathbf{X}$ is the adjacency matrix. The Laplacian matrix has thus diagonal elements equal to the degree $L_{ii} = k_i$ and is the opposite of the adjacency matrix for off diagonal elements $L_{i \neq j} = -x_{ij}$. It is therefore symmetric if the graph is undirected. This matrix is a central concept in spectral graph analysis (Mohar, 1997). For undirected graphs, some important properties appear, such as the following:

- If the graph is an infinite square lattice grid, this definition of the Laplacian can be shown to correspond to the continuous Laplacian.
- The Laplacian matrix L is symmetric; it has real non-negative eigenvalues $0 \leq \lambda_1 \leq \ldots \leq \lambda_N$.
- The multiplicity of 0 as an eigenvalue of L is equal to the number of connected components of the graph.
- The second smallest eigenvalue is called the algebraic connectivity. It is non-zero only if the graph is formed of a single connected component. The magnitude of this value reflects how well connected the overall graph is, and has implications for properties such as synchronizability and clustering.

Appendix 5

Return probability and spectral density

In this appendix, we give a simple derivation of the relation between the eigenvalue distribution $\rho(\lambda)$ (also called the spectral density) of a generic Laplacian operator L and the return probability $p_0(t)$ for the random walk described by the following master equation

$$\partial_t p(i, t|i_0, 0) = - \sum_j L_{ij} p(j, t|i_0, 0) \tag{A5.1}$$

with the initial condition $p(i, 0|i_0, 0) = \delta_{ii_0}$. Here the Laplacian can be either of the form (8.9) or of the form studied in Chapter 7 and Appendix 4, the important point being that $\sum_i L_{ij} = 0$ to ensure that the evolution equation (A5.1) preserves the normalization $\sum_i p(i, t|i_0, 0) = 1$. The spectral density is defined as

$$\rho(\lambda) = \left\langle \frac{1}{N} \sum_{i=1}^{N} \delta(\lambda - \lambda_i) \right\rangle \tag{A5.2}$$

where the λ_i are the eigenvalues of the Laplacian and where the brackets denote the average over different realizations of the random network on which the random walk takes place.

The Laplace transform of $p(i, t|i_0, 0)$ is defined as

$$\tilde{p}_{ii_0}(s) = \int_0^\infty dt\, e^{-st} p(i, t|i_0, 0). \tag{A5.3}$$

The Laplace transform of $\partial_t p(i, t|i_0, 0)$ can be obtained by an integration by parts as

$$\int_0^\infty dt\, e^{-st} \partial_t p(i, t|i_0, 0) = -p(i, 0|i_0, 0) + s\tilde{p}_{ii_0}(s) = s\tilde{p}_{ii_0}(s) - \delta_{ii_0} \tag{A5.4}$$

where $\delta_{ii_0} = 1$ is $i = i_0$, and 0 otherwise. This allows us to rewrite Equation (A5.1) as

$$s\tilde{p}_{ii_0}(s) - \delta_{ii_0} = - \sum_j L_{ij} \tilde{p}_{ji_0}(s) \tag{A5.5}$$

or

$$\sum_j \left(s\delta_{ij} + L_{ij} \right) \tilde{p}_{ji_0}(s) = \delta_{ii_0}. \tag{A5.6}$$

This equality means that the matrix $\tilde{\mathbf{p}}(s)$ is therefore the inverse of the matrix $s\mathbf{I} + \mathbf{L}$, where $\mathbf{I}$ is the unit matrix ($I_{ij} = \delta_{ij}$).

The probability of return is by definition the probability of returning to the initial point i_0, averaged over all initial nodes and over different realizations of the random network

$$p_0(t) = \left\langle \frac{1}{N} \sum_{i_0} p(i_0, t | i_0, 0) \right\rangle. \tag{A5.7}$$

Taking the Laplace transform of this equation leads to

$$\tilde{p}_0(s) = \left\langle \frac{1}{N} \sum_{i_0} \tilde{p}_{i_0 i_0}(s) \right\rangle = \left\langle \frac{1}{N} \mathrm{Tr}\,\tilde{\mathbf{p}}(s) \right\rangle, \tag{A5.8}$$

where Tr denotes the trace operation. The trace of a matrix is moreover equal to the sum of its eigenvalues, and we can use the fact that $\tilde{\mathbf{p}}(s)$ is the inverse of $s\mathbf{I} + \mathbf{L}$, to obtain that its eigenvalues are given by $1/(s + \lambda_i)$, $i = 1, \ldots, N$. We therefore obtain

$$\tilde{p}_0(s) = \left\langle \frac{1}{N} \sum_i \frac{1}{s + \lambda_i} \right\rangle. \tag{A5.9}$$

We can now obtain $p_0(t)$ as the inverse Laplace transform of $\tilde{p}_0(s)$, which is given by the integral in the complex plane

$$p_0(t) = \int_{c-i\infty}^{c+i\infty} \mathrm{d}s\, e^{ts}\, \tilde{p}_0(s), \tag{A5.10}$$

where here $\mathrm{i}^2 = -1$ and c is a constant larger than any singularity of $\tilde{p}_0(s)$. This yields

$$
\begin{aligned}
p_0(t) &= \int_{c-i\infty}^{c+i\infty} \mathrm{d}s\, e^{ts} \left\langle \frac{1}{N} \sum_j \frac{1}{s + \lambda_j} \right\rangle \\
&= \left\langle \frac{1}{N} \sum_j e^{-\lambda_j t} \right\rangle, \tag{A5.11}
\end{aligned}
$$

where we have used the residue theorem in order to obtain the last equality. Using the definition of $\rho(\lambda)$, this can then be rewritten as Equation (8.10)

$$p_0(t) = \int_0^\infty \mathrm{d}\lambda\, e^{-\lambda t}\, \rho(\lambda).$$

References

Abramowitz, M. and Stegun, I. A. (1972), *Handbook of Mathematical Functions*, Dover.

Abramson, G. and Kuperman, M. (2001), Social games in a social network, *Phys. Rev. E* **63**, 030901.

Acebrón, J. A., Bonilla, L. L., Pérez Vicente, C. J., Ritort, F. and Spigler, R. (2005), The Kuramoto model: A simple paradigm for synchronization phenomena, *Rev. Mod. Phys.* **77**, 137–185.

Achard, S., Salvador, R., Whitcher, B., Suckling, J. and Bullmore, E. (2006), A resilient, low-frequency, small-world human brain functional network with highly connected association cortical hubs, *J. Neurosci.* **26**, 63–72.

Adamic, L. A. and Adar, E. (2005), How to search a social network, *Social Networks* **27**, 187–203.

Adamic, L. A. and Huberman, B. A. (2001), The Web's hidden order. In *Communications of the ACM*, Vol. 44, ACM Press, pp. 55–60.

Adamic, L. A., Lukose, R. M. and Huberman, B. A. (2003), Local search in unstructured networks. In S. Bornholdt and H. G. Schuster, eds., *Handbook of Graphs and Networks: From the Genome to the Internet*, Wiley-VCH, pp. 295–317.

Adamic, L. A., Lukose, R. M., Puniyani, A. R. and Huberman, B. A. (2001), Search in power-law networks, *Phys. Rev. E* **64**, 046135.

Ahn, Y.-Y., Jeong, H., Masuda, N. and Noh, J. D. (2006), Epidemic dynamics of two species of interacting particles on scale-free networks, *Phys. Rev. E* **74**, 066113.

Ahuja, R. K., Magnanti, T. L. and Orlin, J. B. (1993), *Network Flows*, Prentice Hall.

Aiello, W., Chung, F. and Lu, L. (2001), A random graph model for power law graphs, *Exp. Math.* **10**, 53–66.

Alava, M., Nukala, P. K. V. V. and Zapperi, S. (2006), Statistical models of fracture, *Adv. Phys.* **55**, 349–476.

Albert, R. (2004), Boolean modeling of genetic regulatory networks. In E. Ben-Naim, H. Frauenfelder and Z. Toroczkai, eds., *Complex Networks*, Springer-Verlag, pp. 459–482.

Albert, R., Albert, I. and Nakarado, G. L. (2004), Structural vulnerability of the North American power grid, *Phys. Rev. E* **69**, 025103.

Albert, R. and Barabási, A.-L. (2000), Dynamics of complex systems: Scaling laws for the period of Boolean networks, *Phys. Rev. Lett.* **84**, 5660–5663.

Albert, R. and Barabási, A.-L. (2002), Statistical mechanics of complex networks, *Rev. Mod. Phys.* **74**, 47–97.

Albert, R., Jeong, H. and Barabási, A.-L. (1999), Internet: Diameter of the World Wide Web, *Nature* **401**, 130–131.

Albert, R., Jeong, H. and Barabási, A.-L. (2000), Error and attack tolerance of complex networks, *Nature* **406**, 378–382.

Albert, R. and Othmer, H. G. (2003), The topology of the regulatory interactions predicts the expression pattern of the segment polarity genes in Drosophila melanogaster, *J. Theor. Biol.* **223**, 1–18.

Aldana, M. (2003), Boolean dynamics of networks with scale-free topology, *Physica D* **185**, 45–66.

Aldana, M. and Cluzel, P. (2003), A natural class of robust networks, *Proc. Natl Acad. Sci. (USA)* **100**, 8710–8714.

Aldana, M., Coppersmith, S. and Kadanoff, L. P. (2003), Boolean dynamics with random couplings. In E. Kaplan, J. E. Marsden and K. R. Sreenivasan, eds., *Perspectives and Problems in Nonlinear Science. A Celebratory Volume in Honor of Lawrence Sirovich*, Springer, pp. 23–89.

Alderson, D. (2004), *Technological and Economic Drivers and Constraints in the Internet's "Last Mile"*. Technical Report CIT-CDS 04-004.

Aleksiejuk, A., Holyst, J. A. and Stauffer, D. (2002), Ferromagnetic phase transition in Barabási–Albert networks, *Physica A* **310**, 260–266.

Allesina, S., Bodini, A. and Bondavalli, C. (2005), Ecological subsystems via graph theory: the role of strongly connected components, *Oikos* **110**, 164–176.

Almaas, E., Kovács, B., Viscek, T., Oltvai, Z. N. and Barabási, A.-L. (2004), Global organization of metabolic fluxes in the bacterium *Escherichia coli, Nature* **427**, 839–843.

Almaas, E., Krapivsky, P. L. and Redner, S. (2005), Statistics of weighted treelike networks, *Phys. Rev. E* **71**, 036124.

Almaas, E., Kulkarni, R. V. and Stroud, D. (2003), Scaling properties of random walks on small-world networks, *Phys. Rev. E* **68**, 056105.

Alon, U. (2003), Biological networks: the tinkerer as an engineer, *Science* **301**, 1866–1867.

Alon, U. (2007a), *An Introduction to Systems Biology: Design Principles of Biological Circuits*, Chapman & Hall/CRC.

Alon, U. (2007b), Network motifs: theory and experimental approaches, *Nat. Rev. Genet.* **8**, 450–461.

Alvarez-Hamelin, J. I. and Schabanel, N. (2004), An Internet graph model based on trade-off optimization, *Eur. Phys. J. B* **38**, 231–237.

Amaral, L. A. N. and Barthélemy, M. (2003), Complex systems: Challenges, successes and tools. In E. Korutcheva and R. Cuerno, eds., *Advances in Condensed Matter and Statistical Physics*, Nova Science Publishers.

Amaral, L. A. N. and Ottino, J. M. (2004), Complex networks: augmenting the framework for the study of complex systems, *Eur. Phys. J. B* **38**, 147–162.

Amaral, L. A. N., Scala, A., Barthélemy, M. and Stanley, H. E. (2000), Classes of small-world networks, *Proc. Natl Acad. Sci. (USA)* **97**, 11149–11152.

Andersen, R., Chung, F. and Lu, L. (2005), Modeling the small-world phenomenon with local network flow, *Internet Math.* **2**, 359–385.

Anderson, R. M. and May, R. M. (1984), Spatial, temporal and genetic heterogeneity in hosts populations and the design of immunization programs, *IMA J. Math. Appl. Med. Biol.* **1**, 233–266.

Anderson, R. M. and May, R. M. (1992), *Infectious Diseases of Humans: Dynamics and Control*, Oxford University Press.

Antal, T., Redner, S. and Sood, V. (2006), Evolutionary dynamics on degree-heterogeneous graphs, *Phys. Rev. Lett.* **96**, 188104.

Arenas, A., Díaz-Guilera, A. and Guimerà, R. (2001), Communication in networks with hierarchical branching, *Phys. Rev. Lett.* **86**, 3196–3199.

Arenas, A., Díaz-Guilera, A. and Pérez-Vicente, C. J. (2006a), Synchronization processes in complex networks, *Physica D* **224**, 27–34.

Arenas, A., Díaz-Guilera, A. and Pérez-Vicente, C. J. (2006b), Synchronization reveals topological scales in complex networks, *Phys. Rev. Lett.* **96**, 114102.

Aron, J. L., Gove, R. A., Azadegan, S. and Schneider, M. C. (2001), The benefits of a notification process in addressing the worsening computer virus problem: results of a survey and a simulation model, *Computers Security* **20**, 693–714.

Axelrod, R. (1997a), *The Complexity of Cooperation*, Princeton University Press.

Axelrod, R. (1997b), The dissemination of culture: A model with local convergence and global polarization, *J. Conflict Resolut.* **41**, 203–226.

Axelrod, R. and Hamilton, W. D. (1981), The evolution of cooperation, *Science* **211**, 1390–1396.

Bader, G. D. and Hogue, C. W. V. (2002), Analyzing yeast protein–protein interaction data obtained from different sources, *Nat. Biotechnol.* **20**, 991–997.

Bailey, N. T. (1975), *The Mathematical Theory of Infectious Diseases*, Griffin.

Balescu, R. (1997), *Statistical Dynamics: Matter Out of Equilibrium*, Imperial College Press.

Banavar, J. R., Maritan, A. and Rinaldo, A. (1999), Size and form in efficient transportation networks, *Nature* **399**, 130–132.

Banks, D., House, L., McMorris, F. R., Arabie, P. and Gaul, W., eds. (2004), *Classification, Clustering, and Data Mining Applications*, Springer.

Banks, D. L. and Carley, K. M. (1996) , Models for network evolution, *J. Math. Sociol.* **21**, 173–196.

Barabási, A.-L. (2005), Taming complexity, *Nat. Phys.* **1**, 68–70.

Barabási, A.-L. and Albert, R. (1999), Emergence of scaling in random networks, *Science* **286**, 509–512.

Barabási, A.-L., Albert, R. and Jeong, H. (1999), Mean-field theory for scale-free random networks, *Physica A* **272**, 173–187.

Barabási, A.-L., de Menezes, M. A., Balensiefer, S. and Brockman, J. (2004), Hot spots and universality in network dynamics, *Eur. Phys. J. B* **38**, 169–175.

Barabási, A.-L. and Oltvai, Z. N. (2004), Network biology: Understanding the cell's functional organization, *Nat. Rev. Genet.* **5**, 101–113.

Barabási, A.-L. and Stanley, H. E. (1995), *Fractal Concepts in Surface Growth*, Cambridge University Press.

Barahona, M. and Pecora, L. M. (2002), Synchronization in small-world systems, *Phys. Rev. Lett.* **89**, 054101.

Barford, P., Bestavros, A., Byers, J. and Crovella, M. (2001), On the marginal utility of deploying measurement infrastructure. In *Proceedings of the ACM SIGCOMM Internet Measurement Workshop 2001, CA*, Boston University.

Baronchelli, A., Felici, M., Loreto, V., Caglioti, E. and Steels, L. (2006), Sharp transition towards shared vocabularies in multi-agent systems, *J. Stat. Mech.* P06014.

Baronchelli, A. and Loreto, V. (2006), Ring structures and mean first passage time in networks, *Phys. Rev. E* **73**, 026103.

Baroyan, O. V., Genchikov, L. A., Rvachev, L. A. and Shashkov, V. A. (1969), An attempt at large-scale influenza epidemic modelling by means of a computer, *Bull. Int. Epidemiol. Assoc.* **18**, 22–31.

Barrat, A., Barthélemy, M., Pastor-Satorras, R. and Vespignani, A. (2004a), The architecture of complex weighted networks, *Proc. Natl Acad. Sci. (USA)* **101**, 3747–3752.

Barrat, A., Barthélemy, M. and Vespignani, A. (2004b), Traffic-driven model of the World Wide Web graph, *Lect. Notes Comp. Sci* **3243**, 56–67.

Barrat, A., Barthélemy, M. and Vespignani, A. (2004c), Weighted evolving networks: coupling topology and weights dynamics, *Phys. Rev. Lett.* **92**, 228701.

Barrat, A., Barthélemy, M. and Vespignani, A. (2005), The effects of spatial constraints on the evolution of weighted complex networks, *J. Stat. Mech.* p. P05003.

Barrat, A. and Pastor-Satorras, R. (2005), Rate equation approach for correlations in growing network models, *Phys. Rev. E* **71**, 036127.

Barrat, A. and Weigt, M. (2000), On the properties of small-world network models, *Eur. Phys. J. B* **13**, 547–560.

Barthélemy, M. (2003), Crossover from scale-free to spatial networks, *Europhys. Lett.* **63**, 915–921.

Barthélemy, M. (2004), Betweenness centrality in large complex networks, *Eur. Phys. J. B* **38**, 163–168.

Barthélemy, M. and Amaral, L. A. N. (1999a), Erratum: Small-world networks: Evidence for a crossover picture, *Phys. Rev. Lett.* **82**, 5180–5180.

Barthélemy, M. and Amaral, L. A. N. (1999b), Small-world networks: Evidence for a crossover picture, *Phys. Rev. Lett.* **82**, 3180–3183.

Barthélemy, M., Barrat, A., Pastor-Satorras, R. and Vespignani, A. (2004), Velocity and hierarchical spread of epidemic outbreaks in complex scale-free networks, *Phys. Rev. Lett.* **92**, 178701.

Barthélemy, M., Barrat, A., Pastor-Satorras, R. and Vespignani, A. (2005), Dynamical patterns of epidemic outbreaks in complex heterogeneous networks, *J. Theor. Biol.* **235**, 275–288.

Barthélemy, M. and Flammini, A. (2006), Optimal traffic networks, *J. Stat. Mech.* p. L07002.

Barthélemy, M., Gondran, B. and Guichard, E. (2003), Spatial structure of the Internet traffic, *Physica A* **319**, 633–642.

Bassett, D. S. and Bullmore, E. (2006), Small-world brain networks, *Neuroscientist* **12**, 512–523.

Battiston, S. and Catanzaro, M. (2004), Statistical properties of corporate board and director networks, *Eur. Phys. J. B* **38**, 345–352.

Baxter, R. J. (1982), *Exactly Solved Models in Statistical Mechanics*, Academic Press.

Ben-Naim, E., Krapivsky, P. L. and Redner, S. (2003), Bifurcations and patterns in compromise processes, *Physica D* **183**, 190–204.

Bender, E. A. and Canfield, E. R. (1978), The asymptotic number of labelled graphs with given degree sequences, *J. Comb. Theory A* **24**, 296–307.

Berg, J. and Lässig, M. (2002), Correlated random networks, *Phys. Rev. Lett.* **89**, 228701.

Bergé, C. (1976), *Graphs and Hypergraphs*, North-Holland.

Bergé, P., Pomeau, Y. and Vidal, C. (1984), *Order Within Chaos: Towards a Deterministic Approach to Turbulence*, Wiley.

Bettencourt, L. M. A., Cintrón-Arias, A., Kaiser, D. I. and Castillo-Chávez, C. (2006), The power of a good idea: quantitative modeling of the spread of ideas from epidemiological models, *Physica A* **364**, 513–536.

Bianconi, G. (2002), Mean field solution of the Ising model on a Barabási–Albert network, *Phys. Lett. A* **303**, 166–168.

Bianconi, G. (2005), Emergence of weight–topology correlations in complex scale-free networks, *Europhys. Lett.* **71**, 1029–1035.

Bianconi, G. and Barabási, A.-L. (2001), Competition and multiscaling in evolving networks, *Europhys. Lett.* **54**, 436–442.

Bianconi, G. and Marsili, M. (2006a), Emergence of large cliques in random scale-free networks, *Europhys. Lett.* **74**, 740–746.

Bianconi, G. and Marsili, M. (2006b), Number of cliques in random scale-free network ensembles, *Physica D* **224**, 1–6.

Binder, K. and Heermann, D. W. (2002), *Monte Carlo Simulation in Statistical Physics*, Springer.

Binney, J. J., Dowrick, N. J., Fisher, A. J. and Newman, M. E. J. (1992), *The Theory of Critical Phenomena*, Oxford University Press.

Blatt, M., Wiseman, S. and Domany, E. (1996), Superparamagnetic clustering of data, *Phys. Rev. Lett.* **76**, 3251–3254.

Blekhman, I. I. (1988), *Synchronization in Science and Technology*, American Society of Mechanical Engineers.

Boccaletti, S., Kurths, J., Osipov, G., Valladares, D. L. and Zhou, C. S. (2002), The synchronization of chaotic systems, *Phys. Rep.* **366**, 1–101.

Boccaletti, S., Latora, V., Moreno, Y., Chavez, M. and Hwang, D.-U. (2006), Complex networks: Structure and dynamics, *Phys. Rep.* **424**, 175–308.

Bochner, S., Buker, E. A. and McLeod, B. M. (1976), Communication patterns in an international student dormitory: A modification of the "small world" method, *J. Appl. Soc. Psychol.* **6**, 275–290.

Boguñá, M., Krioukov, D. and claffy, k. (2007), Navigability of complex networks, `http://arXiv.org/abs/0709.0303`.

Boguñá, M. and Pastor-Satorras, R. (2002), Epidemic spreading in correlated complex networks, *Phys. Rev. E* **66**, 047104.

Boguñá, M. and Pastor-Satorras, R. (2003), Class of correlated random networks with hidden variables, *Phys. Rev. E* **68**, 036112.

Boguñá, M., Pastor-Satorras, R. and Vespignani, A. (2003a), Absence of epidemic threshold in scale-free networks with degree correlations, *Phys. Rev. Lett.* **90**, 028701.

Boguñá, M., Pastor-Satorras, R. and Vespignani, A. (2003b), Epidemic spreading in complex networks with degree correlations, *Lect. Notes Phys.* **625**, 127–147.

Boguñá, M., Pastor-Satorras, R. and Vespignani, A. (2004), Cut-offs and finite size effects in scale-free networks, *Eur. Phys. J. B* **38**, 205–209.

Boguñá, M. and Serrano, M. A. (2005), Generalized percolation in random directed networks, *Phys. Rev. E* **72**, 016106.

Bolker, B. M. and Grenfell, B. (1995), Space persistence and dynamics of measles epidemics, *Phil. Trans.: Biol. Sci.* **348**, 309–320.

Bolker, B. M. and Grenfell, B. T. (1993), Chaos and biological complexity in measles dynamics, *Proc. R. Soc. B* **251**, 75–81.

Bollobás, B. (1981), Degree sequences of random graphs, *Discrete Math.* **33**, 1–19.

Bollobás, B. (1985), *Random Graphs*, Academic Press.

Bollobás, B. (1998), *Modern Graph Theory*, Springer-Verlag.

Bollobás, B. and Riordan, O. (2003), Mathematical results on scale-free random graphs. In S. Bornholdt and H. G. Schuster, eds., *Handbook of Graphs and Networks: From the Genome to the Internet*, Wiley-VCH, pp. 1–34.

Bollobás, B., Riordan, O., Spencer, J. and Tusnády, G. (2001), The degree sequence of a scale-free random graph process, *Random Struct. Algor.* **18**, 279–290.

Bollt, E. M. and ben Avraham, D. (2005), What is special about diffusion on scale-free nets?, *New J. Phys.* **7**, 26.

Börner, K., Dall'Asta, L., Ke, W. and Vespignani, A. (2004), Studying the emerging global brain: analyzing and visualizing the impact of coauthorship teams, *Complexity* **10**, 57–67.

Bornholdt, S. (2001), Modeling genetic networks and their evolution: A complex dynamical systems perspective, *Biol. Chem.* **382**, 1289–1299.

Bornholdt, S. (2005), Less is more in modeling large genetic networks, *Science* **310**, 449–451.

Borrvall, C., Ebenman, B. and Jonsson, T. (2000), Biodiversity lessens the risk of cascading extinction in model food webs, *Ecol. Lett.* **3**, 131–136.

Bota, M., Dong, H.-W. and Swanson, L. W. (2003), From gene networks to brain networks, *Nat. Neurosci.* **6**, 795–799.

Boyer, D. and Miramontes, O. (2003), Interface motion and pinning in small-world networks, *Phys. Rev. E* **67**, 035102.

Brandes, U. (2001), A faster algorithm for betweenness centrality, *J. Math. Sociol.* **25**, 163–177.

Braunewell, S. and Bornholdt, S. (2007), Superstability of the yeast cell-cycle dynamics: Ensuring causality in the presence of biochemical stochasticity, *J. Theor. Biol.* **245**, 638–643.

Bray, A. J. (1994), Theory of phase-ordering kinetics, *Adv. Phys.* **43**, 357–459.

Brin, S. and Page, L. (1998), The anatomy of a large-scale hypertextual Web search engine, *Comput. Netw. ISDN Syst.* **30**, 107–117.

Broder, A. Z., Kumar, S. R., Maghoul, F. *et al.* (2000), Graph structure in the Web, *Comput. Netw.* **33**, 309–320.

Broido, A. and claffy, k. (2001), Internet topology: connectivity of IP graphs. In S. Fahmy and K. Park, eds., *Proceedings of SPIE*, Vol. 4526, *Scalability and Traffic Control in IP Networks*, July 2001, SPIE, pp. 172–187.

Brunel, N. and Hakim, V. (1999), Fast global oscillations in networks of integrate-and-fire neurons with low firing rates, *Neural Comput.* **11**, 1621–1671.

Buhl, J., Gautrais, J., Reeves, N. *et al.* (2006), Topological patterns in street networks of self-organized urban settlements, *Eur. Phys. J. B* **49**, 513–522.

Bunde, A. and Havlin, S. (1991), Percolation. In A. Bunde and S. Havlin, eds., *Fractals and Disordered Systems*, Springer Verlag, pp. 51–95.

Burda, Z., Correia, J. D. and Krzywicki, A. (2001), Statistical ensemble of scale-free random graphs, *Phys. Rev. E* **64**, 046118.

Burda, Z. and Krzywicki, A. (2003), Uncorrelated random networks, *Phys. Rev. E* **67**, 046118.

Burioni, R. and Cassi, D. (2005), Random walks on graphs: ideas, techniques and results, *J. Phys. A: Math. Gen.* **38**, R45–R78.

Caldarelli, G. (2007), *Scale-Free Networks*, Oxford University Press.

Caldarelli, G., Capocci, A., De Los Rios, P. and Muñoz, M. A. (2002), Scale-free networks from varying vertex intrinsic fitness, *Phys. Rev. Lett.* **89**, 258702.

Caldarelli, G., Higgs, P. G. and MacKane, A. J. (1998), Modelling coevolution in multispecies communities, *J. Theor. Biol.* **193**, 345–358.

Callaway, D. S., Newman, M. E. J., Strogatz, S. H. and Watts, D. J. (2000), Network robustness and fragility: Percolation on random graphs, *Phys. Rev. Lett.* **85**, 5468–5471.

Camacho, J., Guimerà, R. and Amaral, L. A. N. (2002a), Analytical solution of a model for complex food webs, *Phys. Rev. E* **65**, 030901.

Camacho, J., Guimerà, R. and Amaral, L. A. N. (2002b), Robust patterns in food web structure, *Phys. Rev. Lett.* **88**, 228102.

Capocci, A., Servedio, V. D. P., Caldarelli, G. and Colaiori, F. (2005), Detecting communities in large networks, *Physica A* **352**, 669–676.

Cardillo, A., Scellato, S., Latora, V. and Porta, S. (2006), Structural properties of planar graphs of urban streets patterns, *Phys. Rev. E* **73**, 066107.

Carmi, S., Cohen, R. and Dolev, D. (2006), Searching complex networks efficiently with minimal information, *Europhys. Lett.* **74**, 1102–1108.

Castellano, C. (2005), Effect of network topology on the ordering dynamics of voter models, *Proceedings of the 8th Granada Seminar on Computational and Statistical Physics* **779**, 114–120.

Castellano, C., Fortunato, S. and Loreto, V. (2007), Statistical physics of social dynamics, *Rev. Mod. Phys.*, http://arxiv:org/abs/0710.3256.

Castellano, C., Loreto, V., Barrat, A., Cecconi, F. and Parisi, D. (2005), Comparison of voter and Glauber ordering dynamics on networks, *Phys. Rev. E* **71**, 066107.

Castellano, C., Marsili, M. and Vespignani, A. (2000), Non-equilibrium phase transition in a model for social influence, *Phys. Rev. Lett.* **85**, 3536–3539.

Castellano, C. and Pastor-Satorras, R. (2006a), Non mean-field behavior of the contact process on scale-free networks, *Phys. Rev. Lett.* **96**, 038701.

Castellano, C. and Pastor-Satorras, R. (2006b), Zero temperature Glauber dynamics on complex networks, *J. Stat. Mech.* p. P05001.

Castellano, C., Vilone, D. and Vespignani, A. (2003), Incomplete ordering of the voter model on small-world networks, *Europhys. Lett.* **63**, 153–158.

Castelló, X., Eguíluz, V. M. and San Miguel, M. (2006), Ordering dynamics with two non-excluding options: bilingualism in language competition, *New J. Phys.* **8**, 308.

Catanzaro, M., Boguñá, M. and Pastor-Satorras, R. (2005), Generation of uncorrelated random scale-free networks, *Phys. Rev. E* **71**, 027103.

Centola, D., González-Avella, J. C., Eguíluz, V. M. and San Miguel, M. (2007), Homophily, cultural drift and the co-evolution of cultural groups, *J. Conflict Resol.* **51**, 905–929.

Chandler, D. (1987), *Introduction to Modern Statistical Mechanics*, Oxford University Press.

Chang, H., Jamin, S. and Willinger, W. (2003), Internet connectivity at the AS-level: an optimization-driven modeling approach. In *Proceedings of the ACM SIGCOMM Workshop on Models, Methods and Tools for Reproducible Network Research,* ACM Press, pp. 33–46.

Chang, H., Jamin, S. and Willinger, W. (2006), To peer or not to peer: Modeling the evolution of the Internet's AS-level topology. In *Proceedings of IEEE INFOCOM 2006,* IEEE, pp. 1–12.

Chartrand, G. and Lesniak, L. (1986), *Graphs & Digraphs*, Wadsworth & Brooks/Cole.

Chavez, M., Hwang, D.-U., Amman, A., Hentschel, H. G. E. and Boccaletti, S. (2005), Synchronization is enhanced in weighted complex networks, *Phys. Rev. Lett.* **94**, 218701.

Chen, P., Xie, H., Maslov, S. and Redner, S. (2007), Finding scientific gems with Google, *J. Informet.* **1**, 8–15.

Chen, X. and Cohen, J. E. (2001), Global stability, local stability and permanence in model food webs, *J. Theor. Biol.* **212**, 223–235.

Chen, Z. Y. and Wang, X. F. (2006), Effects of network structure and routing strategy on network capacity, *Phys. Rev. E* **73**, 036107.

Chowell, G., Hyman, J. M., Eubank, S. and Castillo-Chavez, C. (2003), Scaling laws for the movement of people between locations in a large city, *Phys. Rev. E* **68**, 066102.

Christaller, W. (1966), *Central Places in Southern Germany*, Prentice Hall.

Chung, F. and Lu, L. (2004), The small world phenomenon in hybrid power law graphs. In
    E. Ben-Naim *et al.*, eds., *Complex Networks*, Springer-Verlag, pp. 91–106.
claffy, k. (1999), Internet measurement and data analysis: topology, workload, performance
    and routing statistics. http://www.caida.org/outreach/papers/1999/Nae.
Clark, J. and Holton, D. A. (1991), *A First Look at Graph Theory*, World Scientific.
Clauset, A. and Moore, C. (2005), Accuracy and scaling phenomena in Internet mapping,
    Phys. Rev. Lett. **94**, 018701.
Clerc, J. P., Giraud, G., Laugier, J. M. and Luck, J. M. (1990), The electrical conductivity of
    binary disordered systems, percolation clusters, fractals and related models, *Adv.
    Phys.* **39**, 191–309.
Cohen, J. E., Briand, F. and Newman, C. M. (1990), *Community Food Webs: Data and
    Theory*, Springer.
Cohen, R., Erez, K., ben Avraham, D. and Havlin, S. (2000), Resilience of the Internet to
    random breakdowns, *Phys. Rev. Lett.* **85**, 4626–4628.
Cohen, R., Erez, K., ben Avraham, D. and Havlin, S. (2001), Breakdown of the Internet
    under intentional attack, *Phys. Rev. Lett.* **86**, 3682–3685.
Cohen, R. and Havlin, S. (2003), Scale-free networks are ultrasmall, *Phys. Rev. Lett.*
    90, 058701.
Cohen, R., Havlin, S. and ben Avraham, D. (2003), Efficient immunization strategies for
    computer networks and populations, *Phys. Rev. Lett.* **91**, 247901.
Colizza, V., Banavar, J. R., Maritan, A. and Rinaldo, A. (2004), Network structures from
    selection principles, *Phys. Rev. Lett.* **92**, 198701.
Colizza, V., Barrat, A., Barthélemy, M. and Vespignani, A. (2006a), The role of the airline
    transportation network in the prediction and predictability of global epidemics, *Proc.
    Natl Acad. Sci. (USA)* **103**, 2015–2020.
Colizza, V., Barrat, A., Barthélemy, M. and Vespignani, A. (2007a), Modeling the
    worldwide spread of pandemic influenza: Baseline case and containment
    interventions, *PLoS Med.* **4(1)**, e13.
Colizza, V., Flammini, A., Serrano, M. A. and Vespignani, A. (2006b), Detecting rich-club
    ordering in complex networks, *Nat. Phys.* **2**, 110–115.
Colizza, V., Pastor-Satorras, R. and Vespignani, A. (2007b), Reaction–diffusion processes
    and metapopulation models in heterogeneous networks, *Nat. Phys.* **3**, 276–282.
Colizza, V. and Vespignani, A. (2008), Epidemic modeling in metapopulation systems with
    heterogeneous coupling pattern: theory and simulations, *J. Theor. Biol.* **251**, 450–457.
Cormen, T. H., Leiserson, C. E., Rivest, R. L. and Stein, C. (2003), *Introduction to
    Algorithms*, 2nd Edn, MIT Press.
Correale, L., Leone, M., Pagnani, A., Weigt, M. and Zecchina, R. (2006), Computational
    core and fixed-point organisation in Boolean networks, *J. Stat. Mech.* p. P03002.
Costa, M. Crowford, J., Castro, M. *et al.* (2005), Vigilante: end-to-end containment of
    Internet worms, *ACM SIGOPS Operating Syst. Rev.* **39**, 133–147.
Crépey, P., Alvarez, F. P. and Barthélemy, M. (2006), Epidemic variability in complex
    networks, *Phys. Rev. E* **73**, 046131.
Crovella, M. E. and Bestavros, A. (1997), Self-similarity in World Wide Web traffic:
    evidence and possible causes, *IEEE/ACM Trans. Networking* **5**, 835–846.
Crovella, M. E. and Krishnamurthy, B. (2006), *Internet Measurements: Infrastructure,
    Traffic and Applications*, Wiley.
Crovella, M. E., Lindemann, C. and Reiser, M. (2000), Internet performance modeling: the
    state of the art at the turn of the century, *Performance Eval.*, **42**, 91–108.
Crucitti, P., Latora, V. and Marchiori, M. (2004), Model for cascading failures in complex
    networks, *Phys. Rev. E* **69**, 045104(R).

Crucitti, P., Latora, V. and Porta, S. (2006), Centrality measures in spatial networks of urban streets, *Phys. Rev. E* **73**, 036125.

Csabai, I. (1994), 1/f noise in computer networks traffic, *J. Phys. A: Math. Gen.* **27**, 417–419.

da Fontoura Costa, L. (2004), Reinforcing the resilience of complex networks, *Phys. Rev. E* **69**, 066127.

Daley, D. J. and Gani, J. (2000), *Epidemic Modelling: An Introduction*, Cambridge University Press.

Daley, D. J. and Kendall, D. G. (1964), Epidemics and rumours, *Nature* **204**, 1118–1118.

Daley, D. J. and Kendall, D. G. (1965), Stochastic rumours, *IMA J. Appl. Math.* **1**, 42–55.

Dall'Asta, L. (2005), Inhomogeneous percolation models for spreading phenomena in random graphs, *J. Stat. Mech.* p. P08011.

Dall'Asta, L., Alvarez-Hamelin, I., Barrat, A., Vàzquez, A. and Vespignani, A. (2005), Statistical theory of internet exploration, *Phys. Rev. E* **71**, 036135.

Dall'Asta, L., Alvarez-Hamelin, I., Barrat, A., Vàzquez, A. and Vespignani, A. (2006a), Exploring networks with traceroute-like probes: Theory and simulations, *Theor. Comput. Sci.* **355**, 6–24.

Dall'Asta, L. and Baronchelli, A. (2006), Microscopic activity patterns in the Naming Game, *J. Phys. A: Math. Gen.* **39**, 14851–14867.

Dall'Asta, L., Baronchelli, A., Barrat, A. and Loreto, V. (2006b), Agreement dynamics on small-world networks, *Europhys. Lett.* **73**, 969–975.

Dall'Asta, L., Barrat, A., Barthélemy, M. and Vespignani, A. (2006c), Vulnerability of weighted networks, *J. Stat. Mech.* p. P04006.

Dall'Asta, L. and Castellano, C. (2007), Effective surface-tension in the noise-reduced voter model, *Europhys. Lett.* **77**, 60005.

Danila, B., Yu, Y., Earl, S. *et al.* (2006a), Congestion-gradient driven transport on complex networks, *Phys. Rev. E* **74**, 046114.

Danila, B., Yu, Y., Marsh, J. A. and Bassler, K. E. (2006b), Optimal transport on complex networks, *Phys. Rev. E* **74**, 046106.

Danon, L., Díaz-Guilera, A., Duch, J. and Arenas, A. (2005), Comparing community structure identification, *J. Stat. Mech.* p. P09008.

Davidsen, J., Ebel, H. and Bornholdt, S. (2002), Emergence of a small world from local interactions: Modeling acquaintance networks, *Phys. Rev. Lett.* **88**, 128701.

Davis, G. F., Yoo, M. and Baker, W. E. (2003), The small world of the American corporate elite, 1982–2001, *Strateg. Organ.* **1**, 301–326.

De Angelis, D. (1975), Stability and connectance in food web models, *Ecology* **56**, 238–243.

de Arcangelis, L., Redner, S. and Herrmann, H. J. (1985), A random fuse model for breaking processes, *J. Phys. Lett.* **46**, 585–590.

de Menezes, M. A. and Barabási, A.-L. (2004a), Fluctuations in network dynamics, *Phys. Rev. Lett.* **92**, 028701.

de Menezes, M. A. and Barabási, A.-L. (2004b), Separating internal and external dynamics of complex systems, *Phys. Rev. Lett.* **93**, 068701.

de Solla Price, D. J. (1976), A general theory of bibliometric and other cumulative advantage processes, *J. Am. Soc. Inform. Sci.* **27**, 292–306.

de Solla Price, D. J. (1986), *Little Science, Big Science. . .and Beyond*, Columbia University Press.

Deane, C. M., Salwinski, L., Xenarios, I. and Eisenberg, D. (2002), Protein interactions – two methods for assessment of the reliability of high throughput observations, *Mol. Cell. Proteom.* **1**, 349–356.

Deffuant, G., Neau, D., Amblard, F. and Weisbuch, G. (2000), Mixing beliefs among interacting agents, *Adv. Compl. Syst.* **3**, 87–98.

Denker, M., Timme, M., Diesmann, M., Wolf, F. and Geisel, T. (2004), Breaking synchrony by heterogeneity in complex networks, *Phys. Rev. Lett.* **92**, 074103.

Derrida, B. and Flyvbjerg, H. (1987), Statistical properties of randomly broken objects and of multivalley structures in disordered systems, *J. Phys. A: Math. Gen.* **20**, 5273–5288.

Derrida, B. and Pomeau, Y. (1986), Random networks of automata: a simple annealed approximation, *Europhys. Lett.* **1**, 45–49.

Dezsö, Z. and Barabási, A.-L. (2002), Halting viruses in scale-free networks, *Phys. Rev. E* **65**, 055103(R).

Diekmann, O. and Heesterbeek, J. A. P. (2000), *Mathematical Epidemiology of Infectious Diseases*, Wiley.

Dietz, K. (1967), Epidemics and rumours: A survey, *J. Royal Stat. Soc. A* **130**, 505–528.

Dimitropoulos, X., Riley, G., Krioukov, D. and Sundaram, R. (2005), Towards a topology generator modeling AS relationships, IEEE International Conference on Network Protocols, http://www.caida.org/publications/papers/2005/top_modeling_as .

Dodds, P. S., Muhamad, R. and Watts, D. J. (2003), An experimental study of search in global social networks, *Science* **301**, 827–829.

Dodds, P. S. and Watts, D. J. (2004), Universal behavior in a generalized model of contagion, *Phys. Rev. Lett.* **92**, 218701.

Dodds, P. S. and Watts, D. J. (2005), A generalized model of social and biological contagion, *J. Theor. Biol.* **232**, 587–604.

Dodel, S., Herrmann, J. M. and Geisel, T. (2002), Functional connectivity by cross-correlation clustering, *Neurocomputing* **44**, 1065–1070.

Domany, E. (1999), Superparamagnetic clustering of data – the definitive solution of an ill-posed problem, *Physica A* **263**, 158–169.

Donetti, L., Hurtado, P. I. and Muñoz, M. A. (2005), Entangled networks, synchronization, and optimal network topology, *Phys. Rev. Lett.* **95**, 188701.

Donetti, L. and Muñoz, M. A. (2004), Detecting network communities: a new systematic and efficient algorithm, *J. Stat. Mech.* p. P10012.

Dornic, I., Chaté, H., Chave, J. and Hinrichsen, H. (2001), Critical coarsening without surface tension: The universality class of the voter model, *Phys. Rev. Lett.* **87**, 045701.

Dorogovtsev, S. N., Goltsev, A. V. and Mendes, J. F. F. (2002), Ising model on networks with an arbitrary distribution of connections, *Phys. Rev. E* **66**, 016104.

Dorogovtsev, S. N., Goltsev, A. V. and Mendes, J. F. F. (2004), Potts model on complex networks, *Eur. Phys. J. B* **38**, 177–182.

Dorogovtsev, S. N. and Mendes, J. F. F. (2001), Comment on "breakdown of the Internet under intentional attack", *Phys. Rev. Lett* **87**, 219801.

Dorogovtsev, S. N. and Mendes, J. F. F. (2002), Evolution of networks, *Adv. Phys.* **51**, 1079–1187.

Dorogovtsev, S. N. and Mendes, J. F. F. (2003), *Evolution of Networks: From Biological Nets to the Internet and WWW*, Oxford University Press.

Dorogovtsev, S. N. and Mendes, J. F. F. (2005), Evolving weighted scale-free networks. In *Science of Complex Networks: From Biology to the Internet and WWW: CNET 2004, AIP Conference Proceedings* Vol. 776, pp. 29–36.

Dorogovtsev, S. N., Mendes, J. F. F. and Oliveira, J. G. (2006), Degree-dependent intervertex separation in complex networks, *Phys. Rev. E* **73**, 056122.

Dorogovtsev, S. N., Mendes, J. F. F. and Samukhin, A. N. (2000), Structure of growing networks with preferential linking, *Phys. Rev. Lett.* **85**, 4633–4636.

Dorogovtsev, S. N., Mendes, J. F. F. and Samukhin, A. N. (2001a), Giant strongly connected component of directed networks, *Phys. Rev. E* **64**, 025101.

Dorogovtsev, S. N., Mendes, J. F. F. and Samukhin, A. N. (2001b), Size-dependent degree distribution of a scale-free growing network, *Phys. Rev. E* **63**, 062101.

Dorogovtsev, S. N., Mendes, J. F. F. and Samukhin, A. N. (2003), Principles of statistical mechanics of uncorrelated random networks, *Nucl. Phys. B* **666**, 396–416.

Doyle, J. C., Alderson, D. L., Li, L. *et al.* (2005), The "robust yet fragile" nature of the Internet, *Proc. Natl Acad. Sci. (USA)* **102**, 14497–14502.

Dreyfus, G., ed. (2005), *Neural Networks: Methodology And Applications*, Springer.

Drossel, B. (2007), Random Boolean networks. In H. G. Schuster, ed., *Annual Review of Nonlinear Dynamics and Complexity*, Vol. 1, Wiley-VCH.

Drossel, B., Higgs, P. G. and MacKane, A. J. (2001), The influence of predator–prey population dynamics on the long-term evolution of food web structure, *J. Theor. Biol.* **208**, 91–107.

Drossel, B. and MacKane, A. J. (2003), Modelling food webs. In S. Bornholdt and H. G. Schuster, eds., *Handbook of Graphs and Networks: From the Genome to the Internet*, Wiley-VCH, pp. 218–247.

Duch, J. and Arenas, A. (2006), Scaling of fluctuations in traffic on complex networks, *Phys. Rev. Lett.* **96**, 218702.

Dunne, J. A. (2006), The network structure of food webs. In M. Pascual and J. Dunne, eds., *Ecological Networks: Linking Structure to Dynamics in Food Webs*, Oxford University Press, pp. 27–86.

Dunne, J. A., Williams, R. J. and Martinez, N. D. (2002a), Food-web structure and network theory: the role of connectance and size, *Proc. Natl Acad. Sci. (USA)* **99**, 12917–12922.

Dunne, J. A., Williams, R. J. and Martinez, N. D. (2002b), Network structure and biodiversity loss in food-webs: robustness increases with connectance, *Ecol. Lett.* **5**, 558–567.

Earn, D. J. D., Rohani, P. and Grenfell, B. T. (1998), Persistence, chaos and synchrony in ecology and epidemiology, *Proc. R. Soc. B* **265**, 7–10.

Ebel, H., Mielsch, L.-I. and Bornholdt, S. (2002), Scale-free topology of e-mail networks, *Phys. Rev. E* **66**, 035103.

Echenique, P., Gómez-Gardeñes, J. and Moreno, Y. (2004), Improved routing strategies for Internet traffic delivery, *Phys. Rev. E* **70**, 056105.

Echenique, P., Gómez-Gardeñes, J. and Moreno, Y. (2005), Dynamics of jamming transitions in complex networks, *Europhys. Lett.* **71**, 325–331.

Economou, E. N. (2006), *Green's Functions in Quantum Physics*, Springer.

Eguíluz, V. M., Chialvo, D. R., Cecchi, G. A., Baliki, M. and Vania Apkarian, A. (2005a), Scale-free brain functional networks, *Phys. Rev. Lett.* **94**, 018102.

Eguíluz, V. M., Zimmermann, M. G., Cela-Conde, C. J. and San Miguel, M. (2005b), Cooperation and emergence of role differentiation in the dynamics of social networks, *Am. J. Social.* **110**, 977–1008.

Ehrhardt, G. C. M. A., Marsili, M. and Vega-Redondo, F. (2006a), Diffusion and growth in an evolving network, *Int. J. Game Theory* **34**, 282–297.

Ehrhardt, G. C. M. A., Marsili, M. and Vega-Redondo, F. (2006b), Phenomenological models of socioeconomic network dynamics, *Phys. Rev. E* **74**, 036106.

Emmerling, M., Dauner, M., Ponti, A. *et al.* (2002), Metabolic flux response to pyruvate kinase knockout in *E. coli, J. Bacteriol.* **184**, 152–164.

Erdős, P. and Rényi, A. (1959), On random graphs, *Publ. Math.* **6**, 290–297.

Erdős, P. and Rényi, A. (1960), On the evolution of random graphs, *Publ. Math. Inst. Hung. Acad. Sci.* **5**, 17–60.

Erdős, P. and Rényi, A. (1961), On the strength of connectedness of random graphs, *Acta. Math. Sci. Hung.* **12**, 261–267.

Eubank, S., Guclu, H., Kumar, V. S. A. *et al.* (2004), Modelling disease outbreaks in realistic urban social networks, *Nature* **429**, 180–184.

Euler, L. (1736), Solutio problematis ad geometriam situs pertinentis, *Comment. Acad. Sci. J. Petropol.* **8**, 128–140.

Fabrikant, A., Koutsoupias, E. and Papadimitriou, C. H. (2002), Heuristically optimized trade-offs: A new paradigm for power laws in the Internet. In *Proceedings of the 29th International Colloquium on Automata, Languages, and Programming (ICALP)*, Vol. 2380 of *Lecture Notes in Computer Science*, Springer, pp. 110–122.

Farkas, I., Derenyi, I., Palla, G. and Vicsek, T. (2004), Equilibrium statistical mechanics of network structures, *Lect. Notes Phys.* **650**, 163–187.

Fell, D. (1996), *Understanding the Control of Metabolism*, Portland Press.

Ferguson, N. M., Cummings, D. A. T., Cauchemez, S. *et al.* (2005), Strategies for containing an emerging influenza pandemic in Southeast Asia, *Nature* **437**, 209–214.

Ferguson, N. M., Keeling, M. J., Edmunds, W. J. *et al.* (2003), Planning for smallpox outbreaks, *Nature* **425**, 681–685.

Flahault, A. and Valleron, A.-J. (1991), A method for assessing the global spread of HIV-1 infection based on air-travel, *Math. Popul. Stud.* **3**, 161–171.

Fortunato, S., Boguñá, M., Flammini, A. and Menczer, F. (2005), Approximating PageRank from in-degree, *Internet Math.*, http://arXiv.org/abs/cs/0511016

Fortunato, S., Flammini, A. and Menczer, F. (2006a), Scale-free network growth by ranking, *Phys. Rev. Lett.* **96**, 218701.

Fortunato, S., Flammini, A., Menczer, F. and Vespignani, A. (2006b), Topical interests and the mitigation of search engine bias, *Proc. Natl Acad. Sci. (USA)* **103**, 12684–12689.

Fowler, T. B. (1999), A short tutorial on fractals and Internet traffic, *Telecom. Rev.* **10**, 1–14.

Fox, J. J. and Hill, C. C. (2001), From topology to dynamics in biochemical networks, *Chaos* **11**, 809–815.

Frachebourg, L. and Krapivsky, P. L. (1996), Exact results for kinetics of catalytic reactions, *Phys. Rev. E* **53**, R3009–R3012.

Frank, O. (2004), Network sampling. In P. Carrington, J. Scott and S. Wasserman, eds., *Models and Methods in Social Network Analysis*, Cambridge University Press, pp. 98–162.

Frank, O. and Strauss, D. (1986), Markov graphs, *J. Am. Stat. Assoc.* **81**, 832–842.

Freeman, L. C. (1977), A set of measures of centrality based on betweenness, *Sociometry* **40**, 35–41.

Fujita, M., Krugman, P. and Venables, A. J. (1999), *The Spatial Economy: Cities, Regions, and International Trade*, MIT Press.

Fukuda, K., Takayasu, H. and Takayasu, M. (2000), Origin of critical behavior in Ethernet traffic, *Physica A* **287**, 289–301.

Gade, P. M. and Hu, C.-K. (2000), Synchronous chaos in coupled map lattices with small-world interactions, *Phys. Rev. E* **62**, 6409–6413.

Galam, S. (2002), Minority opinion spreading in random geometry, *Eur. Phys. J. B* **25**(4), 403–406.

Gallos, L. K. (2004), Random walk and trapping processes on scale-free networks, *Phys. Rev. E* **70**, 046116.

Gallos, L. K., Argyrakis, P., Bunde, A., Cohen, R. and Havlin, S. (2004), Tolerance of scale-free networks: from friendly to intentional attack strategies, *Physica A* **344**, 504–509.

Gallos, L. K., Cohen, R., Argyrakis, P., Bunde, A. and Havlin, S. (2005), Stability and topology of scale-free networks under attack and defense strategies, *Phys. Rev. Lett.* **94**, 188701.

Gantmacher, F. R. (1974), *The Theory of Matrices*, Vol. II, Chelsea Publishing Company.

Gardner, M. R. and Ashby, W. R. (1970), The connectance of large dynamics (cybernetic) systems: critical values for stability, *Nature* **228**, 784.

Garlaschelli, D., Caldarelli, G. and Pietronero, L. (2003), Universal scaling relations in food webs, *Nature* **423**, 165–168.

Gastner, M. T. and Newman, M. E. J. (2006), Shape and efficiency in spatial distribution networks, *J. Stat. Mech.* p. P01015.

Gavin, A. C., Bösche, M., Krause, R. *et al.* (2002), Functional organization of the yeast proteome by systematic analysis of protein complexes, *Nature* **415**, 141–147.

Gilbert, E. N. (1959), Random graphs, *Ann. Math. Stat.* **30**, 1141–1144.

Giot, L., Bader, J. S., Brouwer, C. *et al.* (2003), A protein interaction map of *Drosophila melanogaster*, *Science* **302**, 1727–1736.

Girvan, M. and Newman, M. E. J. (2002), Community structure in social and biological networks, *Proc. Natl Acad. Sci. (USA)* **99**, 7821–7826.

Gitterman, M. (2000), Small-world phenomena in physics: the Ising model, *J. Phys. A: Math. Gen.* **33**, 8373–8381.

Glass, L. (2001), Synchronization and rhythmic processes in physiology, *Nature* **410**, 277–284.

Glauche, I., Krause, W., Sollacher, R. and Greiner, M. (2004), Distributive routing and congestion control in wireless multihop ad hoc communication networks, *Physica A* **341**, 677–701.

Gnedenko, B. V. (1962), *The Theory of Probability*, Chelsea Publishing Company.

Goffman, W. (1966), Mathematical approach to the spread of scientific ideas–the history of mast cell research, *Nature* **212**, 449–452.

Goffman, W. and Newill, V. A. (1964), Generalization of epidemic theory: An application to the transmission of ideas, *Nature* **204**, 225–228.

Goh, K.-I., Cusick, M. E., Valle, D. *et al.* (2007), The human disease network, *Proc. Natl Acad. Sci. (USA)* **104**, 8685–8690.

Goh, K.-I., Kahng, B. and Kim, D. (2001), Universal behavior of load distribution in scale-free networks, *Phys. Rev. Lett.* **87**, 278701.

Goh, K.-I., Kahng, B. and Kim, D. (2003), Packet transport and load distribution in scale-free network models, *Physica A* **318**, 72–79.

Goldenberg, J., Shavitt, Y., Shir, E. and Solomon, S. (2005), Distributive immunization of networks against viruses using the "honey-pot" architecture, *Nat. Phys.* **1**, 184–188.

Goltsev, A. V., Dorogovtsev, S. N. and Mendes, J. F. F. (2003), Critical phenomena in networks, *Phys. Rev. E* **67**, 026123.

Gómez-Gardeñes, J., Campillo, M., Floria, L. M. and Moreno, Y. (2007a), Dynamical organization of cooperation in complex topologies, *Phys. Rev. Lett.* **98**, 108103.

Gómez-Gardeñes, J., Echenique, P. and Moreno, Y. (2006), Immunization of real complex communication networks, *Eur. Phys. J. B* **49**, 259–264.

Gómez-Gardeñes, J., Moreno, Y. and Arenas, A. (2007b), Paths to synchronization on complex networks, *Phys. Rev. Lett.* **98**, 034101.

Gómez-Gardeñes, J., Moreno, Y. and Arenas, A. (2007c), Synchronizability determined by coupling strengths and topology on complex networks, *Phys. Rev. E* **75**, 066106.

González-Avella, J. C., Eguíluz, V. M., Cosenza, M. G. *et al.* (2006), Local versus global interactions in nonequilibrium transitions: A model of social dynamics, *Phys. Rev. E.* **73**, 046119.

Grais, R. F., Ellis, J. H. and Glass, G. E. (2003), Assessing the impact of airline travel on the geographic spread of pandemic influenza, *Eur. J. Epidemiol.* **18**, 1065–1072.

Grais, R. F., Ellis, J. H., Kress, A. and Glass, G. E. (2004), Modeling the spread of annual influenza epidemics in the US: The potential role of air travel, *Health Care Manag. Sci.* **7**, 127–134.

Granovetter, M. (1978), Threshold models of collective behavior, *Am. J. Sociol.* **83**, 1420–1443.

Grassberger, P. (1983), On the critical behavior of the general epidemic process and dynamical percolation, *Math. Biosci.* **63**, 157–172.

Grenfell, B. T. and Bolker, B. M. (1998), Cities and villages: infection hierarchies in a measles metapopulation, *Ecol. Lett.* **1**, 63–70.

Grenfell, B. T. and Harwood, J. (1997), (meta)population dynamics of infectious diseases, *Trends Ecol. Evol.* **12**, 395–399.

Guardiola, X., Díaz-Guilera, A., Llas, M. and Pérez, C. J. (2000), Synchronization, diversity, and topology of networks of integrate and fire oscillators, *Phys. Rev. E* **62**, 5565–5570.

Guelzim, N., Bottani, S., Bourgine, P. and Kepes, F. (2002), Topological and causal structure of the yeast transcriptional regulatory network, *Nat. Genet.* **31**, 60–63.

Guimerà, R. and Amaral, L. A. N. (2004), Modeling the world-wide airport network, *Eur. Phys. J. B* **38**, 381–385.

Guimerà, R., Arenas, A., Díaz-Guilera, A. and Giralt, F. (2002a), Dynamical properties of model communication networks, *Phys. Rev. E* **66**, 026704.

Guimerà, R., Díaz-Guilera, A., Vega-Redondo, F., Cabrales, A. and Arenas, A. (2002b), Optimal network topologies for local search with congestion, *Phys. Rev. Lett.* **89**, 248701.

Guimerà, R., Mossa, S., Turtschi, A. and Amaral, L. A. N. (2005), The worldwide air transportation network: Anomalous centrality, community structure, and cities' global roles, *Proc. Natl Acad. Sci. (USA)* **102**, 7794–7799.

Häggström, O. (2002), Zero-temperature dynamics for the ferromagnetic Ising model on random graphs, *Physica A* **310**, 275–284.

Hansel, D. and Sompolinsky, H. (1992), Synchronization and computation in a chaotic neural network, *Phys. Rev. Lett.* **68**, 718–721.

Hanski, I. and Gaggiotti, O. E. (2004), *Ecology, Genetics, and Evolution of Metapopulations*, Elsevier, Academic Press.

Harris, S. E., Sawhill, B. K., Wuensche, A. and Kauffman, S. (2002), A model of transcriptional regulatory networks based on biases in the observed regulation rules, *Complexity* **7**, 23–40.

Harris, T. E. (1974), Contact interactions on a lattice, *Ann. Prob.* **2**, 969–988.

Haykin, S. (2007), *Neural Networks: A Comprehensive Foundation*, Prentice Hall.

Heagy, J. F., Pecora, L. M. and Carroll, T. L. (1995), Short wavelength bifurcations and size instabilities in coupled oscillator systems, *Phys. Rev. Lett.* **74**, 4185–4188.

Hegselmann, R. and Krause, U. (2002), Opinion dynamics and bounded confidence models, analysis and simulation, *J. Artif. Soc. Social Sim.* **5**, paper 2.

Herfindal, O. C. (1959), *Copper Costs and Prices: 1870–1957*, John Hopkins University Press.

Herrero, C. (2002), Ising model in small-world networks, *Phys. Rev. E* **65**, 066110.

Herrmann, H. J. and Roux, S., eds. (1990), *Statistical Models for the Fracture of Disordered Media*, North-Holland.

Hethcote, H. W. (1978), An immunization model for a heterogeneous population, *Theor. Popul. Biol.* **14**, 338–349.

Hethcote, H. W. and Yorke, J. A. (1984), Gonorrhea: transmission and control, *Lect. Notes Biomath.* **56**, 1–105.

Hirschman, A. O. (1964), The paternity of an index, *Am. Econ. Rev.* **54**, 761–762.

Ho, Y., Gruhler, A., Heilbut, A. *et al.* (2002), Systematic identification of protein complexes in *Saccharomyces cerevisiae* by mass spectrometry, *Nature* **415**, 180–183.

Holland, P. W. and Leinhardt, S. (1981), An exponential family of probability distributions for directed graphs, *J. Am. Stat. Assoc.* **76**, 33–50.

Holme, P. (2004), Efficient local strategies for vaccination and network attack, *Europhys. Lett.* **68**, 908–914.

Holme, P. and Kim, B. J. (2002a), Growing scale-free networks with tunable clustering, *Phys. Rev. E* **65**, 026107.

Holme, P. and Kim, B. J. (2002b), Vertex overload breakdown in evolving networks, *Phys. Rev. E* **65**, 066109.

Holme, P., Kim, B. J., Yoon, C. N. and Han, S. K. (2002), Attack vulnerability of complex networks, *Phys. Rev. E* **65**, 056109.

Holme, P. and Newman, M. E. J. (2006), Nonequilibrium phase transition in the coevolution of networks and opinions, *Phys. Rev. E* **74**, 056108.

Holme, P., Trusina, A., Kim, B. J. and Minnhagen, P. (2003), Prisoners' dilemma in real-world acquaintance networks: Spikes and quasiequilibria induced by the interplay between structure and dynamics, *Phys. Rev. E* **68**, 030901.

Hołyst, J. A., Sienkiewicz, J., Fronczak, A., Fronczak, P. and Suchecki, K. (2005), Universal scaling of distances in complex networks, *Phys. Rev. E* **72**, 026108.

Hong, H., Choi, M. Y. and Kim, B. J. (2002a), Synchronization on small-world networks, *Phys. Rev. E* **65**, 026139.

Hong, H., Kim, B. J. and Choi, M. Y. (2002b), Comment on "Ising model on a small world network", *Phys. Rev. E* **66**, 018101.

Hong, H., Kim, B. J., Choi, M. Y. and Park, H. (2004), Factors that predict better synchronizability on complex networks, *Phys. Rev. E* **69**, 067105.

Huang, K. (1987), *Statistical Mechanics*, Wiley.

Huffaker, B., Fomenkov, M., Moore, D., Nemeth, E. and claffy, k. (2000), Measurements of the Internet topology in the Asia–Pacific region. In *INET '00, Yokohama, Japan, 18–21 July 2000*. The Internet Society http://www.caida.org/publications/papers/2000/asia_paper/

Huffaker, B., Fomenkov, M., Plummer, D. Moore, D., and claffy, k. (2002a), Distance metrics in the Internet. In *IEEE International Telecommunications Symposium (ITS), Brazil, Sept. 2002*, IEEE Computer Society Press http://www.caida.org/publications/papers/2002/Distance/index.xml

Huffaker, B., Plummer, D., Moore, D. and claffy, k. (2002b), Topology discovery by active probing. In *Proceedings of 2002 Symposium on Applications and the Internet, (SAINT) Workshops*, pp. 90–96.

Hufnagel, L., Brockmann, D. and Geisel, T. (2004), Forecast and control of epidemics in a globalized world, *Proc. Natl Acad. Sci. (USA)* **101**, 15124–15129.

Hwang, D.-U., Chavez, M., Amman, A. and Boccaletti, S. (2005), Synchronization in complex networks with age ordering, *Phys. Rev. Lett.* **94**, 138701.

Ichinomiya, T. (2004), Frequency synchronization in random oscillator network, *Phys. Rev. E* **70**, 026116.

Ito, T., Chiba, T., Ozawa, R. *et al.* (2001), A comprehensive two-hybrid analysis to explore the yeast protein interactome, *Proc. Natl Acad. Sci. (USA)* **98**, 4569–4574.

Jalan, S. and Amritkar, R. E. (2002), Self-organized and driven phase synchronization in coupled map, *Phys. Rev. Lett.* **90**, 014101.

Jeong, H., Mason, S. P., Barabási, A.-L. and Oltvai, Z. N. (2001), Lethality and centrality in protein networks, *Nature* **411**, 41–42.

Jeong, H., Tombor, B., Albert, R., Oltvai, Z. N. and Barabási, A.-L. (2000), The large scale organization of metabolic networks, *Nature* **407**, 651–654.

Jeong, J. and Berman, P. (2007), Low-cost search in scale-free networks, *Phys. Rev. E* **75**, 036104.

Joo, J. and Lebowitz, J. L. (2004), Behavior of susceptible–infected–susceptible epidemics on heterogeneous networks with saturation, *Phys. Rev. E* **69**, 066105.

Jordán, F. and Molnar, I. (1999), Reliable flows and preferred patterns in food webs, *Evol. Ecol. Res.* **1**, 591–609.

Jordán, F. and Scheuring, I. (2004), Network ecology: topological constraints on ecosystem dynamics, *Phys. Life Rev.* **1**, 139–172.

Jost, J. and Joy, M. P. (2001), Spectral properties and synchronization in coupled map lattices, *Phys. Rev. E* **65**, 016201.

Jungnickel, D. (2004), *Graphs, Networks and Algorithms*, 2nd edn, Springer.

Kalapala, V., Sanwalani, V., Clauset, A. and Moore, C. (2006), Scale invariance in road networks, *Phys. Rev. E* **73**, 026130.

Kanter, I. and Sompolinsky, H. (1987), Mean-field theory of spin-glasses with finite coordination number, *Phys. Rev. Lett.* **58**, 164–167.

Karsai, M., Juhász, R. and Iglói, F. (2006), Nonequilibrium phase transitions and finite-size scaling in weighted scale-free networks, *Phys. Rev. E* **73**, 036116.

Kauffman, S. A. (1969), Metabolic stability and epigenesis in randomly constructed genetic nets, *J. Theor. Biol.* **22**, 437–467.

Kauffman, S. A. (1993), *The Origins of Order: Self-organization and Selection in Evolution*, Oxford University Press.

Kauffman, S., Peterson, C., Samuelsson, B. and Troein, C. (2003), Random Boolean network models and the yeast transcriptional network, *Proc. Natl Acad. Sci. (USA)* **100**, 14796–14799.

Kauffman, S., Peterson, C., Samuelsson, B. and Troein, C. (2004), Genetic networks with canalyzing Boolean rules are always stable, *Proc. Natl Acad. Sci. (USA)* **101**, 17102–17107.

Kaufman, V., Mihaljev, T. and Drossel, B. (2005), Scaling in critical random Boolean networks, *Phys. Rev. E* **72**, 046124.

Keeling, M. J. (2000), Metapopulation moments: coupling, stochasticity and persistence, *J. Animal Ecol.* **69**, 725–736.

Keeling, M. J. and Rohani, P. (2002), Estinating spatial coupling in epidemiological systems: a mechanistic approach, *Ecol. Lett.* **5**, 20–29.

Keeling, M. J., Woolhouse, M. E. J., Shaw, D. J. *et al.* (2001), Dynamics of the 2001 UK foot and mouth epidemic: Stochastic dispersal in a heterogeneous landscape, *Science* **294**, 813–817.

Keller, E. F. (2005), Revisiting "scale-free" networks, *BioEssays* **27**, 1060–1068.

Kephart, J. O., Sorkin, G. B., Chess, D. M. and White, S. R. (1997), Fighting computer viruses, *Sci. Am.* **277**(5), 56–61.

Kephart, J. O., White, S. R. and Chess, D. M. (1993), Computers and epidemiology, *IEEE Spectrum* **30**, 20–26.

Kermarrec, A. M., Massoulie, L. and Ganesh, A. J. (2003), Probabilistic reliable dissemination in large-scale systems, *IEEE Trans. Parall. Distr. Syst.* **14**, 248–258.

Killworth, P. D. and Bernard, H. R. (1978), The reversal small-world experiment, *Soc. Networks* **1**, 159–192.

Kim, B. J., Hong, H., Holme, P. *et al.* (2001), XY model in small-world networks, *Phys. Rev. E* **64**, 056135.

Kim, B. J., Trusina, A., Holme, P. *et al.* (2002a), Dynamic instabilities induced by asymmetric influence: Prisoners' dilemma game in small-world networks, *Phys. Rev. E* **66**, 021907.

Kim, B. J., Yoon, C. N., Han, S. K. and Jeong, H. (2002b), Path finding strategies in scale-free networks, *Phys. Rev. E* **65**, 027103.

Kim, D.-H., Kim, B. J. and Jeong, H. (2005), Universality class of the fiber bundle model on complex networks, *Phys. Rev. Lett.* **94**, 025501.

Kinney, R., Crucitti, P., Albert, R. and Latora, V. (2005), Modeling cascading failures in the North American power grid, *Eur. Phys. J. B* **46**, 101–107.

Kleinberg, J. M. (1998), Authoritative sources in a hyperlinked environment, pp. 668–677.

Kleinberg, J. M. (2000a), Navigation in a small world, *Nature* **406**, 845.

Kleinberg, J. M. (2000b), The small-world phenomenon: an algorithmic perspective, *in* Proceedings of the 32nd ACM symposium on the theory of computing, ACM Press, NY, pp. 163–170.

Kleinberg, J. M. (2001), Small-world phenomena and the dynamics of information, *Adv. Neural Inform. Process. Syst.* **14**, 431–438.

Kleinberg, J. M. (2006), Complex networks and decentralized search algorithms. In *Proceedings of the International Congress of Mathematicians (ICM)* http://www.icm2006.org/proceedings/Vol_III/contents/ICM_Vol_3_50.pdf

Kleinberg, J. M., Kumar, R., Raghavan, P., Rajagopalan, S. and Tomkins, A. S. (1999), The Web as a graph: Measurements, models and methods, *Lect. Notes Comput. Sci.* **1627**, 1–18.

Klemm, K. and Bornholdt, S. (2005), Stable and unstable attractors in Boolean networks, *Phys. Rev. E* **72**, 055101(R).

Klemm, K. and Eguíluz, V. M. (2002a), Growing scale-free networks with small-world behavior, *Phys. Rev. E.* **65**, 057102.

Klemm, K. and Eguíluz, V. M. (2002b), Highly clustered scale-free networks, *Phys. Rev. E.* **65**, 036123.

Klemm, K., Eguíluz, V. M., Toral, R. and San Miguel, M. (2003a), Global culture: a noise-induced transition in finite systems, *Phys. Rev. E.* **67**, 045101.

Klemm, K., Eguíluz, V. M., Toral, R. and San Miguel, M. (2003b), Nonequilibrium transitions in complex networks: a model of social interaction, *Phys. Rev. E.* **67**, 026120.

Klemm, K., Eguíluz, V. M., Toral, R. and San Miguel, M. (2005), Globalization, polarization and cultural drift, *J. Econ. Dynam. Control* **29**, 321–334.

Koch, C. (1999), *Biophysics of Computation*, Oxford University Press.

Korte, C. and Milgram, S. (1970), Acquaintance networks between racial groups: Application of the small world method, *J. Personality Soc. Psychol.* **15**, 101–108.

Kötter, R. and Sommer, F. T. (2000), Global relationship between anatomical connectivity and activity propagation in the cerebral cortex, *Phil. Trans. R. Soc. Lond. B* **355**, 127–134.

Kozma, B. and Barrat, A. (2008), Consensus formation on adaptive networks, *Phys. Rev. E*, **77**, 016102.

Kozma, B., Hastings, M. B. and Korniss, G. (2005), Diffusion processes on power-law small-world networks, *Phys. Rev. Lett.* **95**, 018701.

Kozma, B., Hastings, M. B. and Korniss, G. (2007), Diffusion processes on small-world networks with distance dependent random links, *J. Stat. Mech.* p. P08014.

Krapivsky, P. L. (1992), Kinetics of monomer–monomer surface catalytic reactions, *Phys. Rev. A* **45**, 1067–1072.

Krapivsky, P. L. and Redner, S. (2001), Organization of growing random networks, *Phys. Rev. E* **63**, 066123.

Krapivsky, P. L. and Redner, S. (2003a), Dynamics of majority rule in two-state interacting spin systems, *Phys. Rev. Lett.* **90**, 238701.

Krapivsky, P. L. and Redner, S. (2003b), Rate equation approach for growing networks, *Lect. Notes Phys.* **665**, 3–22.

Krapivsky, P. L. and Redner, S. (2005), Network growth by copying, *Phys. Rev. E* **71**, 036118.

Krause, W., Glauche, I., Sollacher, R. and Greiner, M. (2004), Impact of network structure on the capacity of wireless multihop ad hoc communication, *Physica A* **338**, 633–658.

Krause, W., Scholz, J. and Greiner, M. (2006), Optimized network structure and routing metric in wireless multihop ad hoc communication, *Physica A* **361**, 707–723.

Krauth, W. (2006), *Statistical Mechanics: Algorithms and Computations*, Oxford University Press.

Kumar, R., Raghavan, P., Rajagopalan, S. *et al.* (2000), Stochastic models for the Web graph, *Proceedings of the 41st IEEE Symposium on Foundations of Computer Science (FOCS)*, pp. 57–65.

Kuperman, M. and Abramson, G. (2001), Small world effect in an epidemiological model, *Phys. Rev. Lett.* **86**, 2909–2912.

Kuramoto, Y. (1984), *Chemical Oscillations, Waves, and Turbulence*, Springer.

Labovitz, C., Ahuja, A. and Jahanian, F. (1999), Experimental study of Internet stability and backbone failures. In *Proceedings of the IEEE Symposium on Fault-Tolerant Computing (FTCS99), Wisconsin*, 278–285.

Labovitz, C. and Malan, G. R. and Jahanian, F. (1998), Internet routing instability, *IEEE-ACM Trans. Netw.* **6**, 515–528.

Lago-Fernández, L. F., Huerta, R., Corbacho, F. and Sigüenza, J. A. (2000), Fast response and temporal coherent oscillations in small-world networks, *Phys. Rev. Lett.* **84**, 2758–2761.

Lakhina, A., Byers, J. W., Crovella, M. and Xie, P. (2002), *Sampling Biases in IP Topology Measurements*. Technical report BUCS-TR-2002-021, Department of Computer Sciences, Boston University.

Lässig, M., Bastolla, U., Manrubia, S. C. and Valleriani, A. (2001), Shape of ecological networks, *Phys. Rev. Lett.* **86**, 4418–4421.

Latora, V. and Marchiori, M. (2001), Efficient behavior of small-world networks, *Phys. Rev. Lett.* **87**, 198701.

Latora, V. and Marchiori, M. (2002), Is the Boston subway a small-world network?, *Physica A* **314**, 109–113.

Lawlor, L. R. (1978), A comment on randomly constructed model ecosystems, *Am. Nat.* **112**, 444–447.

Lebhar, E. and Schabanel, N. (2004), Almost optimal decentralized routing in long-range contact networks. In *Proceedings of the 31st International Colloquium on Automata, Languages and Programming (ICALP 2004)*, pp. 894–905.

Lee, C. and Stepanek, J. (2001), On future global grid communication performance. *10th IEEE Heterogeneous Computing Workshop, May 2001*.

Lee, D.-S. (2005), Synchronization transition in scale-free networks: clusters of synchrony, *Phys. Rev. E* **72**, 026208.

Lee, T. I., Rinaldi, N. J., Robert, F. *et al.* (2002), Transcriptional regulatory networks in *Saccharomyces cerevisiae*, *Science* **298**, 799–804.

Lemke, N., Herédia, F., Barcellos, C. K., dos Reis, A. N. and Mombach, J. C. M. (2004), Essentiality and damage in metabolic networks, *Bioinformatics* **20**, 115–119.

Lenaerts, T., Jansen, B., Tuyls, K. and Vylder, B. D. (2005), The evolutionary language game: An orthogonal approach, *J. Theor. Biol.* **235**, 566–582.

Leone, M., Vázquez, A., Vespignani, A. and Zecchina, R. (2002), Ferromagnetic ordering in graphs with arbitrary degree distribution, *Eur. Phys. J. B* **28**, 191–197.

Leskovec, J., Adamic, L. A. and Huberman, B. A. (2006), The dynamics of viral marketing. In *Proceedings of the 7th ACM Conference on Electronic Commerce, Ann Arbor*, ACM Press, pp. 228–237.

Li, F., Long, T., Lu, Y., Ouyang, Q. and Tang, C. (2004a), The yeast cell-cycle network is robustly designed, *Proc. Natl Acad. Sci. (USA)* **101**, 4781–4786.

Li, L., Alderson, D., Willinger, W. and Doyle, J. (2004b), A first-principles approach to understanding the Internet's router-level topology. In *Proceedings of the 2004 Conference on Applications, Technologies, Architectures, and Protocols for Computer Communications (SIGCOMM)*, ACM Press, pp. 3–14.

Liben-Nowell, D., Novak, J., Kumar, R., Raghavan, P. and Tomkins, A. (2005), Geographic routing in social networks, *Proc. Natl Acad. Sci. (USA)* **102**, 11623–11628.

Liljeros, F., Edling, C. R., Amaral, L. A. N., Stanley, H. E. and Åberg, Y. (2001), The web of human sexual contacts, *Nature* **411**, 907–908.

Lin, H. and Wu, C.-X. (2006), Dynamics of congestion transition triggered by multiple walkers on complex networks, *Eur. Phys. J. B* **51**, 543–547.

Lind, P. G., Gallas, J. A. C. and Herrmann, H. J. (2004), Coherence in complex networks of oscillators. In M. Dirickx, ed., *Proceedings of Verhulst 2004*, Springer. cond-mat/0502307.

Liu, Z., Lai, Y.-C. and Ye, N. (2003), Propagation and immunization of infection on general networks with both homogeneous and heterogeneous components, *Phys. Rev. E* **67**, 031911.

Lloyd, A. L. and May, R. M. (1996), Spatial heterogeneity in epidemic models, *J. Theor. Biol.* **179**, 1–11.

Longini, I. M. (1988), A mathematical model for predicting the geographic spread of new infectious agents, *Math. Biosci.* **90**, 367–383.

Longini, I. M., Nizam, A., Xu, S. *et al.* (2005), Containing pandemic influenza at the source, *Science* **309**, 1083–1087.

Lozano, S., Arenas, A. and Sanchez, A. (2006), Mesoscopic structure conditions the emergence of cooperation on social networks, *PLoS ONE* **3**(4), e1892.

Lundberg, C. C. (1975), Patterns of acquaintanceship in society and complex organization: A comparative study of the small world problem, *Pacif. Sociol. Rev.* **18**, 206–222.

Lv, Q., Cao, P., Cohen, E., Li, K. and Shenker, S. (2002), Search and replication in unstructured peer-to-peer networks. In *Proceedings of the 16th International Conference on Supercomputing*, ACM Press, pp. 84–95.

Ma, S. K. (1985), *Statistical Mechanics*, World Scientific.

Ma, S. K. (2000), *Modern Theory of Critical Phenomena*, The Perseus Books Group.

Maass, W. and Bishop, C., eds. (1998), *Pulsed Neural Networks*, MIT Press.

MacArthur, R. H. (1955), Fluctuations of animal populations and a measure of of community stability, *Ecology* **36**, 533–536.

Madar, N., Kalisky, T., Cohen, R., ben Avraham, D. and Havlin, S. (2004), Immunization and epidemic dynamics in complex networks, *Eur. Phys. J. B* **38**, 269–276.

Magnasco, M. O. (2000), The thunder of distant Net storms. http://arXiv.org/abs/nlin/0010051.

Mahadevan, P., Krioukov, D., Fall, K. and Vahdat, A. (2006), Systematic topology analysis and generation using degree correlations, *ACM SIGCOMM Computer Communication Review (Proceedings of the 2006 conference on Applications, Technologies, Architectures, and Protocols for Computer Communications)* **36**, 135–146.

Mahmoudi, H., Pagnani, A., Weigt, M. and Zecchina, R. (2007), Propagation of external regulation and asynchronous dynamics in random Boolean networks, *Chaos* **17**, 026109.

Mahon, T. A. and Bonner, J. T. (1983), *On Size and Life*, Scientific American Library.

Maki, D. P. and Thompson, M. (1973), *Mathematical Models and Applications, with Emphasis on the Social, Life, and Management Sciences*, Prentice-Hall.

Mandelbrot, B. B. (1969), Long run linearity, locally Gaussian processes, H-spectra and infinite variances, *Intern. Econ. Rev.* **10**, 82–113.

Mandelbrot, B. B. (1982), *The Fractal Geometry of Nature*, Freeman.

Manku, G. S., Naor, M. and Wieder, U. (2004), Know thy neighbor's neighbor: the power of lookahead in randomized P2P networks. In *Proceedings of the 36th annual ACM Symposium on Theory of Computing*, ACM Press, pp. 54–63.

Manna, S. S. and Sen, P. (2002), Modulated scale-free network in Euclidean space, *Phys. Rev. E* **66**, 066114.

Mantegna, R. and Stanley, H. E. (1999), *An Introduction to Econophysics: Correlations and Complexity in Finance*, Cambridge University Press.

Marendy, P. (2001), A review of World-Wide-Web searching techniques, focusing on HITS and related algorithms that utilise the link topology of the World-Wide-Web to provide the basis for a structure based search technology. `http://citeseer.ist.psu.edu/559198.html`.

Maritan, A., Colaiori, F., Flammini, A., Cieplak, M. and Banavar, J. R. (1996), Universality classes of optimal channel networks, *Science* **272**, 984–986.

Markram, H. (2006), The blue brain project, *Nat. Rev. Neurosci.* **7**, 153–160.

Marro, J. and Dickman, R. (1999), *Non-equilibrium Phase Transitions in Lattice Models*, Cambridge University Press.

Martinez, N. D., Williams, R. J. and Dunne, J. A. (2006), Diversity, complexity, and persistence in large model ecosystems. In M. Pascual and J. A. Dunne, eds. *Ecological Networks: Linking Structure to Dynamics in Food Webs*, Oxford University Press, pp. 163–185.

Maslov, S. and Sneppen, K. (2002), Specificity and stability in topology of protein networks, *Science* **296**, 910–913.

Maslov, S., Sneppen, K. and Zaliznyak, A. (2004), Detection of topological patterns in complex networks: correlation profile of the Internet, *Physica A* **333**, 529–540.

Masuda, N., Miwa, H. and Konno, N. (2005), Geographical threshold graphs with small-world and scale-free properties, *Phys. Rev. E* **71**, 036108.

Mathias, N. and Gopal, V. (2001), Small-world: how and why, *Phys. Rev. E* **63**, 021117.

May, R. M. (1973), *Stability and Complexity in Model Ecosystems*, Princeton University Press.

May, R. M. and Anderson, R. M. (1984), Spatial heterogeneity and the design of immunization programs, *Math. Biosci.* **72**, 83–111.

May, R. M. and Lloyd, A. L. (2001), Infection dynamics on scale-free networks, *Phys. Rev. E* **64**, 066112.

McCraith, S., Holtzman, T., Moss, B. and Fields, S. (2000), Genome wide analysis of vaccinia virus protein-protein interactions, *Proc. Natl Acad. Sci. (USA)* **97**, 4879–4884.

Medina, A., Matt, I. and Byers, J. (2000), On the origin of power-laws in Internet topology, *Comput. Commun. Rev.* **30**, 18–28.

Medvedyeva, K., Holme, P., Minnhagen, P. and Kim, B. J. (2003), Dynamic critical behaviour of the XY model in small-world networks, *Phys. Rev. E* **67**, 036118.

Mehta, M. L. (1991), *Random Matrices*, Academic Press.

Memmesheimer, R.-M. and Timme, M. (2006), Designing the dynamics of spiking neural networks, *Phys. Rev. Lett.* **97**, 188101.

Memmott, J., Waser, N. M. and Price, M. V. (2004), Tolerance of pollination networks to species extinctions, *Proc. R. Soc. Lond. B* **271**, 2605–2611.

Menczer, F. (2002), Growing and navigating the small world web by local content, *Proc. Natl Acad. Sci. (USA)* **99**, 14014–14019.

Merton, R. K. (1968), The Matthew effect in science, *Science* **159**, 56–63.

Milgram, S. (1967), The small world problem, *Psychology Today* **1**, 61–67.

Milo, R., Shen-Orr, S., Itzkovitz, S. *et al.* (2002), Network motifs: simple building blocks of complex networks, *Science* **298**, 824–827.

Mirkin, B. G. (1996), *Mathematical Classification and Clustering*, Springer.

Mirollo, R. E. and Strogatz, S. H. (1990), Synchronization of pulse-coupled biological oscillators, *SIAM J. Appl. Math.* **50**, 1645–1662.

Mitzenmacher, M. (2006), The future of power-law research, *Internet Math.* **2**, 525–534.

Mohar, B. (1997), Some applications of Laplace eigenvalues of graphs. In G. Hahn and G. Sabidussi, eds., *Graph Symmetry: Algebraic Methods and Applications*, Vol. 497 of *NATO ASI Series C*, Kluwer, pp. 227–275.

Molloy, M. and Reed, B. (1995), A critical point for random graphs with a given degree sequence, *Random Struct. Algor.* **6**, 161–179.

Molloy, M. and Reed, B. (1998), The size of the giant component of a random graph with a given degree distribution, *Comb. Probab. Comput.* **7**, 295–305.

Monasson, R. (1999), Diffusion, localization and dispersion relations on small-world lattices, *Eur. Phys. J. B* **12**, 555–567.

Montis, A. D., Barthélemy, M., Chessa, A. and Vespignani, A. (2007), The structure of inter-urban traffic: a weighted network analysis, *Envir. Planning J. B* **34**, 905–924.

Montoya, J. M., Pimm, S. L. and Solé, R. V. (2006), Ecological networks and their fragility, *Nature* **442**, 259–264.

Montoya, J. M. and Solé, R. V. (2002), Small-world patterns in food-webs, *J. Theor. Biol.* **214**, 405 – 412.

Moore, C. and Newman, M. E. J. (2000), Epidemics and percolation in small-world networks, *Phys. Rev. E* **61**, 5678–5682.

Moran, P .A .P. (1962), *The Statistical Processes of Evolutionary Theory*. Oxford, Clarendon Press.

Moreira, A. A., Andrade, J. S. and Amaral, L. A. N. (2002), Extremum statistics in scale-free network models, *Phys. Rev. Lett.* **89**, 268703.

Moreno, J. L. (1934), *Who Shall Survive? Foundations of Sociometry, Group Psychotherapy, and Sociodram*, Beacon House.

Moreno, Y., Gómez, J. B. and Pacheco, A. F. (2002a), Instability of scale-free networks under node-breaking avalanches, *Europhys. Lett.* **58**, 630–636.

Moreno, Y., Gómez, J. B. and Pacheco, A. F. (2003a), Epidemic incidence in correlated complex networks, *Phys. Rev. E* **68**, 035103.

Moreno, Y., Nekovee, M. and Pacheco, A. F. (2004a), Dynamics of rumor spreading in complex networks, *Phys. Rev. E* **69**, 066130.

Moreno, Y., Nekovee, M. and Vespignani, A. (2004b), Efficiency and reliability of epidemic data dissemination in complex networks, *Phys. Rev. E* **69**, 055101(R).

Moreno, Y. and Pacheco, A. F. (2004), Synchronization of Kuramoto oscillators in scale-free networks, *Europhys. Lett.* **68**, 603–609.

Moreno, Y., Pastor-Satorras, R., Vázquez, A. and Vespignani, A. (2003b), Critical load and congestion instabilities in scale-free networks, *Europhys. Lett.* **62**, 292–298.

Moreno, Y., Pastor-Satorras, R. and Vespignani, A. (2002b), Epidemic outbreaks in complex heterogeneous networks, *Eur. Phys. J. B* **26**, 521–529.

Moreno, Y. and Vázquez, A. (2003), Disease spreading in structured scale-free networks, *Eur. Phys. J. B* **31**, 265–271.

Moreno, Y., Vázquez-Prada, M. and Pacheco, A. F. (2004c), Fitness for synchronization of network motifs, *Physica A* **343**, 279–287.

Mossa, S., Barthélemy, M., Stanley, H. E. and Amaral, L. A. N. (2002), Truncation of power law behavior in "scale-free" network models due to information filtering, *Phys. Rev. Lett.* **88**, 138701.

Motter, A. E. (2004), Cascade control and defense in complex networks, *Phys. Rev. Lett.* **93**, 098701.

Motter, A. E. and Lai, Y.-C. (2002), Cascade-based attacks on complex networks, *Phys. Rev. E* **66**, 065102.

Motter, A. E., Zhou, C. and Kurths, J. (2005a), Network synchronization, diffusion, and the paradox of heterogeneity, *Phys. Rev. E* **71**, 016116.

Motter, A. E., Zhou, C. and Kurths, J. (2005b), Weighted networks are more synchronizable: how and why. In *Science of Complex Networks: From Biology to the Internet and WWW: CNET 2004*, AIP Conference Proceedings Vol. 776, pp. 201–214.

Motter, A. E., Zhou, C. S. and Kurths, J. (2005c), Enhancing complex-network synchronization, *Europhys. Lett.* **69**, 334–340.

Murray, J. D. (2005), *Mathematical Biology*, Springer-Verlag.

Nekovee, M., Moreno, Y., Bianconi, G. and Marsili, M. (2007), Theory of rumour spreading in complex social networks, *Physica A* **374**, 457–470.

Newman, M. E. J. (2001a), Scientific collaboration networks. I. Network construction and fundamental results, *Phys. Rev. E* **64**, 016131.

Newman, M. E. J. (2001b), Scientific collaboration networks. II. Shortest paths, weighted networks, and centrality, *Phys. Rev. E* **64**, 016132.

Newman, M. E. J. (2001c), The structure of scientific collaboration networks, *Proc. Natl Acad. Sci. (USA)* **98**, 404–409.

Newman, M. E. J. (2002a), Assortative mixing in networks, *Phys. Rev. Lett.* **89**, 208701.

Newman, M. E. J. (2002b), Spread of epidemic disease on networks, *Phys. Rev. E* **66**, 016128.

Newman, M. E. J. (2003a), Mixing patterns in networks, *Phys. Rev. E* **67**, 026126.

Newman, M. E. J. (2003b), Random graphs as models of networks. In S. Bornholdt and H. G. Schuster, eds., *Handbook of Graphs and Networks: From the Genome to the Internet*, Wiley-VCH, pp. 35–68.

Newman, M. E. J. (2003c), The structure and function of complex networks, *SIAM Rev.* **45**, 167–256.

Newman, M. E. J. (2004), Detecting community structure in networks, *Eur. Phys. J. B* **38**, 321–330.

Newman, M. E. J. (2005a), A measure of betweenness centrality based on random walks, *Soc. Networks* **27**, 39–54.

Newman, M. E. J. (2005b), Power laws, Pareto distributions and Zipf's law, *Contemp. Phys.* **46**, 323–351.

Newman, M. E. J. (2006), Modularity and community structure in networks, *Proc. Natl Acad. Sci. (USA)* **103**, 8577–8582.

Newman, M. E. J. and Barkema, G. T. (1999), *Monte Carlo Methods in Statistical Physics*, Oxford University Press.

Newman, M. E. J., Forrest, S. and Balthrop, J. (2002), Email networks and the spread of computer viruses, *Phys. Rev. E* **66**, 035101.

Newman, M. E. J. and Girvan, M. (2004), Finding and evaluating community structure in networks, *Phys. Rev. E* **69**, 026113.

Newman, M. E. J., Strogatz, S. H. and Watts, D. J. (2001), Random graphs with arbitrary degree distributions and their applications, *Phys. Rev. E* **64**, 026118.

Nikoletopoulos, T., Coolen, A. C. C., Pérez Castillo, I. *et al.* (2004), Replicated transfer matrix analysis of Ising spin models on "small world" lattices, *J. Phys. A: Math. Gen.* **37**, 6455–6475.

Nishikawa, T. and Motter, A. E. (2006a), Maximum performance at minimum cost in network synchronization, *Physica D* **224**, 77–89.

Nishikawa, T. and Motter, A. E. (2006b), Synchronization is optimal in nondiagonalizable networks, *Phys. Rev. E* **73**, 065106(R).

Nishikawa, T., Motter, A. E., Lai, Y.-C. and Hoppensteadt, F. C. (2003), Heterogeneity in oscillator networks: Are smaller worlds easier to synchronize ?, *Phys. Rev. Lett.* **91**, 014101.

Noh, J. D. and Rieger, H. (2004), Random walks on complex networks, *Phys. Rev. Lett.* **92**, 118701.

Nowak, A., Szamrej, J. and Latané, B. (1990), From private attitude to public opinion: A dynamic theory of social impact, *Psychol. Rev.* **97**, 362–376.

Nowak, M. A. and May, R. M. (1992), Evolutionary games and spatial chaos, *Nature* **359**, 826–829.

Nowak, M. A. and May, R. M. (1993), The spatial dilemmas of evolution, *Int. J. Bifurc. Chaos* **3**, 35–78.

Oh, E., Rho, K., Hong, H. and Kahng, B. (2005), Modular synchronization in complex networks, *Phys. Rev. E* **72**, 047101.

Ohira, T. and Sawatari, R. (1998), Phase transition in a computer network traffic model, *Phys. Rev. E* **58**, 193–195.

Ohono, S. (1970), *Evolution by Gene Duplication*, Springer.

O'Kelly, M. E. (1998), A geographer's analysis of hubs-and-spokes networks, *J. Transp. Geog.* **6**, 171–186.

Olinky, R. and Stone, L. (2004), Unexpected epidemic thresholds in heterogeneous networks: The role of disease transmission, *Phys. Rev. E* **70**, 030902.

Onnela, J.-P., Saramäki, J., Kertész, J. and Kaski, K. (2005), Intensity and coherence of motifs in weighted complex networks, *Phys. Rev. E* **71**, 065103(R).

Onsager, L., Crystal Statistics. I. A two-dimensional model with an order-disorder transition, *Phys. Rev.* **65**, 117–149 (1944)

Oosawa, C. and Savageau, M. A. (2002), Effects of alternative connectivity on behavior of randomly constructed Boolean networks, *Physica D* **170**, 143–161.

Otsuka, K., Kawai, R., Hwong, S.-L., Ko, J.-Y. and Chern, J.-L. (2000), Synchronization of mutually coupled self-mixing modulated lasers, *Phys. Rev. Lett.* **84**, 3049–3052.

Palla, G., Derényi, I., Farkas, I. and Vicsek, T. (2005), Uncovering the overlapping community structure of complex networks in nature and society, *Nature* **435**, 814–818.

Palsson, B. Ø. (2006), *Systems Biology: Properties of Reconstructed Networks*, Cambridge University Press.

Park, A. W., Gubbins, S. and Gilligan, C. A. (2002), Extinction times for closed epidemics: the effects of host spatial structure, *Ecol. Lett.* **5**, 747–755.

Park, J. and Newman, M. E. J. (2003), Origin of degree correlations in the Internet and other networks, *Phys. Rev. E* **68**, 026112.

Park, J. and Newman, M. E. J. (2004), Statistical mechanics of networks, *Phys. Rev. E* **70**, 066117.

Park, K., Kim, G. and Crovella, M. (1997), On the effect of traffic self-similarity on network performance. In *Proceedings of the SPIE International Conference on Performance and Control of Network Systems*, pp. 296–310.

Park, K., Lai, Y.-C., Gupte, S. and Kim, J.-W. (2006), Synchronization in complex networks with a modular structure, *Chaos* **16**, 015105.

Pastor-Satorras, R., Vázquez, A. and Vespignani, A. (2001), Dynamical and correlation properties of the Internet, *Phys. Rev. Lett.* **87**, 258701.

Pastor-Satorras, R. and Vespignani, A. (2001a), Epidemic dynamics and endemic states in complex networks, *Phys. Rev. E* **63**, 066117.

Pastor-Satorras, R. and Vespignani, A. (2001b), Epidemic spreading in scale-free networks, *Phys. Rev. Lett.* **86**, 3200–3203.

Pastor-Satorras, R. and Vespignani, A. (2002a), Epidemic dynamics in finite size scale-free networks, *Phys. Rev. E* **65**, 035108.

Pastor-Satorras, R. and Vespignani, A. (2002b), Immunization of complex networks, *Phys. Rev. E* **65**, 036104.

Pastor-Satorras, R. and Vespignani, A. (2004), *Evolution and Structure of the Internet: A Statistical Physics Approach*, Cambridge University Press.

Pathria, R. K. (1996), *Statistical Mechanics*, 2nd edn, Butterworth-Heinemann.

Paul, G., Tanizawa, T., Havlin, S. and Stanley, H. E. (2004), Optimization of robustness of complex networks, *Eur. Phys. J. B* **38**, 187–191.

Paxson, V. (1997), End-to-end routing behavior in the Internet, *IEEE/ACM Trans. Networking* **5**, 601–615.

Paxson, V. and Floyd, S. (1995), Wide area traffic:the failure of Poisson modeling, *IEEE/ACM Trans. Networking* **3**, 226–244.

Pękalski, A. (2001), Ising model on a small world network, *Phys. Rev. E* **64**, 057104.

Pecora, L. M. and Carroll, T. L. (1990), Synchronization in chaotic systems, *Phys. Rev. Lett.* **64**, 821–824.

Pecora, L. M. and Carroll, T. L. (1998), Master stability functions for synchronized coupled systems, *Phys. Rev. Lett.* **80**, 2109–2112.

Percacci, R. and Vespignani, A. (2003), Scale-free behavior of the Internet global performance, *Eur. Phys. J. B* **32**, 411–414.

Peskin, C. S. (1975), *Mathematical Aspects of Heart Physiology*, Courant Institute of Mathematical Sciences, New York University.

Petermann, T. and De Los Rios, P. (2004a), Exploration of scale-free networks: Do we measure the real exponents?, *Eur. Phys. J. B* **38**, 201–204.

Petermann, T. and De Los Rios, P. (2004b), Role of clustering and gridlike ordering in epidemic spreading, *Phys. Rev. E* **69**, 066116.

Pikovsky, A., Rosenblum, M. and Kurths, J. (2001), *Synchronization: A Universal Concept in Nonlinear Sciences*, Cambridge Nonlinear Science Series, Cambridge University Press.

Pimm, S. L. (1979), Complexity and stability: another look at MacArthur's original hypothesis, *Oikos* **33**, 351–357.

Pimm, S. L. (2002), *Food Webs*, University of Chicago Press.

Qian, C., Chang, H., Govindan, R. *et al.* (2002), The origin of power laws in Internet topologies revisited. In *INFOCOM 2002. 21st Annual Joint Conference of the IEEE Computer and Communications Societies*. Proceedings IEEE, Vol. 2, IEEE Computer Society Press, pp. 608–617.

Quince, C., Higgs, P. G. and McKane, A. J. (2005), Deleting species from model food webs, *Oikos* **110**, 283–296.

Rain, J. C., Selig, L., De Reuse, H. *et al.* (2001), The protein–protein interaction map of *Helicobacter pylori*, *Nature* **409**, 211–215.

Ramasco, J. J., Dorogovtsev, S. N. and Pastor-Satorras, R. (2004), Self-organization of collaboration networks, *Phys. Rev. E* **70**, 036106.

Rapoport, A. (1953), Spread of information through a population with socio-structural bias: I. assumption of transitivity, *Bull. Math. Biol.* **15**, 523–533.

Rapoport, A. and Chammah, A. M. (1965), *Prisoner's Dilemma: A Study in Conflict and Cooperation*, University of Michigan Press.

Ravasz, E. and Barabási, A.-L. (2003), Hierarchical organization in complex networks, *Phys. Rev. E* **67**, 026112.

Ravasz, E., Somera, A. L., Mongru, D. A., Oltvai, Z. N. and Barabási, A.-L. (2002), Hierarchical organization of modularity in metabolic networks, *Science* **297**, 1551–1555.

Redner, S. (1998), How popular is your paper? An empirical study of the citation distribution, *Eur. Phys. J. B* **4**, 131–134.

Redner, S. (2005), Citation statistics from 110 years of *Physical Review*, *Phys. Today* **58**, 49.

Reichardt, J. and Bornholdt, S. (2004), Detecting fuzzy community structures in complex networks with a Potts model, *Phys. Rev. Lett.* **93**, 218701.

Restrepo, J. G., Ott, E. and Hunt, B. R. (2005a), Onset of synchronization in large networks of coupled oscillators, *Phys. Rev. E* **71**, 036151.

Restrepo, J. G., Ott, E. and Hunt, B. R. (2005b), Synchronization in large directed networks of coupled phase oscillators, *Chaos* **16**, 015107.

Rodgers, G. J. and Bray, A. J. (1988), Density of states of a sparse random matrix, *Phys. Rev. B* **37**, 3557–3562.

Rodriguez-Iturbe, I. and Rinaldo, A. (1997), *Fractal River Basins: Chance and Self-Organization*, Cambridge University Press.

Rohani, P., Earn, D. J. D. and Grenfell, B. T. (1999), Opposite patterns of synchrony in sympatric disease metapopulations, *Science* **286**, 968–971.

Rosenblum, M. G., Pikovsky, A. S. and Kurths, J. (1996), Phase synchronization of chaotic oscillators, *Phys. Rev. Lett.* **76**, 1804–1807.

Rosvall, M., Grönlund, A., Minnhagen, P. and Sneppen, K. (2005), Searchability of networks, *Phys. Rev. E* **72**, 046117.

Rosvall, M., Minnhagen, P. and Sneppen, K. (2005), Navigating networks with limited information, *Phys. Rev. E* **71**, 066111.

Rozenfeld, A. F., Cohen, R., ben Avraham, D. and Havlin, S. (2002), Scale-free networks on lattices, *Phys. Rev. Lett.* **89**, 218701.

Rulkov, N. F., Sushchik, M. M., Tsimring, L. S. and Abarbanel, H. D. I. (1995), Generalized synchronization of chaos in directionally coupled chaotic systems, *Phys. Rev. E* **51**, 980–994.

Rvachev, L. A. and Longini, I. M. (1985), A mathematical model for the global spread of influenza, *Math. Biosci.* **75**, 3–22.

Samukhin, A., Dorogovtsev, S. and Mendes, J. (2008), Laplacian spectra of, and random walks on, complex networks: are scale-free architectures really important? *Phys. Rev. E* **77**, 036115.

Sander, L. M., Warren, C. P., Sokolov, I. M., Simon, C. and Koopman, J. (2002), Percolation on heterogeneous networks as a model for epidemics, *Math. Biosci.* **180**, 293–305.

Sanil, A., Banks, D. and Carley, K. (1995), Models for evolving fixed node networks: Model fitting and model testing, *Soc. Networks* **17**, 65–81.

Santos, F. C. and Pacheco, J. M. (2005), Scale-free networks provide a unifying framework for the emergence of cooperation, *Phys. Rev. Lett.* **95**, 098104.

Santos, F. C. and Pacheco, J. M. (2006), A new route to the evolution of cooperation, *Journal of Evolutionary Biology* **19**, 726–733.

Santos, F. C., Pacheco, J. M. and Lenaerts, T. (2006a), Cooperation prevails when individuals adjust their social ties, *PLoS Comput. Biol.* **2**, e140.

Santos, F. C., Pacheco, J. M. and Lenaerts, T. (2006b), Evolutionary dynamics of social dilemmas in structured heterogeneous populations, *Proc. Natl Acad. Sci. (USA)* **103**, 3490–3494.

Santos, F. C., Rodrigues, J. F. and Pacheco, J. M. (2005), Graph topology plays a determinant role in the evolution of cooperation, *Proc. R. Soc. B* **273**, 51–55.

Saramäki, J., Kivelä, M., Onnela, J.-P., Kaski, K. and Kertész, J. (2007), Generalizations of the clustering coefficient to weighted complex networks, *Phys. Rev. E* **75**, 027105.

Sattenspiel, L. and Dietz, K. (1995), A structured epidemic model incorporating geographic mobility among regions, *Math. Biosci.* **128**, 71–91.

Scellato, S., Cardillo, A., Latora, V. and Porta, S. (2006), The backbone of a city, *Eur. Phys. J. B* **50**, 221–225.

Schmidt, B. M. and Chingos, M. M. (2007), Ranking doctoral programs by placement: A new method, *Political Science and Politics* **40**, 523–529.

Schneeberger, A., Mercer, C. H., Gregson, S. A. J. *et al.* (2004), Scale-free networks and sexually transmitted diseases, *Sex. Transm. Dis.* **31**, 380–387.

Schwartz, N., Cohen, R., ben Avraham, D., Barabási, A.-L. and Havlin, S. (2002), Percolation in directed scale-free networks, *Phys. Rev. E* **66**, 015104.

Schwikowski, B., Uetz, P. and Fields, S. (2000), A network of protein–protein interactions in yeast, *Nat. Biotechnol.* **18**, 1257–1261.

Segrè, D., Vitkup, D. and Church, G. M. (2002), Analysis of optimality in natural and perturbed metabolic networks, *Proc. Natl Acad. Sci. (USA)* **99**, 15112–15117.

Sen, P., Dasgupta, S., Chatterjee, A. *et al.* (2003), Small-world properties of the Indian railway network, *Phys. Rev. E* **67**, 036106.

Serrano, M. A. and Boguñá, M. (2005), Tuning clustering in random networks with arbitrary degree distributions, *Phys. Rev. E* **72**, 036133.

Serrano, M. A. and Boguñá, M. (2006), Percolation and epidemic thresholds in clustered networks, *Phys. Rev. Lett.* **97**, 088701.

Serrano, M. A., Boguñá, M. and Díaz-Guilera, A. (2005), Competition and adaptation in an Internet evolution model, *Phys. Rev. Lett.* **94**, 038701.

Serrano, M. A., Boguñá, M. and Pastor-Satorras, R. (2006), Correlations in weighted networks, *Phys. Rev. E* **74**, 055101(R).

Shavitt, Y. and Shir, E. (2005), DIMES: Let the Internet measure itself, *ACM SIGCOMM Comput. Commun. Rev.* **35**, 71–74.

Sheffi, Y. (1985), *Urban Transportation Networks*, Prentice Hall.

Shen-Orr, S., Milo, R., Mangan, S. and Alon, U. (2002), Network motifs in the transcriptional regulation network of *Escherichia coli*, *Nat. Genet.* **31**, 64–68.

Shmulevich, I., Kauffman, S. A. and Aldana, M. (2005), Eukaryotic cells are dynamically ordered or critical but not chaotic, *Proc. Natl Acad. Sci. (USA)* **102**, 13439–13444.

Simard, D., Nadeau, L. and Kröger, H. (2005), Fastest learning in small-world neural networks, *Phys. Lett. A* **336**, 8–15.

Simon, H. A. (1955), On a class of skew distribution functions, *Biometrika* **42**, 425–440.

Şimşek, O. and Jensen, D. (2005), Decentralized search in networks using homophily and degree disparity. In *Proceedings of the 19th International Joint Conference on Artificial Intelligence*, Vol. 19, Lawrence Erlbaum Associates Ltd, pp. 304–310.

Sneppen, K., Trusina, A. and Rosvall, M. (2005), Hide-and-seek on complex networks, *Europhys. Lett.* **69**, 853–859.

Snijders, T. A. B. (2001), The statistical evaluation of social network dynamics. In M. E. Sobel and M. P. Becker, eds., *Sociological Methodology*, Blackwell Ltd., pp. 361–395.

Söderberg, B. (2002), General formalism for inhomogeneous random graphs, *Phys. Rev. E* **66**, 066121.

Solé, R. V. and Montoya, J. M. (2001), Complexity and fragility in ecological networks, *Proc. R. Soc. B* **268**, 2039–2045.

Solé, R. V., Pastor-Satorras, R., Smith, E. and Kepler, T. B. (2002), A model of large-scale proteome evolution, *Adv. Complex Syst.* **5**, 43–54.

Solé, R. V. and Valverde, S. (2001), Information transfer and phase transitions in a model of internet traffic, *Physica A* **289**, 595–605.

Sood, V. and Redner, S. (2005), Voter model on heterogeneous graphs, *Phys. Rev. Lett.* **94**, 178701.

Sood, V., Redner, S. and ben Avraham, D. (2005), First-passage properties of the Erdős–Rényi random graph, *J. Phys. A: Math. Gen.* **38**, 109–123.

Sorrentino, F., di Bernardo, M., and Garofalo, F. (2007), Synchronizability and synchronization dynamics of weighted and unweighted scale-free networks with degree mixing, *Int. J. Bifure. Chaos* **17**, 2419–2434.

Sorrentino, F., di Bernardo, M., Huerta Cuéllar, G. and Boccaletti, S. (2006), Synchronization in weighted scale-free networks with degree–degree correlation, *Physica D* **224**, 123–129.

Spirin, V., Krapivsky, P. L. and Redner, S. (2001), Fate of zero-temperature Ising ferromagnets, *Phys. Rev. E* **63**, 036118.

Spirin, V., Krapivsky, P. L. and Redner, S. (2002), Freezing in Ising ferromagnets, *Phys. Rev. E* **65**, 016119.

Sporns, O. (2003), Network analysis, complexity and brain function, *Complexity* **8**, 56–60.

Sporns, O. (2004), Complex neural dynamics. In V. K. Jirsa and J. A. S. Kelso, eds., *Coordination Dynamics: Issues and Trends*, Springer-Verlag, pp. 197–215.

Sporns, O., Chialvo, D., Kaiser, M. and Hilgetag, C. C. (2004), Organization, development and function of complex brain networks, *Trends Cogn. Sci.* **8**, 418–425.

Sporns, O. and Tononi, G. (2002), Classes of network connectivity and dynamics, *Complexity* **7**, 28–38.

Sporns, O., Tononi, G. and Kötter, R. (2005), The human connectome: A structural description of the human brain, *PLoS Comput. Biol.* **1**, 245–251.

Sporns, O. and Zwi, J. (2004), The small world of the cerebral cortex, *Neuroinformatics* **2**, 145–162.

Sreenivasan, S., Cohen, R., López, E., Toroczkai, Z. and Stanley, H. E. (2007), Structural bottlenecks for communication in networks, *Phys. Rev. E* **75**, 036105.

Stam, C. J. (2004), Functional connectivity patterns of human magnetoencephalographic recordings: a "small-world" network?, *Neurosci. Lett.* **355**, 25–28.

Stam, C. J., Jones, B. F., Nolte, G., Breakspear, M. and Scheltens, P. (2007), Small-world networks and functional connectivity in Alzheimer's disease, *Cereb. Cortex* **17**, 92–99.

Stam, C. J. and Reijneveld, J. C. (2007), Graph theoretical analysis of complex networks in the brain, *Nonlin. Biomed. Phys.* **1**, 3.

Stauffer, D. and Aharony, A. (1992), *Introduction to Percolation Theory*, Taylor & Francis.

Stauffer, D. and Barbosa, V. C. (2006), Dissemination strategy for immunizing scale-free networks, *Phys. Rev. E* **74**, 056105.

Steels, L. (1996), A self-organizing spatial vocabulary, *Artif. Life J.* **2**, 319–332.

Strauss, D. and Ikeda, M. (1990), Pseudolikelihood estimation for social networks, *J. Am. Stat. Assoc.* **85**, 204–212.

Strogatz, S. H. (2000a), Exploring complex networks, *Nature* **410**, 268–276.

Strogatz, S. H. (2000b), From Kuramoto to Crawford: exploring the onset of synchronization in populations of coupled oscillators, *Physica D* **143**, 1–20.

Strogatz, S. H. and Mirollo, R. E. (1988), Phase-locking and critical phenomena in lattices of coupled nonlinear oscillators with random intrinsic frequencies, *Physica D* **31**, 143–168.

Suchecki, K., Eguíluz, V. M. and San Miguel, M. (2005a), Conservation laws for the voter model in complex networks, *Europhys. Lett.* **69**, 228–234.

Suchecki, K., Eguíluz, V. M. and San Miguel, M. (2005b), Voter model dynamics in complex networks: Role of dimensionality, disorder and degree distribution, *Phys. Rev. E* **72**, 036132.

Sudbury, A. (1985), The proportion of the population never hearing a rumour, *J. Appl. Prob.* **22**, 443–446.

Svenson, P. (2001), Freezing in random graph ferromagnets, *Phys. Rev. E* **64**, 036122.

Svenson, P. and Johnston, D. A. (2002), Damage spreading in small world Ising models, *Phys. Rev. E* **65**, 036105.

Szabó, G., Alava, M. and Kertész, J. (2003), Structural transitions in scale-free networks, *Phys. Rev. E* **67**, 056102.

Sznajd-Weron, K. and Sznajd, J. (2000), Opinion evolution in closed community, *Int. J. Mod. Phys. C* **11**, 1157–1165.

Tabah, A. N. (1999), Literature dynamics: Studies on growth, diffusion, and epidemics, *Ann. Rev. Inform. Sci. Technol.* **34**, 249–286.

Tadić, B. and Thurner, S. (2004), Information super-diffusion on structured networks, *Physica A* **332**, 566–584.

Tadić, B., Thurner, S. and Rodgers, G. J. (2004), Traffic on complex networks: Towards understanding global statistical properties from microscopic density fluctuations, *Phys. Rev. E* **69**, 036102.

Takayasu, M., Fukuda, K. and Takayasu, H. (1999), Application of statistical physics to the Internet traffics, *Physica A* **274**, 140–148.

Thadakamalla, H. P., Albert, R. and Kumara, S. R. T. (2005), Search in weighted complex networks, *Phys. Rev. E* **72**, 066128.

Thadakamalla, H. P., Albert, R. and Kumara, S. R. T. (2007), Search in spatial scale-free networks, *New J. Phys.* **9**, 190.

Timme, M. (2006), Does dynamics reflect topology in directed networks?, *Europhys. Lett.* **76**, 367–373.

Timme, M., Wolf, F. and Geisel, T. (2002), Coexistence of regular and irregular dynamics in complex networks of pulse-coupled oscillators, *Phys. Rev. Lett.* **89**, 258701.

Tong, A. H. Y., Lesage, G., Bader, G. D. *et al.* (2004), Global mapping of the yeast genetic interaction network, *Science* **303**, 808 – 813.

Travers, J. and Milgram, S. (1969), An experimental study of the small world phenomenon, *Sociometry* **32**, 425–443.

Trusina, A., Rosvall, M. and Sneppen, K. (2005), Communication boundaries in networks, *Phys. Rev. Lett.* **94**, 238701.

Uetz, P., Dong, Y.-A., Zeretzke, C. *et al.* (2006), Herpesviral protein networks and their interaction with the human proteome, *Science* **311**, 239–242.

Uhlig, S. and Bonaventure, O. (2001), *The Macroscopic Behavior of Internet Traffic: A Comparative Study*. Technical Report Infonet-TR-2001-10, University of Namur.

Urry, J. (2004), Small worlds and the new "social physics", *Glob. Networks* **4**, 109–130.

Valverde, S., Ferrer i Cancho, R. and Solé, R. V. (2002), Scale-free networks from optimal design, *Europhys. Lett.* **60**, 512–517.

Valverde, S. and Solé, R. V. (2002), Self-organized critical traffic in parallel computer networks, *Physica A* **312**, 636–648.

Valverde, S. and Solé, R. V. (2004), Internet's critical path horizon, *Eur. Phys. J. B* **38**, 245–252.

van Vreeswijk, C. and Sompolinsky, H. (1996), Chaos in neuronal networks with balanced excitatory and inhibitory activity, *Science* **274**, 1724–1726.

Varma, A. and Palsson, B. O. (1993), Metabolic capabilities of *Escherichia coli* ii. Optimal growth patterns, *J. Theor. Biol.* **165**, 503–522.

Vázquez, A. (2006a), Polynomial growth in branching processes with diverging reproductive number, *Phys. Rev. Lett.* **96**, 038702.

Vázquez, A. (2006b), Spreading dynamics on heterogeneous populations: Multitype network approach, *Phys. Rev. E* **74**, 066114.

Vázquez, A. (2006c), Spreading dynamics on small-world networks with connectivity fluctuations and correlations, *Phys. Rev. E* **74**, 056101.

Vázquez, A., Flammini, A., Maritan, A. and Vespignani, A. (2003), Modeling of protein interaction networks, *ComPlexUs* **1**, 38–44.

Vázquez, A., Pastor-Satorras, R. and Vespignani, A. (2002), Large-scale topological and dynamical properties of the Internet, *Phys. Rev. E* **65**, 066130.

Viana Lopes, J., Pogorelov, Y. G., Lopes dos Santos, J. M. B. and Toral, R. (2004), Exact solution of Ising model on a small-world network, *Phys. Rev. E* **70**, 026112.

Viger, F., Barrat, A., Dall'Asta, L., Zhang, C.-H. and Kolaczyk, E. (2007), What is the real size of a sampled network? The case of the Internet, *Phys. Rev. E* **75**, 056111.

Vilone, D. and Castellano, C. (2004), Solution of voter model dynamics on annealed small-world networks, *Phys. Rev. E* **69**, 016109.

Vilone, D., Vespignani, A. and Castellano, C. (2002), Ordering phase transition in the one-dimensional Axelrod model, *Eur. Phys. J. B* **30**, 399–406.

Vogels, W., van Renesse, R. and Birman, K. (2003), The power of epidemics: robust communication for large-scale distributed systems, *SIGCOMM Comput. Commun. Rev.* **33**(1), 131–135.

Volz, E. (2004), Random networks with tunable degree distribution and clustering, *Phys. Rev. E* **70**, 056115.

Wagner, A. (2003), How the global structure of protein interaction networks evolves, *Proc. R. Soc. B* **270**, 457–466.

Wagner, A. and Fell, D. A. (2001), The small world inside large metabolic networks, *Proc. R. Soc. B* **268**, 1803–1810.

Walker, D., Xie, H., Yan, K.-K. and Maslov, S. (2007), Ranking scientific publications using a model of network traffic, *J. Stat. Mech.* p. P06010.

Wang, S. and Zhang, C. (2004), Weighted competition scale-free network, *Phys. Rev. E* **70**, 066127.

Wang, W.-X., Wang, B.-H., Hu, B., Yan, G. and Ou, Q. (2005), General dynamics of topology and traffic on weighted technological networks, *Phys. Rev. Lett.* **94**, 188702.

Wang, W.-X., Wang, B.-H., Yin, C.-Y., Xie, Y.-B. and Zhou, T. (2006a), Traffic dynamics based on local routing protocol on a scale-free network, *Phys. Rev. E* **73**, 026111.

Wang, W.-X., Yin, C.-Y., Yan, G. and Wang, B.-H. (2006b), Integrating local static and dynamic information for routing traffic, *Phys. Rev. E* **74**, 016101.

Wardrop, J. G. (1952), Some theoretical aspects of road traffic research, *Proc. Inst. Civil Engi. Part II* **II**, 325–378.

Warren, C. P., Sander, L. M. and Sokolov, I. M. (2002), Geography in a scale-free network model, *Phys. Rev. E* **66**, 056105.

Wasserman, S. and Faust, K. (1994), *Social Network Analysis: Methods and Applications*, Cambridge University Press.

Wasserman, S. and Pattison, P. (1996), Logit models and logistic regressions for social networks, *Psychometrika* **61**, 401–425.

Watts, D. J. (1999), *Small Worlds: The Dynamics of Networks Between Order and Randomness*, Princeton University Press.

Watts, D. J. (2002), A simple model of global cascades on random networks, *Proc. Natl Acad. Sci. (USA)* **99**, 5766–5771.

Watts, D. J. (2004), The new science of networks, *Ann. Rev. Sociol.* **30**, 243–270.

Watts, D. J., Dodds, P. S. and Newman, M. E. J. (2002), Identity and search in social networks, *Science* **296**, 1302–1305.

Watts, D. J., Muhamad, R., Medina, D. C. and Dodds, P. S. (2005), Multiscale, resurgent epidemics in a hierarchical metapopulation model, *Proc. Natl Acad. Sci. (USA)* **102**, 11157–11162.

Watts, D. J. and Strogatz, S. H. (1998), Collective dynamics of "small-world" networks, *Nature* **393**, 440–442.

West, G. B., Brown, J. H. and Enquist, B. J. (1997), A general model for the origin of allometric scaling laws in biology, *Science* **276**, 122–126.

Williams, R. J., Berlow, E. L., Dunne, J. A., Barabási, A.-L. and Martinez, N. D. (2002), Two degrees of separation in complex food webs, *Proc. Natl Acad. Sci. (USA)* **99**, 12913–12916.

Williams, R. J. and Martinez, N. D. (2000), Simple rules yield complex food webs, *Nature* **404**, 180 – 183.

Willinger, W., Govindan, R., Jamin, S., Paxson, V. and Shenker, S. (2002), Scaling phenomena in the Internet: Critically examining criticality, *Proc. Natl Acad. Sci. (USA)* **99**, 2573–2580.

Willinger, W., Taqqu, M. S. and Erramilli, A. (1996), A bibliographic guide to self-similar traffic and performance modeling for modern high-speed networks. In F. Kelly, S. Zachary and I. Ziedins, eds., *Stochastic Networks: Theory and Applications*, Clarendon Press, pp. 82–89.

Wissner-Gross, A. D. (2006), Preparation of topical readings lists from the link structure of Wikipedia. In *Proceedings of the IEEE International Conference on Advanced Learning Technology*. (ICALT '06), IEEE, pp. 825–829.

Wu, A.-C., Xu, X.-J., Wu, Z.-X. and Wang, Y.-H. (2007), Walks on weighted networks, *Chin. Phys. Lett.* **24**, 557.

Wu, F. and Huberman, B. A. (2004), Social structure and opinion formation, *Computational Economics* 0407002, EconWPA, http://ideas.repec.org/p/wpa/wuwpco/0407002.html.

Xulvi-Brunet, R. and Sokolov, I. M. (2002), Evolving networks with disadvantaged long-range connections, *Phys. Rev. E* **66**, 026118.

Yan, G., Zhou, T., Hu, B., Fu, Z.-Q. and Wang, B.-H. (2006), Efficient routing on complex networks, *Phys. Rev. E* **73**, 046108.

Yang, B. and Garcia-Molina, H. (2002), Improving search in peer-to-peer networks. In *Proceedings of the 22nd International Conference on Distributed Computing Systems*, IEEE, pp. 5–14.

Yeomans, J. M. (1992), *Statistical Mechanics of Phase Transitions*, Oxford University Press.

Yerra, B. and Levinson, D. (2005), The emergence of hierarchy in transportation networks, *Ann. Region. Sci.* **39**, 541–553.

Yodzis, P. (1981), The stability of real ecosystems, *Nature* **289**, 674–676.

Yook, S.-H., Jeong, H. and Barabási, A.-L. (2002), Modeling the Internet's large-scale topology, *Proc. Natl Acad. Sci. (USA)* **99**, 13382–13386.

Yook, S.-H., Jeong, H., Barabási, A.-L. and Tu, Y. (2001), Weighted evolving networks, *Phys. Rev. Lett.* **86**, 5835–5838.

Zanette, D. H. (2001), Critical behavior of propagation on small-world networks, *Phys. Rev. E* **64**, 050901.

Zanette, D. H. (2002), Dynamics of rumor propagation on small-world networks, *Phys. Rev. E* **65**, 041908.

Zanette, D. H. and Gil, S. (2006), Opinion spreading and agent segregation on evolving networks, *Physica D* **224**, 156–165.

Zhang, Y., Qian, M., Ouyang, Q. et al. (2006), Stochastic model of yeast cell-cycle network, *Physica D* **219**, 35–39.

Zhao, L., Lai, Y.-C., Park, K. and Ye, N. (2005), Onset of traffic congestion in complex networks, *Phys. Rev. E* **71**, 026125.

Zhou, C. and Kurths, J. (2006), Dynamical weights and enhanced synchronization in adaptive complex networks, *Phys. Rev. Lett.* **96**, 164102.

Zhou, C., Motter, A. E. and Kurths, J. (2006), Universality in the synchronization of weighted random networks, *Phys. Rev. Lett.* **96**, 034101.

Zhou, C., Zemanová, L., Zamora, G., Hilgetag, C. C. and Kurths, J. (2006), Hierarchical organization unveiled by functional connectivity in complex brain networks, *Phys. Rev. Lett.* **97**, 238103.

Zhou, C., Zemanová, L., Zamora-López, G., Hilgetag, C. C. and Kurths, J. (2007), Structure-function relationship in complex brain networks expressed by hierarchical synchronization, *New J. Phys.* **9**, 178.

Zhou, H. and Lipowsky, R. (2005), Dynamic pattern evolution on scale-free networks, *Proc. Natl Acad. Sci. (USA)* **102**, 10052–10057.

Zhou, S. and Mondragon, R. J. (2004), The rich-club phenomenon in the Internet topology, *IEEE Commun. Lett.* **8**, 180–182.

Zhou, T., Liu, J.-G., Bai, W.-J., Chen, G. and Wang, B.-H. (2006), Behaviors of susceptible-infected epidemics on scale-free networks with identical infectivity, *Phys. Rev. E* **74**, 056109.

Zhu, J.-Y. and Zhu, H. (2003), Introducing small-world network effects to critical dynamics, *Phys. Rev. E* **67**, 026125.

Zimmermann, M. G. and Eguíluz, V. M. (2005), Cooperation, social networks, and the emergence of leadership in a prisoner's dilemma with adaptive local interactions, *Phys. Rev. E* **72**, 056118.

Zimmermann, M. G., Eguíluz, V. M. and San Miguel, M. (2001), Cooperation, adaptation and the emergence of leadership. In A. Kirman and J. B. Zimmermann, eds., *Economics with Heterogeneous Interacting Agents*, Springer, pp. 73–86.

Zimmermann, M. G., Eguíluz, V. M. and San Miguel, M. (2004), Coevolution of dynamical states and interactions in dynamic networks, *Phys. Rev. E* **69**, 065102.

Index